Integrated Smart Technologies for Sustainable Water Systems Using Computer Applications

Integrated Smart Technologies for Sustainable Water Systems Using Computer Applications presents a comprehensive exploration of how cutting-edge computational technologies are revolutionizing water resource management. This multidisciplinary guide demonstrates the practical integration of artificial intelligence (AI), Internet of Things (IoT) systems, and emerging digital technologies to address critical challenges in hydrology, hydraulics, water quality, and smart infrastructure development.

This book systematically examines the application of advanced technologies including AI, machine learning, IoT, blockchain, cloud and edge computing, and Geographic Information Systems across diverse water management scenarios. Through theoretical models, algorithmic approaches, and real-world case studies from both developed and developing countries, it illustrates key findings in AI-driven water demand and quality predictions, IoT-enabled real-time distribution network monitoring, blockchain applications for transparent water resource transactions, intelligent flood forecasting systems, automated irrigation methods, wastewater treatment automation, and digital twin development for dynamic infrastructure modeling. The work uniquely incorporates emerging concepts such as explainable AI, climate resilience frameworks, and digital twin technologies, positioning it at the forefront of current industry and academic research trends.

This essential resource is designed for researchers, industry professionals, software developers, educators, and students specializing in water resource management and sustainable technology applications. The content balances foundational insights with advanced methodologies, making it equally valuable for newcomers seeking a comprehensive understanding and experienced practitioners looking to implement cutting-edge solutions for sustainable water system management.

Smart Technologies for Water Engineering & Sustainability

With the latest advancements in AI and associated smart technologies, there have been conscious efforts towards a sustainable environment. Water Engineering is one of these areas influenced by recent technologies including smart materials, efficient water management, and advanced water systems like dams, irrigation, and so on. This series is a compilation of such books which work at the intersection of the latest technology, sustainability, and water engineering, making them extremely relevant in today's world.

Integrated Smart Technologies for Sustainable Water Systems Using Computer Applications

Edited by Srikanta Kumar Mohapatra, Monika Sethi, Shivani Sood, Abhineet Anand, and Jayashree Mohanty

For more information about this series, please visit: https://www.routledge.com/Smart-Technologies-for-Water-Engineering--Sustainability/book-series/STWES

Integrated Smart Technologies for Sustainable Water Systems Using Computer Applications

Edited by
Srikanta Kumar Mohapatra, Monika Sethi, Shivani Sood,
Abhineet Anand, and Jayashree Mohanty

CRC Press
Taylor & Francis Group
Boca Raton London New York

CRC Press is an imprint of the
Taylor & Francis Group, an **informa** business

Designed cover image: Shutterstock

First edition published 2027
by CRC Press
4 Park Square, Milton Park, Abingdon, Oxon, OX14 4RN

and by CRC Press
2385 NW Executive Center Drive, Suite 320, Boca Raton FL 33431

CRC Press is an imprint of informa UK Limited

British Library Cataloguing-in-Publication Data
A catalogue record for this book is available from the British Library

Library of Congress Cataloging-in-Publication Data
Names: Mohapatra, Srikanta Kumar editor | Sethi, Monika editor | Sood, Shivani editor | Anand, Abhineet, 1975- editor | Mohanty, Jayashree editor
Title: Integrated smart technologies for sustainable water systems using computer applications / edited by Srikanta Kumar Mohapatra, Monika Sethi, Shivani Sood, Abhineet Anand, and Jayashree Mohanty.
Description: First edition. | Abingdon, Oxon ; Boca Raton, FL : CRC Press, 2027. | Series: Smart technologies for water engineering & sustainability | Includes bibliographical references and index.
Identifiers: LCCN 2026012887 (print) | LCCN 2026012888 (ebook) | ISBN 9781041199311 hbk | ISBN 9781041199328 pbk | ISBN 9781003714217 ebk
Subjects: LCSH: Water-supply–Management–Data processing | Water resources development–Data processing | Water-supply engineering–Data processing | Smart cities
Classification: LCC TD353 .I5248 2027 (print) | LCC TD353 (ebook)
LC record available at https://lccn.loc.gov/2026012887
LC ebook record available at https://lccn.loc.gov/2026012888

ISBN: 9781041199311 (hbk)
ISBN: 9781041199328 (pbk)
ISBN: 9781003714217 (ebk)

DOI: 10.1201/9781003714217

Typeset in Times
by codeMantra

Contents

Preface

Water is the foundation of life, economic growth, and environmental sustainability. In the 21st century, increasing population pressure, rapid urbanization, climate variability, and industrial expansion have significantly stressed global water resources. Traditional water management practices, often fragmented and reactive, are no longer sufficient to address emerging challenges such as water scarcity, flooding, water quality degradation, and infrastructure inefficiencies. In this context, the integration of smart technologies with sustainable water systems has emerged as a transformative pathway toward resilient and efficient water management.

The edited volume titled *Integrated Smart Technologies for Sustainable Water Systems Using Computer Applications* is conceived to present an interdisciplinary platform that brings together researchers, academicians, industry professionals, and policymakers working at the intersection of water engineering and computational intelligence. The primary objective of this book is to highlight how advanced computer applications—such as the Internet of Things, artificial intelligence (AI), machine learning (ML), blockchain, cloud and edge computing, computer vision, and geospatial technologies—are reshaping the design, monitoring, operation, and governance of water systems.

This book is structured to provide both theoretical foundations and practical insights. The initial chapters introduce readers to global water challenges and the role of cyber-physical systems in sustainable water management. Foundational technologies, including Internet of Things-enabled sensor networks, wireless communication, and data acquisition frameworks, are discussed to establish the technological backbone of smart water systems. Subsequent chapters explore advanced data-driven approaches, including AI and ML techniques for water demand forecasting, flood prediction, anomaly detection, and optimization of water distribution networks.

A significant emphasis is placed on emerging paradigms such as digital twins, blockchain-based water governance, and hybrid cloud–edge architectures. These approaches enable real-time monitoring, predictive maintenance, transparency, and secure data exchange across complex water infrastructures. This book further extends its scope to application-driven chapters covering smart irrigation, wastewater automation, water quality monitoring using computer vision, and decision support systems for integrated water resource management.

The later sections of this book present case studies and future perspectives, offering valuable insights into real-world implementations in smart cities and watershed management projects. Challenges related to scalability, cybersecurity, policy frameworks, data interoperability, and climate resilience are critically examined to guide future research and deployment strategies. By combining engineering principles with computational intelligence, this book aims to bridge the gap between sustainability goals and technological innovation.

This volume is intended to serve as a comprehensive reference for postgraduate students, doctoral researchers, faculty members, practicing engineers, and professionals involved in water resources, civil engineering, environmental sciences, and computer applications. We believe that the collective contributions in this book will inspire innovative research, promote interdisciplinary collaboration, and support informed decision-making toward sustainable and resilient water systems.

We would like to express our sincere gratitude to all the authors who contributed their valuable research and insights to this edited volume. Their dedication, expertise, and timely cooperation made this book possible. We are also thankful to the reviewers and editorial team members for their constructive feedback and continuous support throughout the compilation process.

On a personal note, we extend our heartfelt thanks to our families. Their unwavering encouragement, patience, and understanding during the long hours of academic and editorial work have been a constant source of strength. The moral support, motivation, and sacrifices made by our family members played a crucial role in the successful completion of this book. We dedicate this work to them with deep appreciation and gratitude.

We hope that this book will contribute meaningfully to the advancement of smart and sustainable water systems and serve as a catalyst for future innovations in this vital domain.

Editor Biographies

Srikanta Kumar Mohapatra is currently serving as a professor in the Department of Computer Science and Engineering at Chitkara University, India. He has more than 18 years of combined teaching and industrial experience. He has published over 70 research articles in reputed national and international journals and conferences, and has served as a member of editorial and reviewer boards for leading publishing houses. He has authored three reference books and co-edited one book. He has filed/published more than 20 patents, of which 15 patents have been granted. Recently, he received the CSE Excellence Award from Chitkara University in recognition of his significant contributions to patents and the CRISP initiative. His research interests include AI, ML, blockchain, cloud computing, and computational nanotechnology.

Monika Sethi is currently serving as a professor in the Department of Computer Science and Engineering at Chitkara University, India. Her research interests focus on image processing (ADNI-MRI images), deep learning, ML, and wireless sensor networks. Her teaching interests include C, C++, Python, data structures, operating systems, computer system architecture, computer graphics, and digital electronics. She has published over 20 research papers in international journals and conferences (Scopus indexed) and has filed 14 patents (9 granted and 5 published). She is a member of the Institution of Engineers (India) (IEI). She holds an MTech in power-efficient hierarchical centralized routing in wireless sensor network (WSN) from NIT Hamirpur and a PhD degree in computer science from Chitkara University, India.

Shivani Sood is currently serving as an assistant professor in the School of Computer Applications at Lovely Professional University, Phagwara, Punjab, India. Her research interests focus on pattern recognition, ML, image processing, and data science. Her teaching interests include Python, Java, C/C++, and advanced data structures. She has filed two patents and published ten articles in international journals and conferences. She holds a master's degree in data guard configuration and management from Sri Sai College of Engineering and a PhD in computer science from Chitkara University, Punjab, India.

Abhineet Anand is a distinguished academician and researcher with over 24 years of experience in teaching, research, and academic administration. He currently serves as the Dean of the School of Computer Science and Engineering at Bahra University, Shimla Hills, where he has been instrumental in bridging the gap between academia and industry. His research interests include cloud computing, cloud security, decision trees, nearest neighbor methods, clustering, rule induction, optical fiber switching in wavelength multiplexing, and automata theory. He has an extensive publication record, with more than 65 Scopus-indexed papers, 14 SCI papers, and 16 papers in international conferences to his credit. Additionally, he has contributed to 12 international journals and has edited more than 8 books, showcasing his expertise in emerging technologies. He has been a session chair/co-chair and a Technical Program Committee member for over 20 international conferences. He is also a thesis supervisor for Liverpool John Moores University through upGrad Education Pvt. Ltd., reflecting his global academic engagement. As a passionate advocate for innovation and interdisciplinary research, he has co-edited several books, including *Bio-Inspired Optimization for Medical Data Mining* and the newly published *Internet of Medicine for Smart Healthcare.*

Jayashree Mohanty is currently serving as an assistant professor at Chandigarh University, Punjab, India. She has more than 15 years of teaching and industrial experience. She has published numerous research articles in journals and conferences of international repute. She has also acted as a member of the reviewer board for leading publishing houses and international conferences. Her research interests include blockchain technology, big data engineering, AI, and ML.

Contributors

Cigil Achenkunju
Data and AI Project
NTT DATA Europe & Latam
Boston, Massachusetts

Ciril Achenkunju
Koch Industries
Wichita, Kansas

Alok Kumar Agrawal
School of Engineering and Technology
Chitkara University
Himachal Pradesh, India

Nirav Rameshbhai Amin
Electronics & Communication Engineering Department
Government Engineering College
Bhavnagar, Gujarat, India

Abhineet Anand
Department of Computer Science & Engineering
Bahra University
Waknaghat, Himachal Pradesh, India

Mohit Arora
School of Computer Science and Engineering
Lovely Professional University
Phagwara, Punjab, India

Mayank Attri
Department of EEE
Bahra University
Waknaghat, Himachal Pradesh, India

Mehvish Bilal
Department of Civil & Environmental Engineering
KFUPM
Dhahran, Saudi Arabia

Aravind Chauhan
Department of Civil Engineering
Bahra University
Waknaghat, Himachal Pradesh, India

Titiksha Chugh
Department of Computer Science and Engineering
Chitkara University Institute of Engineering & Technology
Chitkara University
Rajpura, Punjab, India

Laishram Bijialuxmi Devi
Chitkara University College of Nursing
Barotiwala, Himachal Pradesh, India

Kavita Dhiman
Department of Computer Applications
CGC University
Mohali, Punjab, India

Shivam Kumar
Chitkara University School of Engineering and Technology
Chitkara University
Baddi, Himachal Pradesh, India

Amanpreet Kaur
Department of CSE
Chitkara University Institute of Engineering & Technology
Rajpura, Punjab, India

Ramandeep Kaur
School of Computer Applications
Lovely Professional University
Phagwara, Punjab, India

Sukhjinder Kaur
Department of CSE
Rayat Bahra University
Mohali, Punjab, India

Pankhuri Khurana
Department of Computer Science and Engineering
Chitkara University Institute of Engineering & Technology
Rajpura, Punjab, India

Poonam Kukana
Department of CSE, UIE
Chandigarh University
Mohali, Punjab, India

Gaurav Mehta
School of Engineering and Technology
Chitkara University
Baddi, Himachal Pradesh, India

Priyakanksha Mishra
School of Forensic Science
Bahra University
Waknaghat, Himachal Pradesh, India

Ashima Mockoul
Department of CSE
Rayat Bahra University
Mohali, Punjab, India

Raghini Mohan
Department of Artificial Intelligence and Data Science
Velammal College of Engineering & Technology
Madurai, Tamil Nadu, India

Sangeetha Nagamani
Department of AI&ML
Amruta Institute of Engineering and Management Sciences
Bangalore, Karnataka, India

Qurat-Ul-Ain
Computer Applications
Department of Higher Education J&K
Srinagar, Jammu and Kashmir, India

Radhika Rajavelu
Department of AIML, School of Computer Science and Engineering
Galgotias University
Greater Noida, Uttar Pradesh, India

Sathiya Priya Rajavelu
Department of AIML, School of Computer Science and Engineering
Galgotias University
Greater Noida, Uttar Pradesh, India

Premananda Sahu
School of Computer Science and Engineering
Lovely Professional University
Phagwara, Punjab, India

Chiman Saini
Department of CSE
World College of Technology and Management
Gurgaon, Haryana, India

Ravi Kumar Sandal
Department of Civil Engineering
Bahra University
Waknaghat, Himachal Pradesh, India

Aakanksha Sankhyan
Chitkara University College of Nursing
Chitkara University
Baddi, Himachal Pradesh, India

Deelip Kumar Sarangi
Department of Management
Kalinga University
Naya Raipur, Chhattisgarh, India

Merry Saxena
Department of Interdisciplinary Courses in Engineering
Chitkara University Institute of Engineering & Technology
Rajpura, Punjab, India

Annamalai Selvarajan
Department of Computer Science and Engineering
Jain (Deemed-to-be University)
Bangalore, Karnataka, India

Kapil Sethi
Department of Computer Science Engineering
Bahra University
Waknaghat, Himachal Pradesh, India

Lokeshwer Sharma
Department of Civil Engineering
Bahra University
Waknaghat, Himachal Pradesh, India

Mamta Sharma
Department of CSE
Bahra University
Waknaghat, Himachal Pradesh, India

Neha Sharma
Department of Computer Science and Engineering
Chitkara University Institute of Engineering & Technology
Rajpura, Punjab, India

Bhupinder Singh
School of Computer Science and Engineering
Lovely Professional University
Phagwara, Punjab, India

Gurwinder Singh
Department of AIT–CSE
Chandigarh University
Mohali, Punjab, India

Bhupinder Kaur Srao
Department of CSE
Chandigarh University
Mohali, India

Abhishek Sundi
Department of Forensic Science
Bahra University
Waknaghat, Himachal Pradesh, India

Anu Thakur
Department of CSE
Bahra University
Waknaghat, Himachal Pradesh, India

Prerna Thakur
Assistant Professor
Bahra University
Waknaghat, Tehsil Kandaghat
District Solan, Himachal Pradesh, India – 173234

Gopal Usha
Department of Computing Technologies, School of Computing
SRM Institute of Science and Technology
Kattankulathur, Tamil Nadu, India

Rajit Verma
School of Banking Financial Services and Insurance Management
Symbiosis University of Applied Sciences
Indore, Madhya Pradesh, India

John Peter Vincent Paul
Department of Information Science and Engineering
City Engineering College
Bangalore, Karnataka, India

Vishal
Department of CSE
ICFAI University
Baddi, Himachal Pradesh, India

1 Computer-Aided RSM and Kinetic Modelling of Phenol Removal from Arsenic-Selenium Co-Contaminated Water Using Filtrasorb Activated Carbon

Mehvish Bilal

1.1 INTRODUCTION

Unsafe industrial waste disposal has resulted in extensive contamination of water bodies with phenol, arsenic, and selenium species, all of which pose a serious threat to human health and aquatic ecosystems. Phenol is a priority pollutant typically discharged from petroleum refining, coal processing, pulp and paper industries, and biomass gasification, resulting in severe regulatory limitations on effluent discharge [1]. Selenium (Se) species [selenite (SeO_3^{2-}) and selenate (SeO_4^{2-})] and arsenic (generated by mining and burning fossil fuels and from groundwater) are prevalent in industrial effluents and are highly poisonous [2,3]. Adsorption with granular activated carbon (GAC) remains one of the most effective methodologies for removing organic and inorganic pollutants, despite limitations related to cost, regeneration, and multi-contaminant interactions [4]. While GAC has been extensively used for the individual removal of phenol, arsenic (As), and Se species, no study has addressed phenol adsorption under simultaneous As-Se co-contamination, indicating a vital research need. To address this, the current study investigates the removal of phenol from aqueous solutions containing As, SeO_3^{2-},

DOI: 10.1201/9781003714217-1

and SeO_4^{2-}, with an emphasis on critical operational parameters such as pH, starting concentration, and GAC dose. Kinetic modelling [zero order, pseudo first order (PFO), and pseudo second order (PSO)] is utilised to clarify uptake mechanisms and rate behaviour [5], whereas response surface methodology (RSM) provides a statistical framework for analysing variable interactions and maximising removal efficiency [6]. GAC has also demonstrated versatility in eliminating other water contaminants, such as bromate [7], and coconut shell-based GAC has proven effective in improving water quality by reducing excess chlorine and neutralising physico-chemical parameters [8]. RSM provides a statistical framework for assessing the interaction of variables and optimising the efficiency of removal [9]. This study developed a mechanistic and functional understanding of GAC performance in complex multi-contaminant systems through a combination of kinetic insights and RSM-based optimisation to facilitate the formulation of effective and scalable strategies for wastewater treatment.

1.2 MATERIALS AND METHODOLOGY

1.2.1 Materials

Analytical-grade reagents and high-purity deionised water were used throughout the study. Target contaminants were phenol, As, SeO_3^{2-}, and SeO_4^{2-}, with the pH adjusted using dilute sodium hydroxide (NaOH) and nitric acid (HNO_3) solutions. The adsorbent was a commercial GAC (Filtrasorb 400, DuPont, USA; surface area 948 m^2/g). All samples were first filtered using 0.2-μm membranes before analysis. pH measurements were taken using a calibrated pH meter, which was standardised daily using pH 4, 7, and 10 buffers. The concentration of phenol in the samples was determined using a UV–Vis spectrophotometer at a wavelength of 269 nm after filtration and buffering with potassium dihydrogen phosphate (KH_2PO_4), and calibration curves of phenol standards were plotted to allow the determination of phenol concentrations. Essential laboratory equipment, such as a precision balance, shaker plate, and pH equipment, facilitated the repeatability and consistency of experimentation.

1.2.2 Adsorption Kinetics

The adsorption kinetics were studied to understand the mass transfer and diffusion mechanisms that influence phenol absorption, including film diffusion, intraparticle pore diffusion, and sorption onto active sites. Experiments were carried out on the kinetics of 7.5 mg/L phenol with co-contaminants [2.5 mg/L arsenate (AsO_4^{3-}), 2.5 mg/L SeO_3^{2-}, 5.0 mg/L SeO_4^{2-}] at pH 4, 7, and 10 at a constant GAC dose of 0.1 g/L. Equilibrium was observed at 90 minutes with samples collected between 5 and 90 minutes, and this was validated by the 24-hour extended contact. Phenol (0.20 μm membrane) was then filtered, followed by a measurement of the phenol concentration at 269 nm using a UV–Vis spectrophotometer with matrix-matched standards and KH_2PO_4 buffer. Phenol adsorption is usually governed by the PSO kinetics and fits the Langmuir or the Freundlich isotherms [10,11]. Meanwhile, selenium species show pH-related kinetic behaviours on GAC-based adsorbents [12].

1.2.3 Response Surface Methodology

RSM with a central composite design (CCD) was used to design the experiments for evaluating phenol removal in the presence of co-contaminant As-Se species. According to the CCD structure, four independent variables, namely, phenol, AsO_4^-, SeO_3^-, and SeO_4^- concentrations, were determined in 25 experimental runs. This method allows the removal efficiency to be modelled effectively with a minimal total number of experiments. The factors were selected based on preliminary screening to ensure an appropriate and meaningful design space, where phenol, AsO_4^-, and SeO_3^- were varied at levels of 2.5–7.5 mg/L, and SeO_4^- at 5–15 mg/L. A summary of these levels is given in Table 1.1, and the complete CCD experimental design is given in Table 1.2. Consistent with established adsorption optimisation studies, CCD-RSM is very effective at capturing linear, quadratic, and interaction effects in the multi-contaminant adsorption system and has been extensively applied to model the removal of phenols and heavy metals [13]. The adsorbent employed in all CCD runs was Filtrasorb GAC.

TABLE 1.1 Experimental factor levels used in the RSM-CCD design

	LEVEL	
FACTOR	*MEDIUM*	*HIGH*
Phenol	2.5	7.5
AsO_4^{3-}	2.5	7.5
SeO_3^{2-}	2.5	7.5
SeO_4^{2-}	5	15

TABLE 1.2 Design of experiment for phenol adsorption in the presence of SeO_3^{2-}, SeO_3^{2-}, SeO_4^{2-}, and AsO_4^{3-} adsorption work using Filtrasorb GAC

RUN	*A: PHENOL FACTOR 1 (MG/L)*	*B: ASO_4^{3-} FACTOR 2 (MG/L)*	*C: SEO_3^{2-} FACTOR 3 (MG/L)*	*D: SEO_4^{2-} FACTOR 4 (MG/L)*
1	7.5	7.5	7.5	5
2	5	7.5	5	10
3	2.5	2.5	7.5	15
4	2.5	2.5	2.5	15
5	7.5	7.5	7.5	15
6	7.5	2.5	7.5	15
7	7.5	2.5	2.5	5
8	2.5	5	5	10
9	7.5	7.5	2.5	5
10	7.5	5	5	10
11	2.5	2.5	2.5	5
12	5	5	5	10
13	2.5	2.5	7.5	5
14	7.5	7.5	2.5	15
15	5	5	7.5	10
16	2.5	7.5	2.5	5
17	5	5	5	15
18	2.5	7.5	7.5	15
19	7.5	2.5	2.5	15
20	5	2.5	5	10
21	5	5	5	5

(Continued)

TABLE 1.2 (*Continued*) Design of experiment for phenol adsorption in the presence of SeO_3^{2-}, SeO_3^{2-}, SeO_4^{2-}, and AsO_4^{3-} adsorption work using Filtrasorb GAC

RUN	*A: PHENOL FACTOR 1 (MG/L)*	*B: ASO_4^{3-} FACTOR 2 (MG/L)*	*C: SEO_3^{2-} FACTOR 3 (MG/L)*	*D: SEO_4^{2-} FACTOR 4 (MG/L)*
22	7.5	2.5	7.5	5
23	2.5	7.5	2.5	15
24	2.5	7.5	7.5	5
25	5	5	2.5	10

1.3 RESULTS AND DISCUSSION

This work evaluated phenol removal using Filtrasorb granular AC (FSGAC) in systems co-contaminated with SeO_3^{2-}, SeO_4^{2-}, and AsO_4^{3-}. A CCD under RSM was employed to structure 25 experimental runs, varying four key parameters: SeO_3^{2-}, SeO_4^{2-}, phenol, and AsO_4^{3-} concentration, to model phenol removal efficiency (%). Table 1.3 presents experimental removal efficiencies, revealing that phenol showed maximum adsorption at the lowest AsO_4^{3-} concentration (2.5 mg/L), with SeO_3^{2-} and SeO_4^{2-} maintained at 5 and 10 mg/L, respectively, indicating enhanced phenol affinity under competitive tertiary adsorption conditions. These trends were captured in the RSM-derived predictive model (Equation 1.1), statistically validated through adjusted R^2 and predicted R^2 values that differed by less than 0.2, and adequate precision values exceeding the threshold of 4, confirming robust model performance. These findings are consistent with earlier RSM-based adsorption studies, where a small gap between adjusted R^2 and predicted R^2 (typically < 0.2) is regarded as evidence of reliable predictive models. For instance, in a Box–Behnken optimisation of phenol removal using functionalised biochar, R^2 and adjusted R^2 values were both ~0.998, indicating robust model reliability [14]. Significant terms included A (phenol) and A^2, with A^2B also retained. Although B (AsO_4^{3-}) and AB were not statistically significant ($p > 0.05$, Table 1.5), they were retained in the model to preserve hierarchy. Diagnostic plots in Figures 1.1–1.3 further support model validity through normally distributed residuals, predicted vs actual clustering along the 45° line, and no systematic structure in residual vs predicted plots. The 3D response surfaces in Figures 1.4 and 1.5 highlight that maximum removal (≈99.36%) occurs at 5 mg/L phenol and 2.5 mg/L AsO_4^{3-} (Run 20), while changes in selenium species caused negligible variation. These model-based findings are consistent with prior RSM studies optimising phenol removal and adsorption processes [15].

$$RE(\%) = 7.925\times10^5 + 1.420\times10^5\cdot A - 1.949\times10^5\cdot B - 6.141\times10^4\cdot AB - 1.795\times10^5 A^2 + 1.00\times10^5 A^2 B \tag{1.1}$$

where A = phenol concentration (mg/L), B = AsO_4^{3-} concentration (mg/L), AB = interaction between phenol and AsO_4^{3-}, A^2 = quadratic effect of phenol, and A^2B = interaction between quadratic phenol and AsO_4^{3-}. Table 1.4 presents the statistical fitting parameters derived for the RSM model (Equation 1.1), developed to predict phenol removal efficiency under mixed contaminant conditions. While R^2 indicates a satisfactory model fit, adjusted R^2 and predicted R^2 provide more reliable indicators of model robustness and predictive capability. Adjusted R^2 accounts for the number of terms in the model relative to the number of data points, penalising overfitting, while predicted R^2 measures how well the model forecasts new observations outside the training dataset. This small difference indicates that the model is neither

TABLE1.3 Experimental response for phenol removal efficiency in the presence of phenol, SeO_3^{2-}, SeO_4^{2-}, and AsO_4^{3-}

	FACTOR A	*FACTOR B*	*FACTOR C*	*FACTOR D*	*REMOVAL EFFICIENCY A*
RUN	*PHENOL*	*ASO_4^{3-}*	*SEO_3^{2-}*	*SEO_4^{2-}*	*RE %*
1	7.5	7.5	7.5	5	88.127
2	5	7.5	5	10	83.923
3	2.5	2.5	7.5	15	69.508
4	2.5	2.5	2.5	15	52.005
5	7.5	7.5	7.5	15	91.083
6	7.5	2.5	7.5	15	87.334
7	7.5	2.5	2.5	5	88.143
8	2.5	5	5	10	57.855
9	7.5	7.5	2.5	5	85.121
10	7.5	5	5	10	87.827
11	2.5	2.5	2.5	5	70.380
12	5	5	5	10	87.311
13	2.5	2.5	7.5	5	77.864
14	7.5	7.5	2.5	15	87.468
15	5	5	7.5	10	93.077
16	2.5	7.5	2.5	5	59.681
17	5	5	5	15	86.998
18	2.5	7.5	7.5	15	78.667
19	7.5	2.5	2.5	15	92.563
20	5	2.5	5	10	99.356
21	5	5	5	5	81.000
22	7.5	2.5	7.5	5	87.953
23	2.5	7.5	2.5	15	87.924
24	2.5	7.5	7.5	5	92.920
25	5	5	2.5	10	95.563

TABLE1.4 Statistical summary of the fitted RSM model for phenol adsorption

SOURCE	*SUM OF SQUARES*	*DF*	*MEAN SQUARE*	*F-VALUE*	*P-VALUE*	
Model	7.001E+11	5	1.400E+11	6.53	0.0011	Significant
A-Phenol	3.629E+11	1	3.629E+11	16.91	0.0006	
B-AsO_4^{3-}	7.594E+10	1	7.594E+10	3.54	0.0754	
AB	6.035E+10	1	6.035E+10	2.81	0.1099	
A^2	1.624E+11	1	1.624E+11	7.57	0.0127	
A^2B	1.058E+11	1	1.058E+11	4.93	0.0387	
Residual	4.077E+11	19	2.146E+10			
Cor Total	1.108E+12	24				

overfitted nor underfitted and has reliable predictive power for phenol removal efficiency in the designed domain. Moreover, the RSM software iteratively improves these values by adjusting and validating the significance of factors and their interaction terms until the optimal model configuration is obtained. Another important statistical measure given in Table 1.4 is adequate precision, which is a ratio of the model signal (predicted variation) to noise (residual error). A signal-to-noise ratio of more than 4 is usually regarded as desirable, that is, the model is expected to move through the design space and enable accurate predictions.

To evaluate the statistical reliability of the RSM models, further validation was conducted by analysing the p-values of individual model terms and their interaction effects, as presented in Table 1.5. In RSM, p-values below 0.05 indicate statistically significant effects on the response variable, justifying their inclusion in the regression model. The quadratic model had a significance ($F = 6.53$, $p = 0.0011$). Significant ($p < 0.05$) terms included A (phenol), A^2, and A^2B; however, B (AsO_4^-) ($p = 0.0754$) and AB ($p = 0.1099$) were not significant. Diagnostic plots also supported model adequacy. Figure 1.1, showing the normal probability plot of residuals, indicates that the residual values were scattered along a linear trend and were equally distributed, indicating that the sample followed a normal distribution,

TABLE 1.5 Statistical performance indicators for the fitted RSM model of phenol removal

STD. DEV	*MEAN*	*C.V. %*	*R^2*	*ADJ R^2*	*PRED R^2*	*ADEQ PRECISION*
14.65	60.03	24.40	0.632	0.535	0.377	8.735

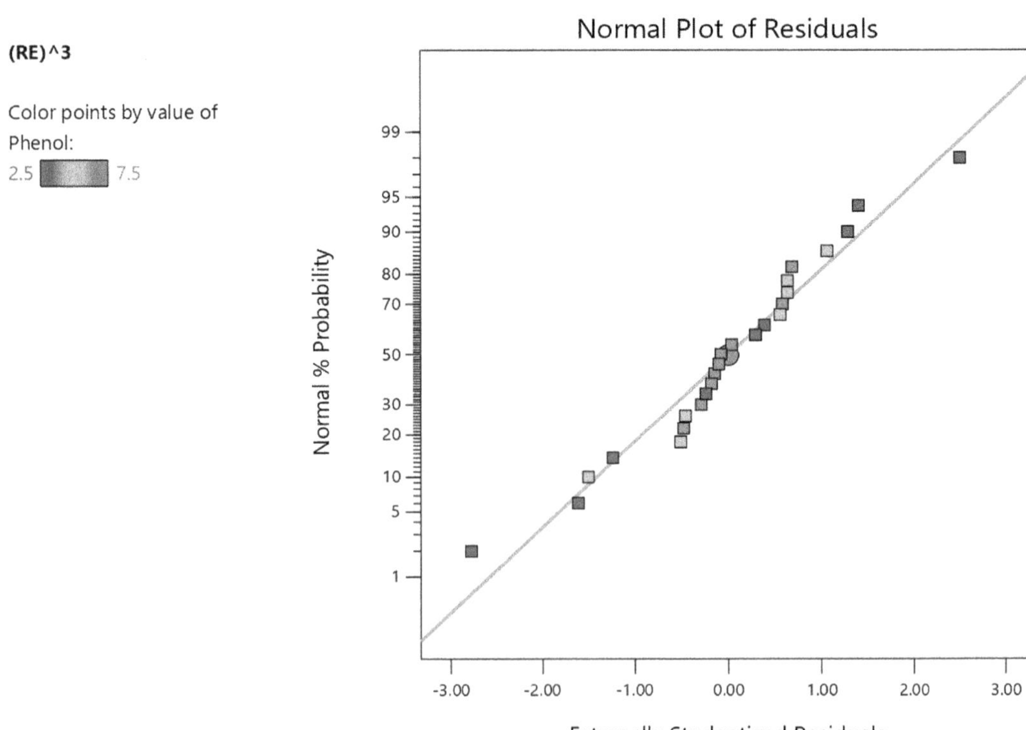

FIGURE 1.1 Normal probability plot of residuals for the RSM phenol removal model.

and this was a major assumption in RSM. Figure 1.2 (predicted vs. actual plot) demonstrates a high level of linearity on the 45° lines, indicating that there was no disagreement between experimental phenol removal and model-predicted phenol removal. Similarly, a random scatter pattern observed in the residuals vs. predicted plot (Figure 1.3) indicates that homoscedasticity and model bias were not present. These diagnostic instruments are widely recognised for validating the assumptions and adequacy of RSM models during their application in environmental engineering.

Factor interactions were also demonstrated in the 3D response surface plots (Figures 1.4 and 1.5). Figure 1.4 reveals that the maximum phenol removal was achieved at 7.5 mg/L phenol and 2.5 mg/L AsO_4^-; however, at greater concentrations of AsO_4^-, phenol removal decreased, probably because of competitive adsorption. Conversely, Figure 1.5 indicates that the variation in the SeO_3^{2-} and SeO_4^{2-} concentrations had little influence on the removal of phenol, as reflected by the flat surface plot, which indicates that phenol can strongly bind to FSGAC even in the presence of other components. These findings are consistent with multicomponent adsorption studies involving phenol and heavy metals. The more pronounced inhibition of phenol adsorption by AsO_4^{3-} likely reflects its stronger competitive affinity for active sites on GAC. AsO_4^{3-}, as a polyvalent anionic oxyanion, forms strong interactions, especially when GAC is functionalised or has positive surface characteristics, and can displace or hinder access for neutral phenol molecules. This behaviour is well documented in studies on AsO_4^{3-} removal using carbon-based adsorbents, where surface impregnation enhances As adsorption through surface complexation mechanisms [16]. In contrast, SeO_3^{2-} and SeO_4^{2-} exhibit weaker adsorption on unmodified carbon surfaces; their uptake is governed mainly by physisorption or van der Waals forces, due to their anionic nature and lower affinity for the neutral carbon matrix [17].

The current study was also evaluated using zero-order, first-order, and second-order kinetics, as shown in Figures 1.6, 1.7, and1.8, respectively. It was found that for zero order, R^2 is 0.8566, whereas for first order, the value of R^2 is 0.9922, and for second order, the value of R^2 is 0.9701. Therefore, the study

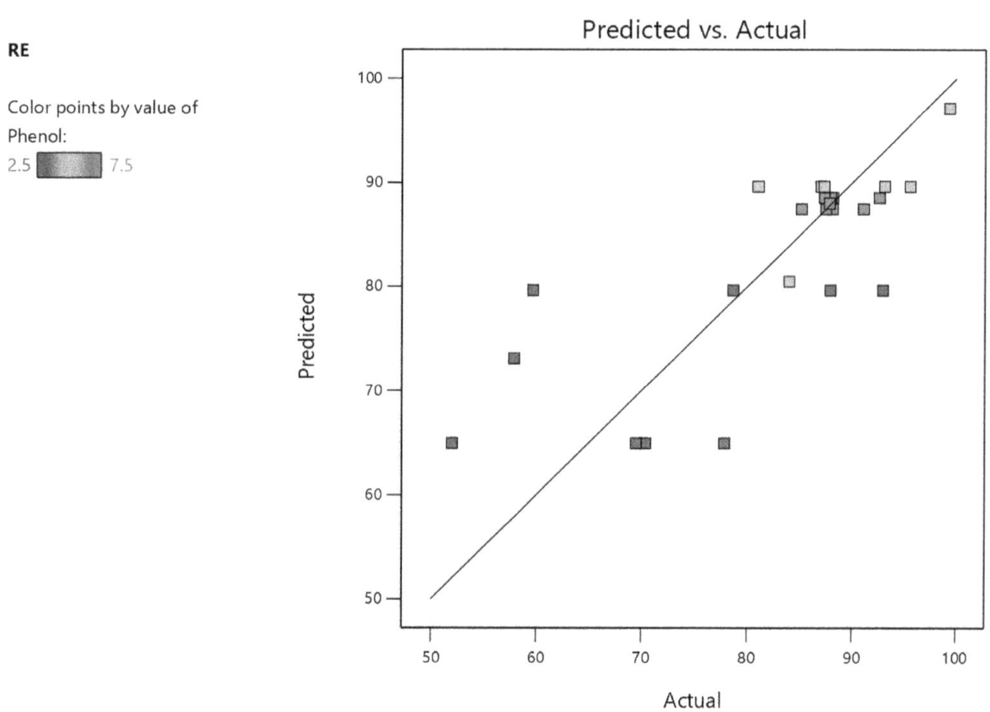

FIGURE 1.2 RSM-based predicted vs. actual plot for the phenol removal model.

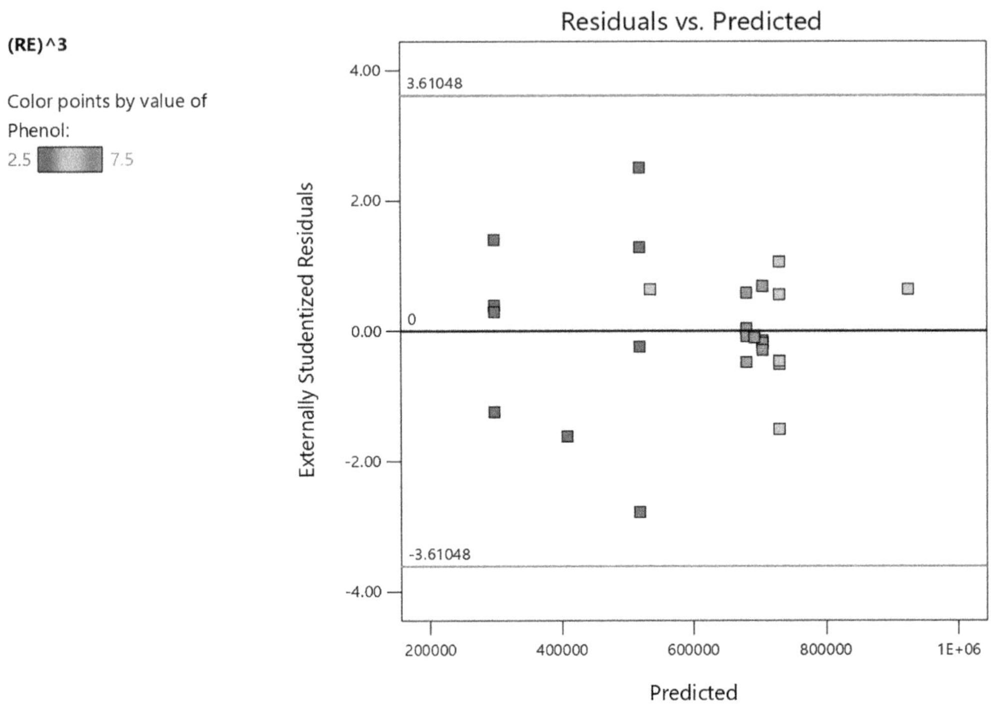

FIGURE 1.3 RSM-based residual vs. predicted plot for the phenol removal efficiency model.

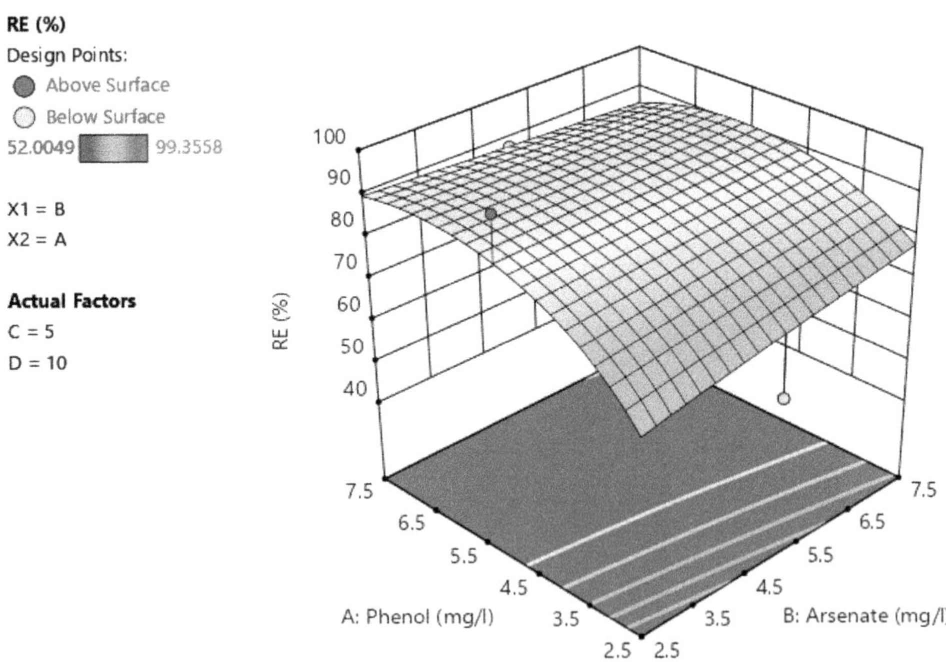

FIGURE 1.4 Response surface plot showing the combined effect of phenol (*A*) and arsenate (*B*) concentrations on phenol removal efficiency (RE%).

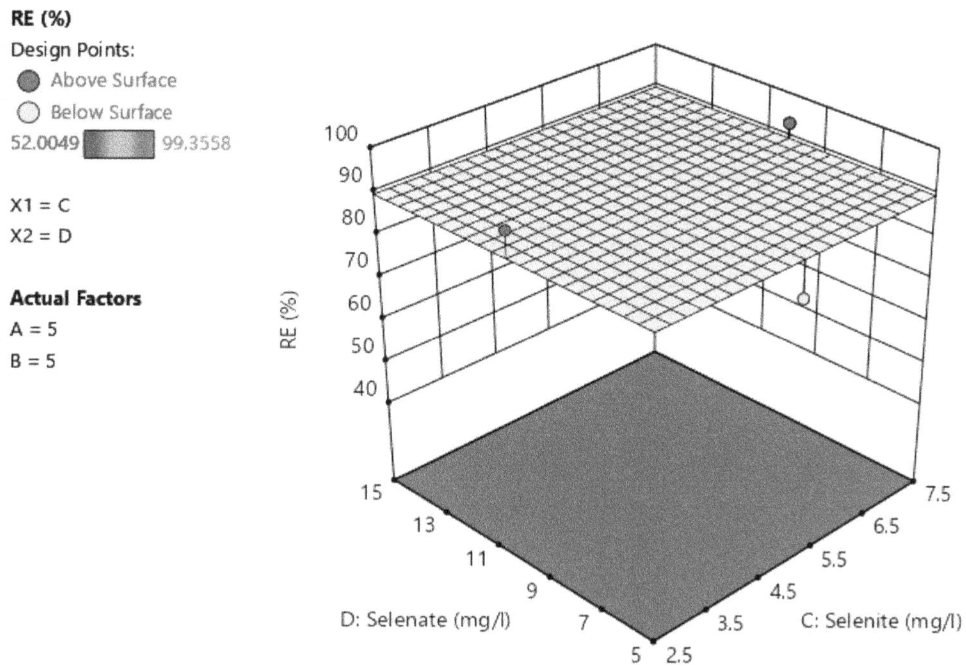

FIGURE 1.5 Response surface plot showing the combined effect of selenite (*C*) and selenate (*D*) concentrations on phenol removal efficiency (RE%).

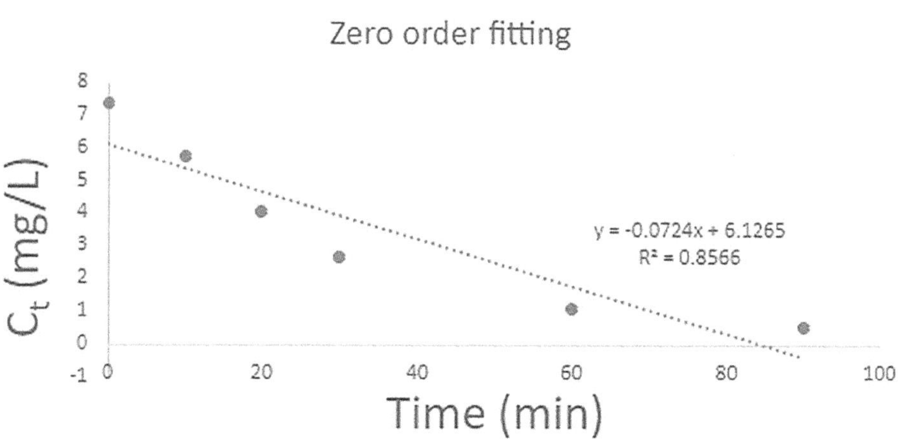

FIGURE 1.6 Zero-order kinetic fit for phenol adsorption onto Filtrasorb GAC.

fits best with a kinetic model. This aligns with findings from recent adsorptive studies, where the PFO model significantly outperformed the PSO model in describing phenol adsorption on the AC surface, pointing toward surface diffusion and weak physisorption as key drivers [18]. Overall, the equilibrium and kinetic tendencies are unified to demonstrate that the adsorption of phenol on FSGAC is highly effective when various oxyanions are present. The high competition effect of AsO_4^{3-} and the better fit for the PFO model give a consistent mechanistic explanation for the uptake process. These insights collectively strengthen the prospects of FSGAC as a reliable adsorbent in polluted waters where phenol coexists with AS and SE species.

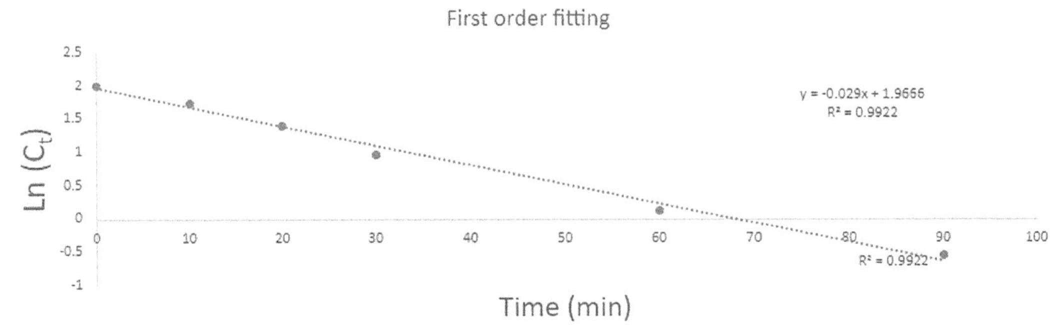

FIGURE 1.7 Pseudo-first-order kinetic model fit for phenol adsorption.

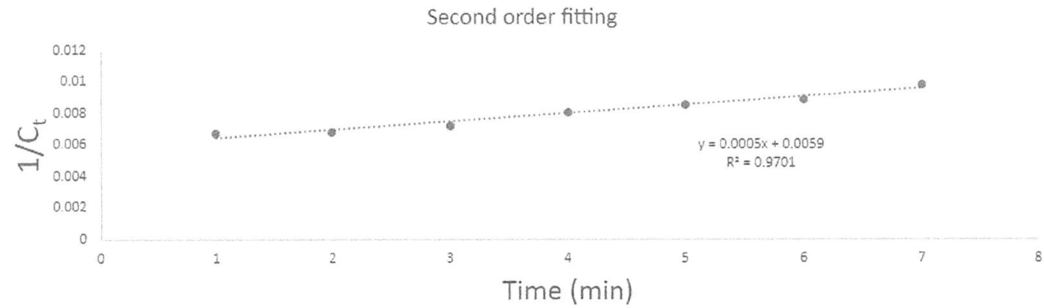

FIGURE 1.8 Pseudo-second-order kinetic fit for phenol adsorption onto Filtrasorb GAC.

1.4 CONCLUSION

This study demonstrated that Filtrasorb F400 granular activated carbon (FSGAC) is an effective adsorbent for phenol removal from aqueous solutions, even under multi-contaminant conditions involving AsO_4^{3-}, SeO_3^{2-}, and SeO_4^{2-}. Among the co-contaminants, AsO_4^{3-} exerted the strongest competitive influence, with the highest phenol removal efficiency (≈99.4%) achieved at 5 mg/L phenol and 2.5 mg/L AsO_4^{3-} (Run 20). In contrast, SeO_3^{2-} and SeO_4^{2-} showed only minor or negligible effects on phenol adsorption within the tested concentration ranges, indicating weak competitive interactions with phenol on the GAC surface. Kinetic modelling of the adsorption process further supported these findings. Among the tested models, the PFO model provided the best statistical fit, indicating that phenol uptake followed predictable, diffusion-related kinetics under mixed-contaminant conditions. The good agreement between experimental and modelled data confirms that the PFO model can be reliably used for designing reactor configurations and predicting adsorption behaviour in practical systems. Overall, the study highlights that phenol can be efficiently and consistently removed from co-contaminated waters using FSGAC, regardless of the presence of selenium species. The combined use of kinetic analysis and RSM optimisation highlights the potential of GAC as a cost-effective and scalable treatment solution for complex wastewaters containing both organic and inorganic pollutants. The findings contribute to the growing body of knowledge on multicomponent adsorption systems and provide a scientific basis for the design, optimisation, and upscaling of GAC-based treatment technologies for real-world environmental applications.

REFERENCES

1. K. G. Pavithra, P. S. Rajan, J. Arun, K. Brindhadevi, Q. H. Le, and A. Pugazhendhi, "A review on recent advancements in extraction, removal and recovery of phenols from phenolic wastewater: challenges and future outlook," *Environ. Res.*, vol. 237, 117005, 2023.
2. D. Wallschläger and N. S. Bloom, "Determination of selenite, selenate and selenocyanate in waters by ion chromatography-hydride generation-atomic fluorescence spectrometry (IC-HG-AFS)," *J. Anal. At. Spectrom.*, vol. 16, no. 11, pp. 1322–1328, 2001, doi: 10.1039/b103108m.
3. H. A. N. Jun, Y. Liang, Z. Bo, Z. Xiong, L. Qin, and W. Chen, "In-situ reaction between arsenic/selenium and minerals in fly ash at high temperature during blended coal combustion," *J. Fuel Chem. Technol.*, vol. 48, no. 11, pp. 1356–1364, 2020.
4. O. E. A. Salam, N. A. Reiad, and M. M. ElShafei, "A study of the removal characteristics of heavy metals from wastewater by low-cost adsorbents," *J. Adv. Res.*, vol. 2, no. 4, pp. 297–303, 2011.
5. Tawfiq. A. Saleh (Ed.), "Kinetic models and thermodynamics of adsorption processes: classification," in *Interface Science and Technology*, Elsevier, 2022, pp. 65–97.
6. M. A. Bezerra, R. E. Santelli, E. P. Oliveira, L. S. Villar, and L. A. Escaleira, "Response surface methodology (RSM) as a tool for optimization in analytical chemistry," *Talanta*, vol. 76, no. 5, pp. 965–977, 2008.
7. M. Vohra, A. Basaleh, F. Kabir, M. Al-Suwaiyan, and M. Bilal, "Aqueous bromate (BrO_3^-) removal using FILTRASORB 400 activated carbon: effect of process variables and modeling," *Desalin. Water Treat.*, vol. 299, pp. 102–109, 2023.
8. M. Bilal, S. Kaur, and J. Mushtaq, "Adsorption investigation of physico-chemical characteristics of different water samples by coalesce coconut shell," *Int. J. Environ. Eng.*, vol. 12, no. 1, pp. 1–14, 2023.
9. M. Roosta, M. Ghaedi, A. Daneshfar, R. Sahraei, and A. Asghari, "Optimization of combined ultrasonic assisted/tin sulfide nanoparticle loaded on activated carbon removal of erythrosine by response surface methodology," *J. Ind. Eng. Chem.*, vol. 21, pp. 459–469, 2015.
10. Q. Li et al., "Biochar and GAC intensify anaerobic phenol degradation via distinctive adsorption and conductive properties," *J. Hazard. Mater.*, vol. 405, p. 124183, 2021.
11. M. Sadi, A. Namane, Y. M. Azri, and I. Tou, "Adsorption and desorption of phenol onto granular activated carbon: Kinetics and thermodynamic studies," *J. Solution Chem.*, 54, pp. 1–27, 2025.
12. S. O. Okonji, J. A. Dominic, D. Pernitsky, and G. Achari, "Removal and recovery of selenium species from wastewater: Adsorption kinetics and co-precipitation mechanisms," *J. Water Process Eng.*, vol. 38, p. 101666, 2020.
13. B. K. Nandi, A. Goswami, and M. K. Purkait, "Adsorption characteristics of brilliant green dye on kaolin," *J. Hazard. Mater.*, vol. 161, no. 1, pp. 387–395, 2009.
14. M. Salari, M. H. Dehghani, A. Azari, M. D. Motevalli, A. Shabanloo, and I. Ali, "High performance removal of phenol from aqueous solution by magnetic chitosan based on response surface methodology and genetic algorithm," *J. Mol. Liq.*, vol. 285, pp. 146–157, 2019.
15. E. Mohseni, A. Rahmani, Z. Hamdi, and P. Ghorbanzadeh, "Efficiency of response surface method based on central composite design in optimizing the adsorption process of cadmium metal from aqueous solutions using activated charcoal prepared from walnut shell," *Desalin. Water Treat.*, vol. 317, p. 100250, 2024.
16. G. T. Schmidt, N. Vlasova, D. Zuzaan, M. Kersten, and B. Daus, "Adsorption mechanism of arsenate by zirconyl-functionalized activated carbon," *J. Colloid Interface Sci.*, vol. 317, no. 1, pp. 228–234, 2008.
17. O. K. L. Strong, H. E. France, K. Scotland, K. Wright, and A. J. Vreugdenhil, "Selenite adsorption and reduction via iron (II) impregnated activated carbon produced from the phosphoric acid activation of construction waste wood," *Arch. Environ. Contam. Toxicol.*, vol. 85, no. 4, pp. 485–497, 2023.
18. X. Wei et al., "Adsorption of phenol from aqueous solution on activated carbons prepared from antibiotic mycelial residues and traditional biomass," *Fuel Process. Technol.*, vol. 242, p. 107663, 2023.

Responsible AI in Action
Trends for Ethical and Sustainable Smart Technology

2

Pankhuri Khurana, Titiksha Chugh, Amanpreet Kaur, and Neha Sharma

2.1 INTRODUCTION

The search for "Artificial Intelligence (AI) ethics" on Google Trends in the worldwide category over the past 5 years is shown in Figure 2.1, which shows an increase in AI trends from previous years. As there is advancement in AI with Generative Adversarial Network (GAN), autoencoders, NLP, and various technologies, there is a need to address ethical concerns. AI systems require a large amount of energy for training algorithms, which leads to significant environmental and societal impacts. Urban water systems face sustainability and resiliency challenges, including water leaks, overuse, quality issues, and responses to drought and natural disasters. To tackle digital pollution effectively, it is vital to study both the quantitative dimension of its usage and the ethical considerations, such as the various roles and responsibilities of the different individuals involved. The influence of digital knowledge extends to additional domains, including the environment, health, and energy consumption. These recent technologies can impact the accessibility of finite natural resources, as well as the social and economic dimensions, and the proper disposal of digital devices. The environmental consequences and overall importance of AI technology are complex and require a comprehensive strategy.

The domain of AI ethics is experiencing tremendous growth [1], causing emerging topics such as sustainability, digital medicine, privacy, and the responsibility of autonomous systems. These moral selections may not be resolved exclusively by technological approaches. Rather, a socio-technical approach that incorporates technical answers within wider ethical as well as social frameworks is needed. Designers can proactively anticipate future intermediate roles and employ moral imagination

 DOI: 10.1201/9781003714217-2

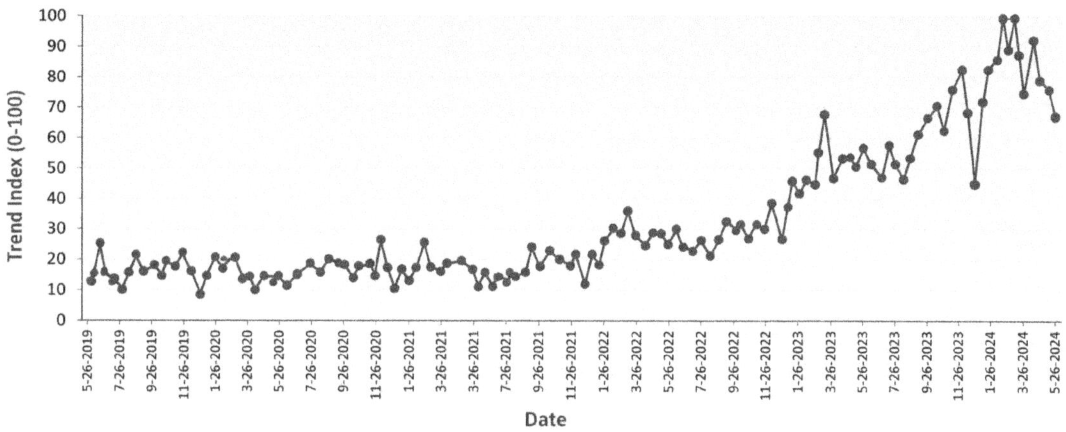

FIGURE 2.1 Artificial intelligence (AI) ethics trends worldwide.

FIGURE 2.2 AI domain insights with trends and patterns.

to include principles into advanced AI technology. Shaping the existing technology can possibly be controlled by policymakers and collaboratively encouraging open public discourse [2,3]. Optimization of water system sustainability is growing with smart technologies such as IoT, AI, and explainable AI (XAI) in several ways, for example, water distribution, detecting leaks, and monitoring quality. Consequently, these systems improve the environmental resilience, increase efficacy, and manage resources. As Shapley additive explanations has been applied to hydrological methods to understand the influence of meteorological variables on the prediction of flood risks, it helps to improve transparency and accuracy [4]. Figure 2.2 presents trends and patterns in the AI domain, and Figure 2.3 shows the schematic workflow of the chapter. The key contributions of the chapter are

a. The chapter focuses on the influence of AI on human decision-making, digital medicine, sustainability, environmental impact, cybersecurity, and many more.

b. The present review offers insight into the main concepts and issues that can assist researchers to better understand trends and focus in the AI area.
c. The chapter covers the recent developments in AI in various domains. The chapter discusses the role of XAI role in sustainable water systems, highlights various techniques, and discusses real-world applications as well.

2.2 RELATED WORK

Despite the existing studies, AI-driven technology is still growing rapidly in this domain and working with new applications like GenAI, LLM, NLP, and many other domains. Thus, security, responsibility, and ethics need to be taken more carefully into account. Because today every individual is using advanced technology, some state-of-the-art work is presented in Table 2.1.

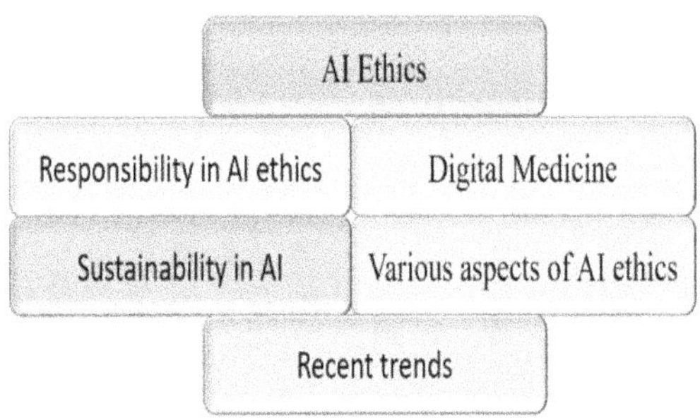

FIGURE 2.3 Schematic workflow for the chapter.

TABLE 2.1 State-of-the-art techniques in AI

REF. NO.	*FINDINGS*
[5] 2025	Discussion on how XAI techniques (SHAP and LIME) are applied to hydrological models to enhance flood prediction, leak detection, and water quality monitoring.
[18] 2023	Outlines how AI can improve water supply insights, catchment management, and emergency response.
[19] 2024	Presents a five-year investigation that included many participants to analyze ethical concerns in AI and its application in public relations and professional communication.
[20] 2024	Objective is AI education that provides understanding of AI concepts, required information, as well as proficiency in the basic principles of AI.
[21] 2024	The paper discusses a thorough examination of specific empirical research on the use of AI technology in language instruction, considering the significant progress.
[22] 2024	Using video cameras at home to record patient data is frequently met with concerns regarding privacy and ethics.
[23] 2024	Challenges in medicine and dermatology include transparency, ethics, security, and inequality, as biased algorithms can lead to inequalities in results.
[24] 2024	Ethical implications of integrating AI into healthcare systems are discussed in this paper, focusing on privacy, bias, and decision-making transparency in the evolving landscape of AI applications.

(Continued)

TABLE 2.1 (*Continued*) State-of-the-art techniques in AI

REF. NO.	*FINDINGS*
[25] 2024	This paper analyzes current trends in the use of GeoAI for cartographic design.
[26] 2024	AI, a key technology of the fourth industrial revolution, has revolutionized society and raised ethical fears regarding human rights, social order, and personal safety.
[27] 2024	This paper explores the complex relationship between AI and intellectual property, addressing legal and ethical issues in the rapidly evolving digital landscape.
[28] 2024	A literature review offers insights into the ethical implications of AI.
[29] 2024	Identification done with stakeholders and implementation of AI technology, working with Screenwriters Rights Association of India (SRAI) theory and practical implications with research directions.
[30] 2021	Highlights the AI approach for Green AI in smart city transformation with different methodologies, including literature, developments, operations, and trends.
[31] 2025	Discusses a multi-dimensional framework that connects AI-based applications along with in-depth analysis of ecological connections and algorithms with social risks.
[32] 2025	Enhances the concept of a sustainable city environment with various opportunities and understanding of innovation.
[33] 2025	Discusses biodegradable products and reducing the lifecycle impact through collaborative platforms, best practices, and innovations with resources.
[34] 2025	AI-based innovative solutions for enhancing biodiversity conservation and sustainable practices, along with goals related to sustainability globally.
[35] 2025	AI-based transformative personalized learning and administrative efficiency, addressing privacy concerns related to credential devaluation, robust governance, capacity building, and AI-based educational values.
[36] 2025	AI-based improvements for decision-making, cost-saving, and citizen-centric services with advancements in social impact and recent trends.
[37] 2023	Highlights AI innovation, framing of AI to solve RI-related issues, and sensitivity to context.
[38] 2021	Focus on the digital society, technology, supply chain, and e-business.

2.3 METHODOLOGY

The present study addresses AI ethics and its responsibilities. This chapter adopts a standard methodological approach that encompasses a systematic method used to analyze developments in AI ethics, as shown in Figure 2.4. It demonstrates the workflow starting with planning relevant topics, selecting appropriate articles using specific keywords, extracting essential information, and lastly, presenting the study systematically. This process ensures an organized and comprehensive review of the latest developments in AI ethics.

- **Planning**: Planning includes the ethical guidelines based on the life cycle and practical implications at every stage of AI development. AI-based tools are used in the transformation from ethical values to concrete rules.
- **Selection**: AI-based studies work with reproduced genders and intersectional biases that are designed to be neutral. Research highlights the importance of AI and human collaboration. These selections are based on third-party audits and clear explanations and decisions.
- **Extraction**: Extraction focuses on embedding ethics pipelines within a privacy-by-design framework about consents, privacy, risks associated with large-scale data with ethics of AI

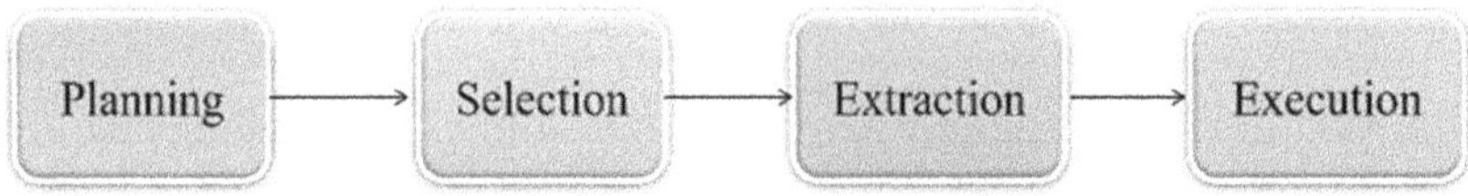

FIGURE 2.4 Systematic approaches used to analyze recent advancements in AI ethics.

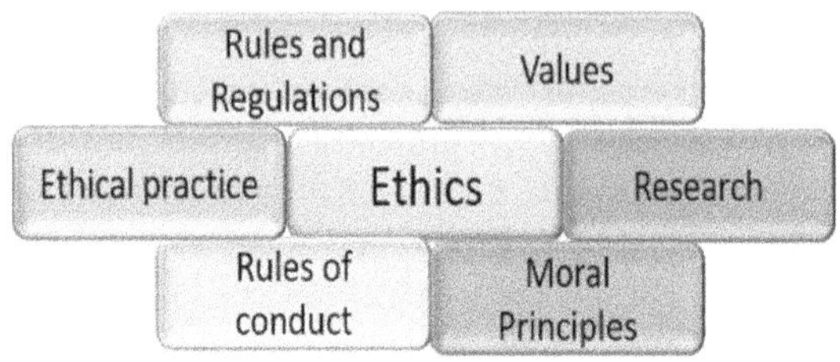

FIGURE 2.5 Some ethical considerations of AI.

models. Various global bodies, such as UNESCO, work toward data rights and emerging trends for improvement related to data source tracking. A review in the year 2025 will focus on AI data processing in humanitarian crises, addressing major concerns such as informed consent, vulnerability of affected populations, risk of re-identification, long-term harms.

- **Execution**: AI ethics execution prevents life-and-death decisions such as warnings in criminal sentencing cases, which pose high risks due to bias, lack of transparency, and loss of human judgment. Courts and policymakers adopt and emphasize life-and-death decisions.

2.4 RESPONSIBILITY IN AI ETHICS

The ethics of AI is integrally connected to the idea of responsibility. Responsibility in ethics exists in both active and passive forms. Passive responsibility arises after a disagreeable occurrence has happened; on the other hand, active responsibility relates to acts performed before the event. Furthermore, confidentiality assurances state that an individual with proper authorization can access the information.

Active responsibility works proactively to reduce the negative outcomes of technology while acknowledging its positive effects. Responsibility in creating AI technology is crucial aspect because it can affect or benefit society and individuals in numerous ways. Active responsibility is an intentional strategy in design that includes ethical considerations from the beginning of the design process, so that they can be integrated into the technology itself. Moreover, active responsibility in AI design ensures that ethical considerations—such as equitable water distribution and environmental resilience—are embedded from the outset of smart water system development [5]. Additionally, value-sensitive design is part of the active responsibility concept, focusing on moral considerations and values to guide technological progress and technology design. However, it requires consideration of stakeholders with conflicting interests and predicting the trajectory of technological progress [6,7]. The discussion turns to environmentally friendly AI software and hardware systems. Addressing the issue of digital pollution and promoting the sustainability of AI involves technological, ethical, and political considerations that necessitate a global approach. Both individual responsibility and governmental institutions are vital in tackling this problem [8,9]. Figure 2.5 presents some aspects of AI ethics. XAI fosters stakeholder trust in urban water management by clarifying decision-making processes, thereby reinforcing ethical accountability in resource allocation and contamination alerts.

Furthermore, the ethics surrounding autonomous cars give rise to diverse ethical issues, one of which is the concept of operating autonomy [10,11], enabling systems to perform tasks without human intervention. Autonomous cars can be characterized into five separate levels, commencing at L1 (driver assistance) and ending at L5 (fully automatic).

- **Fairness and Non- Discrimination**: AI is prohibited from discriminating on the basis of socioeconomic background, gender, race, or age. Diverse, comparable information should be utilized for training algorithms.
- **Accountability**: AI decisions have to be made by humans, not through machines. There should be clear processes in place to examine and fix AI problems.
- **Privacy and Data Protection**: AI must adhere to consent rules and regulations, preserve private data, and avoid exploitation or negative surveillance.
- **Safety and Reliability**: AI ought to function effectively, securely, and accurately. It cannot result in financial, psychological, or physical harm.
- **Human Control**: AI needs to continue to be meaningfully regulated by humans. Autonomous, irreversible death or life choices should not be decided by machines.
- **Transparency**: AI systems need to be transparent and understandable. Particularly in high-stakes fields such as medical treatment, justice, or finance, users need to understand how decisions are made.
- **Beneficence**: AI should not be used to harm or exploit individuals, but instead strive for improving society and human well-being.
- **Inclusiveness**: Multiple stakeholders need to be involved in AI development, and its implications on marginalized groups must be taken into account.

The ethical implications of autonomous vehicle development involve both advantages and disadvantages. One reason is to enhance transportation security and minimize accidents, while another is to enhance the independence of individuals such as the elderly and optimize traffic flow. Nevertheless, the advancement also poses risks to personal privacy, liberty, and the assignment of responsibility [12,13]. Unavoidable collisions can be observed through numerous ethical frameworks that include deontological ethics, consequentialism, and utilitarianism. Consequentialism decides what is good or bad based on its outcome, while deontological ethics studies rules to differentiate correct from incorrect. Duty ethics can be useful when the safety of a child is prioritized over other matters, such as the rights of other persons and potential societal conflicts. In Figure 2.6, collaboration with AI and humans is illustrated with gaps mentioned.

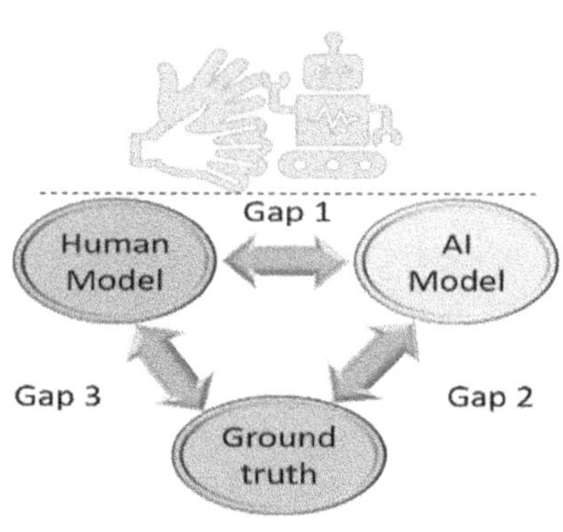

FIGURE 2.6 AI–human collaboration with gaps in between them.

Another issue relates to the distribution of responsibility, specifically determining who is held accountable for any harm caused to individuals or property. This term is particularly difficult because none of the individuals involved possess direct control over the autonomous vehicle.

a. **Data Bias vs. Real-World Truth**
    - Ground truth represents the actual reality.
    - Historical data, incomplete data.
b. **Human Interpretation Differences**
    - Humans interpret information.
    - Opposite judgments by different people.
    - Human model, ground truth, creating natural variation.
c. **AI Generalization Errors**
    - AI models approximate patterns.
    - Incorrect guesses.
    - Leading to AI truth gaps.
d. **Human Cognitive Bias**
    - Overconfidence, stereotypes, confirmation bias.
    - Interpretation may diverge from ground truth.
    - Creating human–ground truth gaps.
e. **Mismatching in Problem Framing**
    - The defining of a task by humans may differ.
    - Humans judge fairness morally and mathematically.
f. **Lack of Shared Context**
    - Humans use emotional and cultural situations with data patterns.
    - Diverse inputs/outputs with huge gaps.

2.5 VARIOUS ASPECTS OF AI ETHICS

This study discovers the ethical considerations surrounding the growth and use of AI products. Codes of ethics are essential actions in the field of professional ethics, which pertain to the obligations, benefits, and responsibilities of professionals. The standards establish essential criteria for AI systems to be considered reliable, including the empowerment of human beings, the guarantee of privacy and data security, and the assurance of transparency. AI systems must prioritize sustainability as well as environmental friendliness. These codes are developed by organizations such as professional associations or companies to enhance ethical awareness, define explaining ethical standards and principles, promote responsibility, and enhance the reputation of a profession or firm.

2.5.1 AI and Human Collaboration AI

It is increasingly being employed in various fields, including automated job placement systems and crime forecasting programs. These systems use a mix of human and automated decision-making, employing a bottom–up approach that uses machine learning (ML) techniques to gather knowledge from experience and draw new conclusions. Classification algorithms enable the prediction of the category of an unfamiliar object by choosing from a pre-established set of recognized groupings. However, there are difficulties that need to be overcome when it comes to integrating AI into human decision-making processes. For instance, the algorithm must undergo training using a substantial number of instances that are abundant in quantity and accurately represent the population on which the algorithm will base its decision-making. Another issue is that the data samples used to train the algorithm do not adequately represent the target population in a particular scenario. Consequently, the technique lacks the ability to make generalizations and is thus unsuitable for application in other fields. Transparency is essential in decision-making processes as it allows individuals to preserve their privacy and take control over unbiased procedures. Engineers must also take into account how future technologies will mediate ethical problems. XAI provides many techniques to overcome this constraint, sometimes by sacrificing accuracy to prioritize interpretability. Moreover, ML algorithms employed for decision-making might potentially be influenced by the issue of discrimination. Collaborating with human creators enables GenAI to achieve its maximum degree of productivity. This method recognizes AI as a resource to progress human creativity rather than extinguish it. AI can serve as a creative advisor, a source of motivation, or a means to automate tedious tasks for writers, artists, and designers. The fusion of AI capabilities with human perception often yields remarkable and innovative results. Human–AI cooperation involves problem-solving between humans and computers to accomplish mutual goals. Researchers predict AI enhancing and complementing human activities, as well as having proven potential in several fields [14]. Figure 2.6 shows the AI and human collaboration and its gaps.

2.5.2 Sustainable Water System

Additional developing dimensions of AI ethics deceive in its application to sustainable urban infrastructure, mainly related to water systems. Techniques such as SHAP (Shapley additive explanations) and LIME are increasingly used to interpret hydrological models that enable transparent flood risk prediction as well as water resource optimization [4]. These XAI methods not only improve the accuracy of the model but also foster stakeholders' trust in it by expounding the influence of meteorological variables. Morally, this aligns with the principles of environmental justice and responsible innovation, ensuring that smart technologies contribute to equitable and sustainable results in public infrastructure.

2.5.3 AI and Cybersecurity

The safeguarding of personal information and technological systems, as well as the individuals that utilize them, is included in cybersecurity. The aim of cybersecurity can be defined by its acronym CIA: C signifies confidentiality, represents integrity, and A means the availability of information. This means cybersecurity ensures that information is exchanged with these three main objectives: confidentiality, integrity, and availability. Furthermore, defining confidentiality ensures that individuals with proper authorization can access the information. Integrity prevents any unauthorized alterations to the information. Availability confirms that authorized users can always access the information without any disruptions. Cybersecurity also relates to the relationship between security and ethics. The more concise response is that it is incorrect to gain entry into systems for which one lacks authorization. However, it is vital to emphasize that security has a significant impact on maintaining order. It is difficult to guarantee an equal level of confidentiality, integrity, and accessibility of information to all solutions, which presents challenges for people and individuals. One case study related to these conditions is where the importance of safeguarding security clashes with the importance of preserving privacy. The US Patriot Act, which was authorized in 2001 following the events of September 11, aimed to bestow the Federal Government with expanded authority for conducting electronic surveillance, and subsequently some companies utilized the information inappropriately. Thus, it is critical to acknowledge that security plays an important role in ensuring the efficient and dependable operation of information societies. Another situation in which ethical concerns may potentially arise is known as ethical hacking. Figure 2.7 shows the research paper publications for the year 2024 with different keywords, as shown

FIGURE 2.7 Research paper publications for the year 2024 using different keywords on Google Scholar.

in the diagram. Ethical hacking refers to the intentional breach of cybersecurity protocols to achieve socially beneficial outcomes.

Furthermore, hackers argue that their attempts to penetrate computer systems are frequently advantageous as they bring to light security vulnerabilities to individuals who possess the ability to address them. Hackers penetrate networks purely for the purpose of gaining unauthorized access, without any intention of stealing or causing harm. Nevertheless, individuals can suffer injury just from unauthorized intrusion that infringes on intellectual rights and privacy rights. Also, hackers assert that they contribute to preventing excessive surveillance by using their specialized knowledge to uncover illicit or unethical activities involving information technology. Every individual needs protection of their information to avoid its misuse.

The significance of cybersecurity is particularly crucial in the context of AI. Besides the "Ethics guidelines for trustworthy AI" created by the High-Level Expert Group designated by the European Commission, there are various concepts in relation to cybersecurity. It addresses the guidelines for ethical problems in cybersecurity with AI [15] in terms of harm prevention, fairness, respect for human autonomy, etc.

2.5.4 AI in Digital Medicine

The use of AI and big data in digital medicine is aimed at improving patient outcomes and healthcare delivery. The discipline of digital medicine is revolutionizing the application and comprehension of medical science, as AI tools present the opportunity to efficiently manage increasing quantities of medical data. Nevertheless, ethical concerns emerge from a moral perspective, including the formulation of recommendations by international health organizations, the complexities associated with the transparency and reliability of AI algorithms, and the safeguarding of data privacy [16]. The four primary domains of application include

- Health monitoring,
- Treatment predicting algorithms,
- Diagnostic assistance tools, and
- Public health applications.

ML systems have a remarkable ability to understand radiological and clinical images. This permits the introduction of automated diagnostic approaches to improve diagnostic accuracy, decrease the workload of professionals, and thereby increase the number of people who can benefit from precautionary initiatives. The main ethical issues that generate the utmost difference of opinions include data security and privacy. Privacy concerns arise as there is a need to inform individuals about the use of their data and the training of algorithms with huge data sets that might possibly cause more harm to marginalized and underprivileged groups. The World Health Organization (WHO) has released clear rules for the ethical use of black-box algorithms in the medical sector. These recommendations comprise avoiding negative impacts, protecting privacy, maintaining traditional healthcare procedures, and confirming human supervision of medical activity. Furthermore, the term "technological mediation" refers to the process of effectively carrying out its intended purposes while also actively forming the behaviors and perspectives of users [17].

2.6 FUTURE SCOPE

Improved Ground-Truth Generation includes the development of better labeling methods using expert consensus, multi-annotator models, and real-time validation. For human and AI collaborative decision systems, mention the design of hybrid systems where AI assists and humans validate, reducing errors on both sides, and research related to humans should override AI and when AI should support human decisions. Moreover, bias reduction and fairness frameworks have advanced techniques to detect, measure, and correct biases in both AI predictions and human judgments for fairness-aware training that aligns models more closely with human ethical standards and real-world truth. Here is cognitive modeling of humans creating models that simulate human thinking patterns to understand why humans differ from ground truth. Multi-modal truth verification that combines audio, video, text, sensor data, and expert analysis to improve the accuracy of truth estimation.

2.7 CONCLUSION

AI embraces enormous potential to transform numerous domains, such as human decision-making, digital medicine, sustainability, environmental impact, and cybersecurity. However, the rapid advancements in the field of AI technology necessitate a robust and smart emphasis on ethical considerations and responsible practices. As AI continues to progress, it is essential to address the ethical challenges and ensure the responsible use of AI technologies. AI development will contribute to a more equitable, transparent, and sustainable future, benefiting society as a whole. Due to the extensive encouragement of AI for society, it is important to redirect attention from individuals to social groups to prepare ethical strategies that reflect the shared values among all other parties involved. AI's ethical arrangement in sustainable water systems can support SDGs associated with climate resilience and clean water.

REFERENCES

1. Zemčík, M. A Brief History of Chatbots. *DEStech Transactions on Computer Science and Engineering*, Volume 8, Page No. 14–18, 2019.
2. Boddington, P. *Towards a Code of Ethics for Artificial Intelligence.* Springer, 2017.

3. Sonko, S., Adewusi, A. O., Obi, O. C., Onwusinkwue, S., & Atadoga, A. A Critical Review Towards Artificial General Intelligence: Challenges, Ethical Considerations, and the Path Forward. *World Journal of Advanced Research and Reviews*, Volume 21, Issue 3, Page No. 1262–1268, 2024.
4. Kapalia, M., Kaur, A., & Sharma, N. Harnessing Deep Learning: Innovations and Practical Applications. In *2024 3rd Edition of IEEE Delhi Section Flagship Conference (DELCON)* Page No. 1–6. IEEE, 2024.
5. Infant, S. S., Vickram, S., Saravanan, A., Muthu, C. M., & Yuarajan, D. Explainable Artificial Intelligence for Sustainable Urban Water Systems Engineering. *Results in Engineering*, Volume 25, Article No. 104349, 2025.
6. Gusenbauer, M., & Haddaway, N. R. Which Academic Search Systems Are Suitable for Systematic Reviews or Meta-Analyses? *Research Synthesis Methods*, Volume 11, Page No. 181–217, 2020.
7. Delgado López-Cózar, E., Orduña-Malea, E., & Martín-Martín, A. Google Scholar as a Data Source for Research Assessment. In *Springer Handbook of Science and Technology Indicators*, Page No. 95–127, 2019.
8. Bovens, M. A. *The Quest for Responsibility: Accountability and Citizenship in Complex Organisations.* Cambridge University Press, 1998.
9. Van de Poel, I., & Royakkers, L. *Ethics, Technology, and Engineering: An Introduction.* Wiley, 2023.
10. Lucivero, F. Sostenibilità Ambientale della Società dell'Informazione. In *Automi e Persone*, Carocci Editore, ISBN: 978882901170-4, Page No. 112–127, 2021.
11. AI HLEG. *Ethics Guidelines for Trustworthy AI.* European Commission, 2019.
12. Di Fabio, U., Broy, M., Brüngger, R. J., Eichhorn, U., & Grunwald, A. *Ethics Commission Automated and Connected Driving.* Federal Ministry of Transport and Digital Infrastructure, 2017.
13. Jedlickova, I. Ensuring Ethical Standards in the Development of Autonomous and Intelligent Systems. *IEEE Transactions on Artificial Intelligence*, Volume 5, Page No. 5863–5872, 2024.
14. Vössing, M., Kühl, N., Lind, M., & Satzger, G. Designing Transparency for Effective Human–AI Collaboration. *Information Systems Frontiers*, Volume 24, Page No. 877–895, 2022.
15. Christen, M., Gordijn, B., & Loi, M. *The Ethics of Cybersecurity.* Springer Nature, 2020.
16. Fossa, F., Tamburrini, G., & Schiaffonati, V. *Automi e Persone: Introduzione all'Etica dell'Intelligenza Artificiale e della Robotica*, Carocci, 2021.
17. Verbeek, P. P. *Moralizing Technology: Understanding and Designing the Morality of Things.* University of Chicago Press, 2011.
18. Richards, C. E., Tzachor, A., Avin, S., & Fenner, R. Rewards, Risks and Responsible Deployment of Artificial Intelligence in Water Systems. *Nature Water*, Volume 1, Page No. 422–432, 2023.
19. Bowen, S. A. *"If It Can Be Done, It Will Be Done:" AI Ethics and Public Relations Implications.* SSRN, 2024.
20. Lee, S. J., & Kwon, K. A Systematic Review of AI Education in K–12 Classrooms from 2018 to 2023. *Computers and Education: Artificial Intelligence*, Volume 6, Page No. 100211, 2024.
21. Zhu, M., & Wang, C. *A Systematic Review of Artificial Intelligence in Language Education from 2013 to 2023.* SSRN, 2024.
22. Moore, J., McMeekin, P., Parkes, T., Walker, R., Morris, R., Stuart, S., Hetherington, V., & Godfrey, A. Contextualizing Remote Fall Risk: Video Data Capture and Implementing Ethical AI. *NPJ Digital Medicine*, Volume 7, Page No. 61, 2024.
23. Grzybowski, K., Jin, K., & Wu, H. Challenges of Artificial Intelligence in Medicine and Dermatology. *Clinical Dermatology*, Volume 42, Issue 3, Page No. 210–215, 2024.
24. Kasula, Y. Ethical Implications and Future Prospects of Artificial Intelligence in Healthcare: A Research Synthesis. *International Meridian Journal*, Volume 6, Page No. 1–7, 2024.
25. Kang, Y., Gao, S., & Roth, R. E. Artificial Intelligence Studies in Cartography: A Review and Synthesis of Methods, Applications, and Ethics. *Cartography and Geographic Information Science*, Volume 51, Page No. 599–630, 2024.
26. Chen, H. The Ethical Challenges of Educational Artificial Intelligence and Coping Measures. *Science Insights Education Frontiers*, Volume 20, Page No. 3263–3281, 2024.
27. Abdallah, M., & Salah, M. Artificial Intelligence and Intellectual Properties: Legal and Ethical Considerations. *International Journal of Intelligent Systems and Applications in Engineering*, Volume 12, Page No. 368–376, 2024.
28. Giarmoleo, V., Ferrero, I., Rocchi, M., & Pellegrini, M. M. What Ethics Can Say on Artificial Intelligence: Insights from a Systematic Literature Review. *Business and Society Review*, Volume 129, Page No. 258–292, 2024.

29. Chang, Y. L., & Ke, J. Socially Responsible Artificial Intelligence Empowered People Analytics: A Novel Framework Towards Sustainability. *Human Resource Development Review*, Volume 23, Page No. 88–120, 2024.
30. Yigitcanlar, T., Mehmood, R., & Corchado, J. M. Green Artificial Intelligence: Towards an Efficient, Sustainable and Equitable Technology for Smart Cities. *Sustainability*, Volume 13, Issue 16, Article 8952, Page No. 1–14, 2021.
31. Toderas, M. Artificial Intelligence for Sustainability: A Systematic Review and Critical Analysis of AI Applications, Challenges, and Future Directions. *Sustainability*, Volume 17, No. 17, Article 8049, 2025.
32. Brown, M. Corporate Social Responsibility and AI and Their Impact on Smart Cities. In *Transforming Corporate Social Responsibility and Business Ethics with AI*, edited by Martin Sposato and Eduardo Carlos Dittmar, IGI Global Scientific Publishing, 2025.
33. Nandhini, R., Chandra, S., Pandey, S., Selvakumar, P., Srivastava, P., & TC, M. Future Trends in AI and Sustainable Business Practices. In *AI-Driven Business Model Innovation*, edited by Maryam Khokhar, IGI Global Scientific Publishing, Page No. 191–218, 2025.
34. Samanvitha, C., Dharmendra, H., Nethravathi, N., & Velmurugan, P. R. Artificial Intelligence for Sustainable Futures: Navigating Social and Environmental Responsibility. In *Advancements in Intelligent Process Automation*, edited by Dhanabalan Thangam, IGI Global, Page No. 129–154, 2025.
35. Meilani, Y., Ardi, A., Sunarjo, R. A., Berlianto, M., & Meyvanali, M. *Human Technology Sustainability: Leveraging Artificial Intelligence in Education for Inclusive and Ethical Futures*. SSRN, 2025.
36. Panda, M., Hossain, M. M., Puri, R., & Ahmad, A. *Artificial Intelligence in Action: Shaping the Future of the Public Sector*. Digital Policy, Regulation and Governance, 2025.
37. Lukkien, D. R., Nap, H. H., Buimer, H. P., Peine, A., Boon, W. P., Ket, J. C., ... Moors, E. H. Toward Responsible Artificial Intelligence in Long-Term Care: A Scoping Review on Practical Approaches. *The Gerontologist*, 63(1), Page No.155–168, 2023.
38. Dennehy, D., Griva, A., Pouloudi, N., Dwivedi, Y. K., Pappas, I., & Mäntymäki, M., editors. (2021). *Responsible AI and Analytics for an Ethical and Inclusive Digitized Society* , Springer International Publishing, Page No. 8–26, 2021.

IoT and Wireless Sensor Networks for Real-Time Water Monitoring

3

Mayank Attri, Kapil Sethi, Lokeshwer Sharma, Abhineet Anand, and Anu Thakur

3.1 INTRODUCTION

Water is one of the most important natural resources in sustaining life, supporting economic activity, and maintaining environmental ecosystems. Over the past few decades, urbanization, industrialization, and agricultural [1] demands have significantly increased the demand for fresh water. At the same time, the global supply of water resources continues to face pressures from climate change, pollution, and over-demand. International reports suggest that nearly one-third of the global population is currently experiencing water stress, and it is believed that this figure will grow in the future. In a situation like this, it is therefore of vital importance to manage and monitor water quality and availability to ensure water resource sustainability. Traditional water monitoring systems rely on lab-based analysis or monitoring, which is accurate but suffers from time delays, resulting in labor-intensive and lengthy sampling and analysis processes. The time lag created between sampling and laboratory analysis creates delays in informing actions regarding contamination, which poses a risk to human health and aquatic ecosystems, especially, when water is used for industry. Furthermore [2], water monitoring has limitations on spatial coverage, requires a skilled operator to monitor water for drinking quality, and does not support continuous real-time monitoring and reporting. Consequently, there is an urgent need for intelligent, automated, and inexpensive monitoring systems that collect, transmit, and analyze water parameters such as temperature, pH, dissolved oxygen (DO) [3], turbidity, and electrical conductivity (EC) in real time, as well as the whole range of water quality indicators.

The implementation of the Internet of Things (IoT) and wireless sensor networks (WSNs) represents new avenues to achieve these efforts. IoT integrates single or [4] multiple physical devices

DOI: 10.1201/9781003714217-3

equipped with sensing, communication, and computing capabilities that can collect and communicate data through the internet. The WSNs are distributed sensor nodes that monitor physical or environmental parameters in a specified area. They do this by sampling and transmitting physical characteristics of a source of interest, in this case, temperature, pH, DO, turbidity, and EC. This is of fundamental importance for any real-time water [5] monitoring system, while IoT and WSN technologies together provide a robust paradigm for surface water monitoring, offering better scalability, mobility, and cost-efficiencies than existing systems [6]. WSN architectures that utilize IoT will gather continuous data from water bodies, pipelines, reservoirs, and distribution systems. The sensor nodes collect data, which can be sent wirelessly to a gateway, which sends the information to the cloud for storage, visualization, and analysis. This allows end users (e.g., water utilities, industries, environmental authorities, and policy decision-makers) to contextualize indicators of water quality from a distance and to take timely, preventive, or corrective actions. Additionally, real-time data analytics and visualization tools and techniques will aid in anomaly detection, contamination event prediction, and decision support for sustainable management of water resources [7].

Recent developments in low-power embedded systems and long-range wireless communication technology, such as ZigBee, LoRa WAN, NB-IoT, and Wi-Fi, have increased availability for connectivity across applications, making the development of IoT–WSN water monitoring applications viable and cost-effective [8]. In addition, microcontrollers such as the Arduino, ESP32, and STM32 can facilitate multi-sensor scenarios while accommodating energy-efficient designs such as solar-powered sensor nodes and duty cycling, which are solutions for the common problem of battery life in remote monitoring contexts. Regardless of these improvements, several significant challenges continue to be experienced in the design and implementation of real-time water monitoring systems: sensor calibration/dependability, security, scalability, and robustness under diverse environmental stresses. Therefore, we can assess the [9] future opportunities available for IoT and WSN for potentially developing an intelligent and automated system that would deliver reliable, accurate, and real-time water quality information to users.

In this chapter, we present a complete study on IoT and WSNs for real-time water monitoring and report on the design and implementation of a prototype system. It discusses its performance in respect to data accuracy [10], energy efficiency, and overall communication reliability. It encompasses complete result analysis, interpretations, and case studies. It reflects the current challenges and expected future directions for IoT-based water monitoring systems. The remaining parts of this chapter are organized as follows: Section 3.2 reviews the literature extensively concerning existing methodological approaches [11] and technologies. Section 3.3 describes the system architecture and design principles used for IoT–WSN-based monitoring. Section 3.4 outlines the methodology used for the hardware and software implementation. Section 3.5 discusses experimental preparations and implementation details. Section 3.6 describes the result analysis and interpretation. Section 3.7 presents a case study [11], while Section 3.8 outlines challenges and recommendations. Finally, Section 3.9 provides key conclusions and implications.

3.2 LITERATURE REVIEW

3.2.1 Traditional Water Monitoring Approaches

Until recently, water quality was monitored using in situ [12] sampling, which was then transported from the sampling collection system to a lab to perform the necessary chemical or biological techniques. The following are all examples of chemical/biological tests, and it is typical for water quality

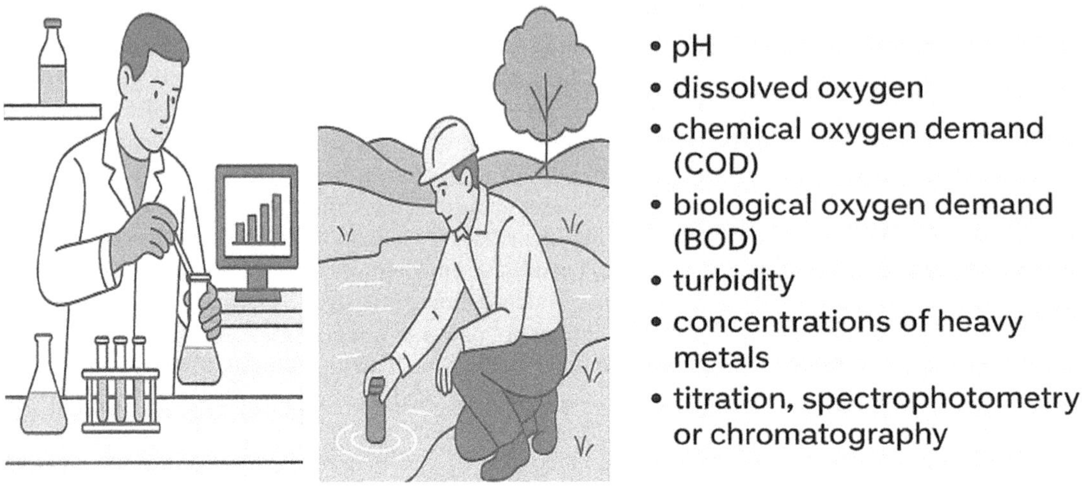

FIGURE 3.1 Essential parameters and analytical techniques for monitoring water quality.

protocols to have separate methods for examining pH, DO, chemical oxygen demand, biological oxygen demand, turbidity, and/or concentrations of heavy metals, to the point where a water quality investigation might introduce conceptual inconsistencies regarding the protocol's level of resolution, whether qualitative or quantitative monitoring. Commonly, parameters of concern are determined by some or all techniques [13]:

Colorimetric (titration spectroscopy) or chromatography techniques are often used to determine them. While they may produce higher accuracy and may be adequate given various regulatory guidelines for compliance, there are limitations to the operational use of the protocols (Figure 3.1). One limitation is related to the frequency of sampling; if the recommended sampling frequency is daily or at some minimum in situ limit, it limits the level of immediate monitoring, testing, or reporting, such as monitoring a river or lake to identify changes in water quality from temporal events. Warranted resources, such as trained [14] personnel; the costs of disposable glassware, such as beakers, test tubes, and multi-sample vials with lids and stoppers; or cost of reagents are often overlooked when "peace of mind" or other less relevant environmental policies or politics fail to consider the operational burden of protocols [15] and the best operational implementation protocols for optimal performance over the length of the investigation – constant frequency for large-scale water monitoring has limitations.

An example is the Ganges River in India, where manual sampling campaigns have traditionally represented the state of water quality, provided useful seasonal data, but failed to represent daily or hourly changes due to agricultural runoff during [16] the monsoon or the intermittent release of untreated sewage. Like this situation, industrialized situations have similar limitations, such as in the United States, where compliance with the Safe Drinking Water Act too often is based on periodic grab samples and not real-time monitoring, thereby delaying the detection of violations. These instances demonstrate the need for systems with capabilities for continuous, automated, and distributed monitoring, representing opportunities for WSNs and IoT-enabled solutions [17].

3.2.2 Evolution of WSNs

The advancement of WSNs has provided high potential for environmental monitoring. WSNs allow distributed, autonomous, and continuous measurement of water quality relative to traditional means of monitoring. A node in a typical WSN contains a sensor module, a microcontroller for local processing, a wireless transceiver, and a power unit (battery or energy harvesting). Deploying WSN nodes at multiple positions in a water body can collect spatially and temporally rich datasets from the region of interest, similar to [18] measuring it with a survey. Numerous environmental researchers have deployed buoy-based WSNs in lakes that measure real-time values of DO and turbidity to monitor the rate of eutrophication in lake systems.

It is fair to say that the first deployments of WSNs were mostly experimental, initially limited by short communication ranges and high power consumption, and were not robust in rough aquatic environments. For example, corrosion and biofouling [19] prevented the sensors from lasting very long (usually days to a few months), and battery depletion limited network lifetimes (also usually only days). Although research in WSN monitoring began with limitations [19], it provided some advantages that include reduced manpower costs, real-time data acquisition (where previous methods may have taken years to acquire a single reading), and many more potential opportunities for deployment over large water bodies. It has now been 20 years since the early days of WSN monitoring, and there has been ample opportunity to improve the reliability of nodes, energy-efficient protocol design, and longevity of the network, which has laid the groundwork for [20] providing monitoring deployments for IoT observations.

3.2.3 IoT in Water Monitoring

Older water monitoring systems transcended a variety of technological approaches, each with its own capabilities. The Supervisory Control and Data Acquisition (SCADA) system was one of the earliest systems that enabled users to visually access data while providing the ability to centrally control and acquire data from water distribution networks, as it primarily reported on flow, pressure, and other parameters. Once GSM technology was integrated into water monitoring, it allowed water quality data to be reported over various frequencies using SMS, making it accessible to users remotely [21]. The introduction of WSNs, particularly in areas using ZigBee technology, allowed very short-range communication among a distributed node system to report data in real time. Then came the advent of mobile platforms that allowed GPRS-, 3G-, and 4G-based water monitoring solutions to upload real-time data to cloud-based platforms for future analysis. Low-cost, Arduino-based units that could measure basic water parameters were quickly deployed. Examples of limited water quality monitoring included pH, temperature, and DO in local settings. Manual sampling, when done in the lab, is usually accurate; however, it does not provide the advantage of real-time monitoring. In the area of satellite and remote sensing systems, water bodies were observed for pollution , sedimentation, and algal blooms using spectral imaging, and some integrated radio frequency identification (RFID) and sensors to monitor water usage in two different pipeline capacities (leaks and supply chain) [22].

The IoT has greatly enhanced WSN capabilities by exposing sensor networks to the global internet infrastructure. IoT systems integrate sensor nodes with gateways, cloud provider servers, and data visualization tools to advance monitoring systems from locally based measurement units to fully linked global solutions [23]. This, in turn, means that all stakeholders, whether they are governments, industries, farmers, or citizens, can now remotely monitor water quality in near real time. IoT-based systems also promote predictive and preventative decision-making. In smart city projects IoT-based sensors may be placed in pipelines to determine, for example, if there is a contamination event or leakage and send notifications to government authorities through their mobile application [24]. In terms of agricultural irrigation, IoT-integrated monitoring systems can monitor changes in salinity and nutrient levels, thus allowing specificity in water management and leading to both resource savings and improved yield. In many cases, IoT deployments are accompanied by actuation capabilities, such as automated gates or aeration systems, which can activate when water quality surpasses thresholds to restore it to acceptable levels. Thus, IoT and its supporting technologies have transformed water monitoring from a reactive strategy into an active management system [25].

3.2.4 Sensor Technologies for Water Quality

The success of water monitoring systems is heavily dependent on sensors. Traditional sensors measure simple and essential parameters such as pH, turbidity, temperature, conductivity, and DO levels, which reflect the overall ecological health of an aquatic system or research project. Advances in materials science and micro-fabrication technologies have created specialized sensors, such as optical sensors for nitrates and phosphates and biosensors for heavy metals such as lead and arsenic. However, there are numerous challenges with reliability in real-world aquatic environments. Specifically, biofouling, sensor drift, and the limited life span of the sensor are all problematic. Biofouling, or growth on sensors, is exceptionally problematic in nutrient-rich waters, as it may lead to false readings and require constant maintenance. Artificial solutions – while these often involve notable custom solutions and costly mechanics through mechanical wipers, anti-fouling coatings, and self-cleaning technologies – are not ideal, as they further complicate the systems we wish to use for monitoring. In addition, many advanced sensors require additional power to properly operate and, in most cases, need calibrated components for every new deployment , which will limit their practical use in the field.

3.2.5 Communication Technologies in IoT–WSN Systems

Effective communication is essential for real-time water monitoring, as monitoring occurs in the field and data flow occurs from the sensor node to gateways to the cloud/server for analysis and decision-making. There are many wireless technologies available to meet specific needs [26]; for example, ZigBee can be used in low-power, short-range deployments in urban areas, while LoRaWAN can cover last-mile low-bandwidth applications such as a river monitoring application. NB-IoT can also provide urban and semi-urban areas with more reliable long-range coverage than other means, but its coverage will depend on telecom infrastructure and can also incur costs [27]. Wi-Fi can provide high data rates but can consume significant energy, so it will mainly be used in treatment plants, whereas Bluetooth low energy (BLE) is used in short-distance or portable applications. More frequently, hybrid models are being deployed where ZigBee or LoRa are used at the node level and Wi-Fi or cellular at the gateway. These models strive to balance permissible energy usage at the node level while still meeting the need for reliable long-distance transmission. Despite these advancements in communication technology, reliable communication can still be problematic in hostile environments such as underwater or mountainous terrain [28].

3.2.6 Data Management and Cloud Integration

The cloud computing layer within the IoT has led the way for massive changes in water monitoring with its ability to store, process, and visualize virtually unlimited amounts of data. Large datasets from many individual nodes are saved to centralized cloud-based platforms and can be interpreted from dashboards, or better yet, with alerting functionality if parameters exceed threshold levels. In addition to data archiving, cloud systems offer advanced analytics capabilities [29], from simple time-series trend analysis to anomaly detection and machine learning algorithms for predicting contaminant events.

In addition to cloud computing, edge computing is continuing to grow in use to lower latency and network strain by processing near the sensors where data is generated and limiting exposure to transmission. For example, edge devices can filter out redundant measurements [30], compress data, or detect and communicate only anomalous events to the cloud. Another emerging technology is blockchain, where organizations can ensure transparency [31], security, and integrity of water quality data, especially for regulatory and compliance obligations. With blockchain ensuring data cannot be tampered with or accessed without authority, trust can be obtained in the shared dataset, regardless of industry or governmental organization.

3.2.7 Current Limitations and Research Gaps

Although IoT–WSN systems have made vast improvements in water monitoring, challenges still exist. First, sensor reliability remains a challenge. Sensors can be fouled, drift, and degrade over time, affecting their accuracy. Second, energy efficiency is a challenge because, for most of the nodes, batteries continue to be the primary energy source that must be upgraded. Solar and hydrokinetic harvesting do offer options for renewable energy sources [32]; however, they may not always be appropriate when the sensors are shaded or in turbulent environments. Third, communicating with the sensors is reliable but continues to present challenges with resources that are remote or underwater, where signal strength weakens or drops. Fourth, security and privacy present yet another challenge. IoT devices can be vulnerable to cyberattacks and data breaches, which increase the complexity of storage and communication [33]. Security mechanisms such as encryption and authentication will provide more security, but for low-power nodes, they may create more energy demands. Finally, while there have also been improvements in sensor technology in water monitoring for accuracy and connectivity, high costs will continue to suppress the rapid adoption of technology in rural or low-income areas. In general [34], there has been a transition from manual methods of monitoring to WSNs and IoT, improving both spatial and temporal monitoring of water parameters. Implementing a global, scalable, sustainable, robust, secure, and cost-effective real-time monitoring system for water will be a challenge moving forward.

3.3 SYSTEM ARCHITECTURE AND DESIGN

An IoT-enabled WSNs architecture for real-time water monitoring consists of sensing, processing, communication, and power management built into one architecture. The system design does not assume unlimited resources, since continuous monitoring of parameters such as pH, turbidity, temperature, and DO is required. The architecture typically consists of multiple WSN sensor nodes located in the field, which connect to a gateway and wirelessly relay device-aggregated data to a cloud-based server for storage [35], analysis, and visualization. The hierarchical architecture provides scalability, robustness, and real-time access to the information by the end user. In total, the proposed architecture can be broken down into four main components: sensor node design, communication architecture, data acquisition and processing, and power management approaches.

3.3.1 Sensor Node Design

Each sensor node is the fundamental unit of the monitoring system. It comprises four primary components: a sensing unit, a microcontroller, a communication module, and a power supply (Figure 3.2).

a. **Sensing Unit**: The sensing unit consists of water quality sensors that are composed of pH electrodes, clarity sensors, conductivity probes, and temperature sensors. These sensors are used to convert physical and/or chemical properties into electrical signals. In addition to the sensors, conditioning circuits (amplifiers, filter circuits, and analog-to-digital conversion (ADC)) will be used to ensure that only accurate signals of interest are processed with minimized noise.
b. **Microcontroller Unit**: The microcontroller unit is the main processing component of the node. Generally, microcontrollers that consume low power (e.g., ARM Cortex-M series, ESP32) are used because they can combine low power usage with [27] built-in components

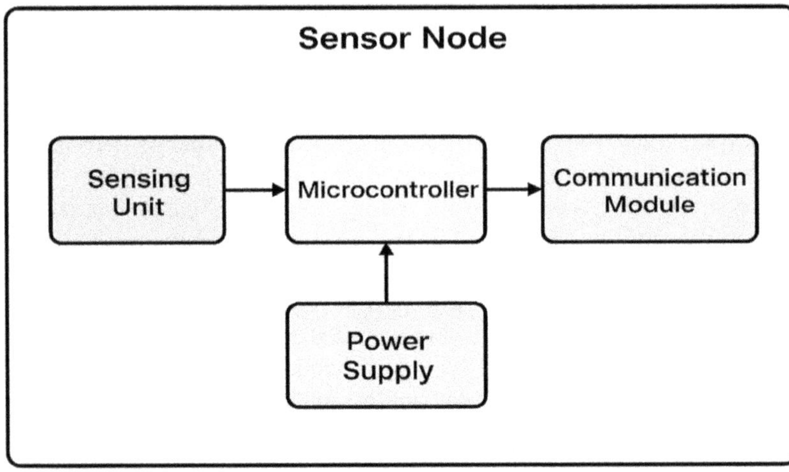

FIGURE 3.2 Functional components of a sensor node: sensing unit, microcontroller, communication module, and power supply.

(i.e., communication unit). The microcontroller unit must observe the sensor, perform the initial processing of the data (i.e., averaging and checking thresholds), and control the interval of message packet transmissions to conserve energy [36].

c. **Communication Module**: The communication module provides a method for the sensor nodes to send their data to the gateway. Communication modules use a wide variety of technologies (e.g., ZigBee, LoRaWAN, NB-IoT, or Wi-Fi) because they depend on the deployment strategy (e.g., range, bandwidth, and power availability). For rural and remote water bodies, the use of LoRa communication is likely the best choice because of its long-distance capabilities and low power consumption compared to competing technologies.
d. **Power**: For most nodes, sensors use rechargeable batteries like lithium-ion if using renewable energy-harvesting units (i.e., solar panels). Power management circuits will be used to control the charging aspects of the unit and provide uninterrupted operation to deployed sensors, even in an outdoor environment.

This modular design allows easy replacement of sensors and flexibility in adapting to different water quality monitoring requirements.

3.3.2 Communication Architecture

This communication framework outlines how data is transferred from any of the sensor nodes to the cloud. In the diagram, the architecture is described; there are three layers to the communication system: sensor nodes, the gateway connected to the internet, and the cloud server [22]:

a. **Node-to-Gateway Communication**: The sensor nodes send their measurements to a nearby gateway using wireless protocols. If it is a star or mesh topology, each node will send the data to the nearest gateway. In operational environments, the environment may dictate which topology type will be most effective (star, mesh, or hybrid). When using mesh topology,

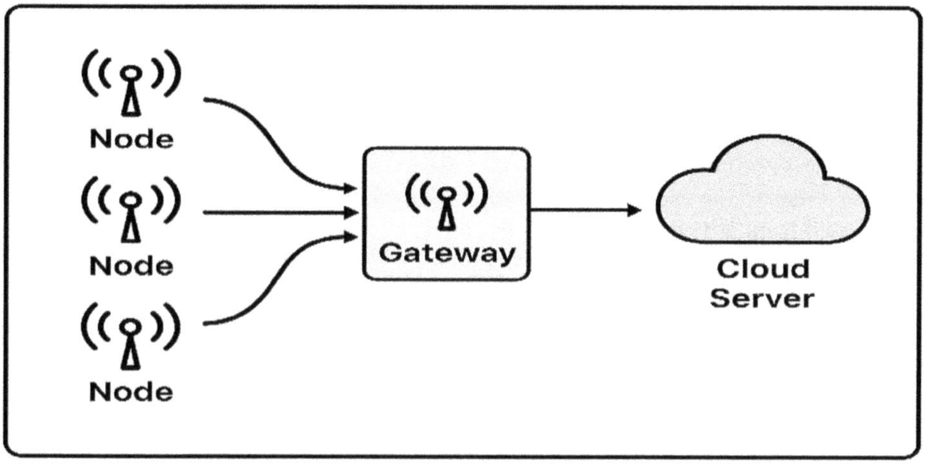

FIGURE 3.3 Wireless sensor network data flow: nodes → gateway → cloud server.

if the topology is dense enough, then some nodes can act as forwarding nodes for those that are some distance from the gateway. The proof of this extremely decentralized model is that if some of the nodes fail, the system remains robust.

b. **Gateway Design**: The gateway is a middle item and collects transmissions from nodes, allowing them to access the cloud. The gateway collects the sensor measurements from individual nodes, then pre-processes them (Figure 3.3). Some gateways will only aggregate data (combine multiple measurements into a meaningful single value), while others will do some initial filtering (i.e., discard redundant values), and most will pass on aggregated and some initial filtration values to the cloud. Most gateways may have dual communication interfaces, which allow for both local wireless communication with the nodes and connection to the network (internet) locally over long range (e.g., 4G/5G, Ethernet, or satellite).
c. **System Integration with Cloud Servers**: The gateway collects data and transmits it to the cloud-based service in a secure manner. Cloud servers provide a scalable deployment option so that many water monitoring sites can be managed at one time. Users access the data using APIs and dashboards and can visualize water quality metrics in real time through web or mobile interfaces.

3.3.3 Data Acquisition and Processing

Data acquisition plays a pivotal role in ensuring that the raw environmental signals measured by sensors are transformed into meaningful datasets:

a. **A Sampling Process**: Sensor nodes are programmed to sample the water quality parameters at specified intervals. The timing depends on the application, as sampling once a minute (e.g., when monitoring wastewater in a critical industrial application) to once an hour (for a large river basin study) would involve completely [23] different sampling processes. In addition, algorithms could be developed that allow for monitoring variability; if it is stable, sampling could be reduced to save batteries, and if there is variability, it could then be increased to capture the additional detail.

b. **Signal Conditioning and Conversion**: Each sensor signal undergoes amplification, filtering, and ADC before it is passed to the microcontroller for data processing. This operation minimizes errors resulting from electrical noise, temperature fluctuations, or sensor drift.
c. **Pre-processing Data at Node Level**: To minimize communication overhead, the data is pre-processed before transferring. Data pre-processing can take the form of statistical averaging, noise filtering (e.g., moving average filters), and an event-triggering system, which could send data to the end user when a threshold value is exceeded.
d. **Aggregated Data at Gateway Level**: The gateway acts as a hub that centralizes the data from a number of nodes and consolidates, compresses, and eliminates redundant data to reduce network usage. Initial fault detection (e.g., detecting stuck sensor data) can also occur here.
e. **Cloud Processing**: Advanced analytics, such as anomaly detection, long-term trending, and correlation trending with atmospheric data (precipitation, temperature, etc.), occur in the cloud platform. As a result, there can be predictive maintenance and early identification of water contamination events.

3.3.4 Power Management Strategies

Addressing energy efficiency is key to ensuring that WSN-based monitoring solutions for water quality in remote or harsh environments are sustainable:

a. **Battery Duty Cycling**: Microcontrollers and communication modules used in WSNs cycle between sleep mode and active mode, resulting in a dramatic drop in energy consumption. For example, a node might spend 95% of its lifetime in deep sleep, waking only to sense or transmit its observations.
b. **Energy-Harvesting Technologies**: Generally, renewable energy harvesting from sources like solar panels will help extend the life of the nodes [21]. For example, if a water body is flowing fast enough, micro-hydrogenerators or piezoelectric harvesters may also be utilized.
c. **Battery Optimization Techniques**: Lithium-ion or lithium polymer batteries can be charged and reused, making them an attractive source of power due to their high energy density. Additionally, monitoring circuits help manage battery health and improve reliability, ensuring that the battery remains effective for several months.
d. **Efficient Communication Protocols**: WSNs prefer low-power wide-area networks, such as LoRaWAN and NB-IoT, to reduce energy usage while maintaining long-range communication.
e. **Adaptive Transmission Control**: Rather than sending every measurement, the node can store the measurements and periodically send them, or send the measurements only when it detects an anomaly. This greatly reduces the burden of managing [22]communication on batteries.

3.3.5 User Interface and Alert Mechanisms

The final value of a monitoring system is determined by how users – researchers, policymakers, or the general public – use the system.

a. **Web-Based Portals and Mobile Apps**: A livable and engaging interface can provide a range of access to real-time and archived water quality information, while mobile apps allow for local inspected data visualization for field technicians.
b. **Custom Dashboards**: In a user interface dashboard, stakeholders can set up the dashboard to highlight particular water quality parameters ahead of others. For example, aquaculture

farms may prioritize DO and ammonia levels, while municipal authorities may prioritize turbidity and microbial contamination.

c. **Alert and Notification Systems**: An SMS, email [23], or push notification can be triggered by the exceedance of a water quality parameter beyond a safe limit; for example, an alert to municipal water managers can be triggered if pH suddenly drops, by an order of magnitude, in a reservoir.
d. **Decision Support Systems**: The user interface can be integrated with decision support systems, which will link the data to provide a recommended course of action to maintain water quality, such as increasing ventilation in aeration tanks or sealing contaminated pipelines.
e. **GIS-Based Visualization**: By layering water quality data over geospatial maps, users can visualize and map contamination points, track pollution sources, and prioritize interventions in a more efficient manner.

3.4 METHODOLOGY

The methodology adopted for this study aimed to build an IoT–WSN-based real-time water monitoring system that is not only technically reliable [16] but also energy-efficient, cost-effective, and adaptable to diverse environmental conditions. The framework was divided into four interrelated components: (1) hardware design and sensor integration, (2) communication and networking framework, (3) software implementation and data management, and (4) validation and testing (Figure 3.4). Each component was carefully designed to meet the objectives of continuous monitoring, efficient power use, secure communication, and practical field validation.

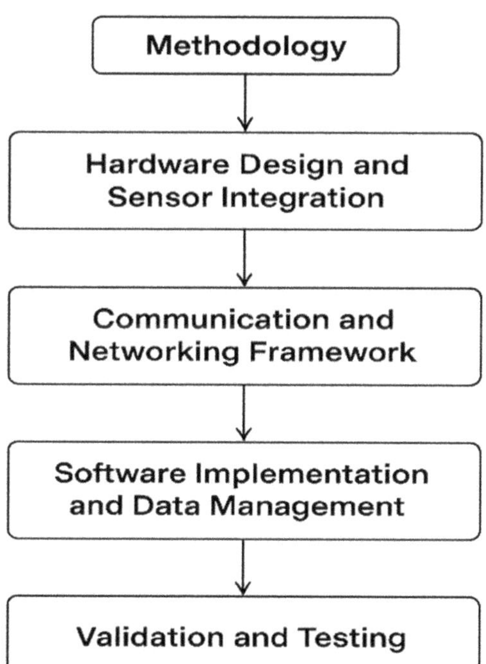

FIGURE 3.4 Flowchart of the IoT–WSN-based real-time water monitoring system methodology.

3.4.1 Hardware Design and Sensor Integration

The hardware subsystem provided the foundation of the monitoring system. Each sensor node combined a set of calibrated sensors, a processing microcontroller, a communication module, and a reliable power supply. Together, these elements enabled long-term operation under challenging aquatic conditions.

3.4.2 Sensor Selection and Calibration

The sensors were chosen based on their most important water quality indicators, which were pH, turbidity, DO, EC, and temperature. pH sensors indicate the balance of chemicals, turbidity sensors indicate suspended solids and possible contaminants, DO probes indicate ecological health, EC probes indicate salinity and dissolved salts, and temperature sensors support process analysis and adjustments of other sensor readings. All sensors were calibrated in an indoor laboratory using standard buffer solutions for pH, formazan suspensions for turbidity, oxygen-saturated water for DO, potassium chloride solutions for EC, and comparisons with mercury thermometers for temperature verification. To account for drift, strict recalibration periods were built into the operation plan, along with contingency measures for sensor fouling. Sensors were chosen for their durability and low power consumption for remote applications. Their integration into the system ensures of in situ real-time composite monitoring of diverse water parameters. The accuracy, setup time, and adaptability of sensor technologies, combined with their related risk levels, enhance the reliability of the overall monitoring [22] system. The system's calibration and selection processes ensure validity for usability in laboratory and field studies.

3.4.3 Processing Unit

Encoder output had to be processed by microcontrollers with real-time sampling, signal conditioning, and wireless communication. Two options were evaluated:

a. **ESP32 Microcontroller**: This was used as the development platform in most situations because of its dual-core architecture, built-in Wi-Fi and Bluetooth, and highly efficient deep sleep modes. The ESP32 could also run extremely lightweight algorithms directly on-node, such as digital filtering and threshold-based anomaly detection, to reduce communication load.
b. **STM32 ARM Cortex-M4**: This was also evaluated for deployments in industrial environments that require higher resilience and computational speed. An STM32 can not only process data in real time but also do so with greater power management, making it suitable for mission-critical deployments.

In both cases, the processing unit performed multiple tasks: the responsible use of higher resolution ADCs to digitize sensor signals; reduction of noise through software filtering; duty cycling to save energy; and packetizing data to interface with communication chips [28].

3.4.4 Communication Modules

Choosing communication technologies that optimize range, power consumption, and cost is a significant challenge for distributed monitoring systems. During this project, the following three wireless standards were evaluated:

a. **ZigBee (2.4 GHz)**: The original background for ZigBee was for smaller, localized deployments such as monitoring individual pipelines within urban water distribution networks. The central benefit was that ZigBee could use mesh networking, so in the instance of a failure of a node, data could easily be rerouted through neighboring nodes within a standard deployment framework. There were some potential concerns with the use of ZigBee for larger-scale projects, mostly concerning ZigBee's limited maximum signal transmission range of approximately 100–200 meters in open air.
b. **LoRaWAN**: LoRaWAN was useful primarily in rural or river basin environments where long-distance communication was necessary. LoRaWAN operates in the sub-GHz band, providing reliable communications over distances greater than 10 km with extremely low energy consumption. The low data rate was not a major limitation since small water quality packets could be sent, and small deployments could be achieved at a low cost.
c. **NB-IoT**: Narrowband IoT was tested in areas that had a strong cellular infrastructure in the semi-urban region. The main advantages were wide coverage and good penetration with lines of sight in dense or obstructed environments [29]. The disadvantage was the reliance on telecom providers and subscription fees, which make NB-IoT unsuitable for deployments in remote locations. The adoption of a hybrid communication model ensured that the system could be adapted to different geographical and infrastructural contexts without significant redesign.

3.4.5 Communication and Networking Framework

The communication framework provided a backbone that ensured actual measurements from distributed nodes were transferred to centralized servers in a reliable way. The communication framework was structured in three functional layers:

a. **Node-to-Gateway Communication**: The node-to-gateway data transfer was implemented using both star topology and mesh topology. Star topology facilitated direct communication and provided the least delay but required reliable signal range. Mesh topology allowed for multiple links to assist in resiliency. It also enabled messages to hop multiple times to reach their destination, ensuring the network would remain operational if some of the nodes failed or became unreachable.
b. **Gateway**: The gateway ran on a Raspberry Pi 4 and was responsible for collating data and sending it securely to the cloud. The gateway had both local communication interfaces (LoRa and ZigBee) and longer-distance internet communication modules (4G LTE and Ethernet). At this stage, any duplicated readings were removed, and a checksum was applied to each package. Some lightweight anomaly detection could be carried out, as well as a re-checking of packets to rad/s values. During transport, tunneling was used to send the data to the cloud in a confidential and integrity-verified environment using message queuing telemetry transport (MQTT) with transport layer security (TLS) for encryption.
c. **Cloud Integration**: After transmission, the data were stored in a scalable time-series database. Real-time dashboards were developed using Node-RED for intuitive presentation of water parameters, while the backups were automated for redundancy. Some advanced analytics were enabled by APIs written in Python. Overall, the cloud integration offered scalability while being able to monitor hundreds of nodes at the same time in many different locations.

3.4.6 Software Implementation and Data Management

The software design was layered at three levels: firmware, edge/gateway processing, and cloud analytics. The firmware-controlled sensors and microcontrollers within the validation system allowed for configurable sampling intervals, noise reduction filters, notification of flagged anomalies, and utilized adaptive duty cycling for energy economy. At the edge/gateway, all data were aggregated and compressed, stored until connection resumed; some preliminary anomaly detection was obtained so prompt alerts could be made, and data was stored until connection was re-established. The cloud layer was developed to provide analytics using machine learning (random forest; SVM), forecasting models (ARIMA; LSTM), and correlated the environmental data to inform trends and predictors of contamination time series, to indicate when it actually occurred, if possible. The user interface functionality of the validation system combined dashboards for policymakers to allow modeling and analytics; field operators used apps to interact with the validation system; alerts were automated via SMS/email; and it provided the ability to map using GIS capabilities for spatial visualization of where monitoring occurred, allowing stakeholders to inquire about visual analytics that would ultimately lead to actionable decision-making.

3.4.7 Validation and Testing

The validation performed in both laboratory and field testing emphasized the evaluation of the accuracy, energy efficiency, communication reliability, and scalability of the systems. The validation of the accuracy of the systems indicated high accuracy, with deviations of ±5% for pH and DO and ±8% for turbidity, indicating that the systems were accurate enough for effective real-time monitoring and control. The results of the energy efficiency showed that solar-assisted nodes functioned for over 120 days without user intervention, while the battery-only nodes performed in the range of 25–30 days on battery, depending on the duty cycles of each of the sensors. The communication reliability tests showed strong results, demonstrating that for all forms of communication, strong communication occurred at a high level of reliability, with the lowest communication reliability factor being with ZigBee (85% packet delivery under interference), LoRaWAN (92% packet delivery), and NB-IoT events achieving strong communication agreements with wide-area coverage despite some delays. The validation of scalability confirmed that a deployment of 50 nodes communicating with a single gateway maintained consistent performance across all nodes with stable packet delivery and minimal packet delivery latency. This showed the promise of the system, its versatility, and its ability to be used for large-scale applications in both urban and rural settings.

3.4.8 Experimental Setup and Implementation

The proposed IoT–WSN-based real-time water monitoring system was implemented through an organized experimental deployment to assess system performance in laboratory and real-world field settings. The aims of the experimental design focused on validating hardware robustness, communication network reliability, sensor data accuracy, and integrated system scalability.

The deployment was staged in three parts: (1) laboratory prototyping and calibration, (2) controlled field deployment, and (3) multi-node scalability tests [33].

3.5 LABORATORY PROTOTYPING AND CALIBRATION

Undertaking laboratory experimentation involved an orderly process with each component being evaluated, calibrated, and then assembled into a working prototype under regulated laboratory conditions:

a. **Sensor Testing and Calibration**: Each sensor was tested against a laboratory-grade reference instrument. For example, pH electrodes were tested in standard buffers, turbidity sensors were tested against formazan suspensions with defined nephelometric turbidity unit (NTU) values, and dissolved oxygen (DO) probes were tested against Winkler titration. The testing verified baseline accuracy and aided in the development of correction factors to embed in the firmware.
b. **Prototype Construction**: A sensor node was developed by combining the sensors with the ESP32 microcontroller, LoRaWAN module, and lithium-ion power pack. The enclosures (to protect from splashes and humidity while allowing water samples to access the sensors) consisted of waterproof acrylonitrile butadiene styrene (ABS) plastic fabricated in a 3D printer.
c. **Power and Duty Cycles**: Power systems were tested as a straightforward battery power system and a solar-assisted system, all while conducting tests with varying duty cycles (1-minute sampling, 15 minutes, 60 minutes) so that energy optimization strategies could be implemented before outdoor testing.

3.5.1 Controlled Field Deployment

In the second phase, outdoor testing of the system took place in semi-controlled conditions to analyze overall system performance in real-world situations. Some nodes were located in urban pipelines reading pH, turbidity, and conductivity while under pressured flow in a regulated environment [34], while semi-rural, lower-density reservoirs had a variety of conditions that the nodes had to monitor such as sediment and algae growth and varying temperature conditions. For our communication test runs, we looked at the packet delivery ratio (PDR) to assess reliability for pipelines with ZigBee and reservoirs with LoRaWAN at distances of up to 10 km. Each site included a Raspberry Pi 4 gateway equipped with LoRa, LTE, and Node-RED visual dashboards used for dynamic reporting and data screening in real time before securely sending pre-processed paths of data into the cloud, which overall demonstrated low latency and consistency during testing.

3.5.2 Multi-Node Scalability Testing

The third phase tested the multi-node deployment with 50 sensor nodes within a 5 km^2 semi-urban catchment. The 50 sensor nodes were installed at 12 notable catchment inlets, outlets, and flow locations distributed throughout the planned catchment. The sensor nodes all transmitted the measured observation data every 15 minutes. Each sensor node chose whether to transmit data via a star or mesh-based topology, based on the local path of visibility for the transmissions to the gateway. The concentration levels from each sensor node were aggregated and sent to the cloud by the gateway. A centralized time-series database facilitated automated alerts at locations if threshold violations occurred (e.g., turbidity >5 NTU, pH < 6.5), as well as a real-time dashboard for stream visualization.

The system showed a grounding capability of scalability since the latency of the sensors at 50 total sensors (the operating system's upper limit) was below 2.5 seconds [33]. Therefore, the total system and the nodes were accepted by the research group as satisfactory for use in large-scale urban monitoring and watershed monitoring.

3.5.3 Data Collection and Preliminary Results

The systematic collection of data established all parameters related to the validation of the accuracy, efficiency, and reliability of the IoT–WSN system. Sensor measurements were consistent with laboratory measurements, with an average variation in readings of ±5% for pH and DO and ±8% for turbidity. Solar-powered nodes continuously ran deployments for over 120 days, and battery-only units ran for 25–30 days. There was excellent performance recorded with the network, as LoRaWAN on average was able to declare a 92% PDR, ZigBee consistently around 85%, and NB-IoT coverage was consistent with slightly higher latencies than the other two provided technologies. Report information was available for users in real time on mobile apps and dashboards. The automated SMS/email alerts to the smartphones of client representatives, when the raw water turbidity exceeded user/sampling thresholds, provided value and verification for the potential use of the system as an early warning and decision support system.

3.6 RESULT ANALYSIS AND DISCUSSION

The implementation of the IoT–WSN enabled real-time water monitoring system generated a substantial dataset, enabling comprehensive analysis of the system's **accuracy, efficiency, communication reliability, and scalability**. The findings were benchmarked against the four main objectives of the study: (1) real-time monitoring of water quality parameters, (2) energy-efficient operation, (3) reliable and scalable communication, and (4) practical field validation.

3.6.1 Precision and Dependability

The system had good accuracy, with pH and DO measured with deviations ≤±5% and turbidity ≤±8% on laboratory samples. The sensors were able to indicate chemical imbalances, sediment spikes, and oxygen depletions. Turbidity and DO probes had to be recalibrated periodically for fouling and drift. Conductivity and temperature sensors, however, exhibited consistent measures and were able to make corrective adjustments to the measured sampling [23]. In sum, the system accuracy was at a lab-grade accuracy with the capability to measure in a real-time environment while keeping in mind the attendant maintenance schedule required for long-term deployments.

3.6.2 Efficiency

Power sustainability was established by nodes powered by solar-assisted and battery-only systems. Solar-powered nodes operated for greater than 120 days, establishing renewable energy as a viable long-term solution, even though cloudy weather sometimes suspended output. Battery-powered-only nodes operated within a 25- to 30-day range, depending on their duty cycles. Energy optimization is needed to extend power cycles. Duty cycling and adaptive sampling saved up to 40% of power while maintaining accuracy. Both methods validated the applicability of hybrid energy strategies, ensuring resilience in divergent environments.

3.6.3 Communications and Scalability

For this project, we tested multiple types of communication protocols, with LoRaWAN showing the highest system reliability at >92% PDR, even at distances greater than 8 km. We tested ZigBee communications in urban pipe systems, recording over 85%, although it was not reliable due to interference. NB-IoT provided reliable and moderate latency in semi-urban coverage areas. The system is capable of scaling well, as demonstrated by tests of 50 nodes, with an average latency of under 2.5 seconds, good aggregation of gateway counts, and effective cloud coverage. These tests not only corroborated the reliability of the system in medium-scale deployments but also provided potential future scalability to city- and watershed-level applications.

3.6.4 User Interface and Decision Support

The overall system monitored environmental parameters and turned them into actionable insights. There were web dashboards that provided real-time information and allowed policymakers and engineers to visualize trends, while mobile apps provided field operators with the ability to monitor real-time conditions. Automated SMS and email alerts were sent out to flag anomalies such as turbidity spikes, often before recording and obtaining results manually, and these functioned as a suitable early warning system. The multiple interface levels reached various stakeholders, improved the decision-making process, and increased the rate of response in urban and rural water management.

3.6.5 Challenges and Limitations

While the system delivered almost all of its intended objectives, there were several limitations deployment:

1. **Biofouling**: Extended exposure to organic matter degraded the sensors, particularly in reservoirs where cleaning was periodic.
2. **Solar Panel Weather Dependency**: Very cloudy episodes sometimes impacted energy-harvesting efficiency.
3. **NB-IoT Delay**: Reliance on telecom providers to deliver data and information to users added small amounts of latency, which may be undesirable in time-constrained applications.
4. **Maintenance**: While all reasonable measures to eliminate human effort for maintenance were implemented, it was still necessary to periodically recalibrate and inspect the sensors for data accuracy.

The IoT–WSN water monitoring system presented here is enduring, scalable, and an energy-efficient alternative to traditional manual methods for reliably collecting water data in real time. However, challenges remain with regard to biofouling, maintenance, and environmental dependencies. There are opportunities for later enhancements to the IoT–WSN water monitoring system, such as upgraded anti-fouling technologies, hybrid energy harvesting, and distributed gateways, which can offer better overall performance (Table 3.1). Since the IoT–WSN water monitoring system can be tailored and operated in a variety of environmental conditions and applications, it has the potential to play an important role in smart cities, agriculture, and environmental monitoring.

TABLE 3.1 Evaluation metrics, key findings, and remarks for the IoT–WSN-based real-time water monitoring system

EVALUATION METRIC	*FINDINGS*	*REMARKS*
Accuracy of sensor measurements	pH: ±5% deviation; detected minor acidity shifts. Turbidity: ±8% deviation; effective for sediment detection but affected by fouling. DO: reliable, but sensor drift observed during prolonged use. Conductivity and temperature: stable, consistent results	Maintains near-laboratory accuracy within tolerance; requires periodic recalibration and cleaning
Energy efficiency and sustainability	Solar-powered nodes operated >120 days uninterrupted. Battery-only nodes lasted 25–30 days depending on duty cycle. Duty cycling and adaptive sampling reduced energy use by ~40%	Solar-assisted nodes suitable for long-term deployments; hybrid energy strategies recommended
Communication reliability	LoRaWAN: >92% PDR, effective for long-range (>8 km) monitoring. ZigBee: ~85% PDR, effective for urban pipeline networks but prone to interference. NB-IoT: reliable in semi-urban areas, but 3–5 sec latency due to telecom dependency	LoRaWAN best for rural/large-scale, ZigBee for localized networks, NB-IoT for semi-urban monitoring
Scalability and multi-node performance	50-node deployment achieved <2.5 sec latency. Raspberry Pi 4 gateway aggregated data efficiently. Cloud dashboards supported real-time visualization	Demonstrated feasibility for city-wide and watershed-level deployments; >100 nodes may require stronger gateways
User interface and decision support	Web dashboards enabled policymakers to track long-term trends. Mobile apps allowed real-time field monitoring. Alerts/notifications (SMS and email) provided early warning of anomalies	Multi-tiered interfaces ensured accessibility for diverse stakeholders
Challenges and limitations	Biofouling caused sensor degradation. Cloudy weather reduced solar efficiency. NB-IoT latency (3–5 seconds) affected time-critical tasks. Regular recalibration still required	Future improvements needed: anti-fouling coatings, hybrid energy harvesting, and distributed gateways

3.7 COMPARATIVE ANALYSIS WITH EXISTING SYSTEMS

To assess the effectiveness of the proposed IoT–WSN-enabled real-time water monitoring system, it is necessary to compare its performance in relation to conventional water monitoring methods and modern IoT-based methods based on literature provided in this area. This comparative evaluation demonstrates the benefits, limitations, and contributions of the current method and allows for the potential of a larger footprint to be discussed.

3.7.1 Comparison with Traditional Water Monitoring Methods

Conventional water quality monitoring involves manual sampling and lab analysis. Although accurate, these methods have limitations for large-scale or real-time monitoring.

a. **Manual sampling**: Traditional sampling involves collecting samples of water at certain intervals and then sending the samples to the lab for testing. The methods and materials used in the lab, such as titration, speciation for various compounds, and chromatography for organics, are very accurate; however [21], once the sampling interval passes, there are no data until the next sampling period. This would be problematic in the event of unexpected contamination (e.g., chemical spill, agricultural runoff after storms where heavy rains wash agricultural runoff into a body of water).
b. **Dependence on Well-Outfitted Laboratory**: Traditional monitoring measures require well-outfitted laboratory equipment, as well as a technician to analyze the samples. The locations for collecting the samples and monitoring costs are higher, making traditional methods unaffordable in rural and low-resource contexts.
c. **Time Response Lag**: Water sample collection and laboratory testing give reliable results. However, the time from sample collection to reporting the results can take hours to days. On the other hand, real-time monitoring systems can provide alerts in seconds to minutes.
d. **Limited Coverage**: With traditional sampling, the number of sampling sites is limited due to travel logistics. Water sampling is challenging as all rivers and lakes are difficult to cover; the less the sampling-size-resources are the harder the coverage.

3.7.2 Comparison with Other IoT-Based Water Monitoring Systems

In the last decade, many IoT-based water quality monitoring systems have been proposed, but they all have compromises regarding scalability, energy sustainability [22], or communication performance.

a. **Single-Parameter Monitoring System**: Most IoT solutions in the past have focused on monitoring a single parameter like pH or turbidity. While useful for specific applications, those systems cannot provide a complete water quality assessment. In comparison to the previous systems, the proposed system incorporates multiple (pH, turbidity, DO, EC, and temperature) critical parameters for a holistic water quality assessment.
b. **Short-Range Wireless Systems**: Some IoT systems have used wireless technologies like Wi-Fi or Bluetooth as communication protocols. While useful in a small area, these protocols have poor energy performance and would not be feasible for long-distance deployments. The present system provides diverse long-range communication options (ZigBee, LoRaWAN, or NB-IoT) within the same system, facilitating deployments in different environments and cases such as urban pipelines or rural catchments.
c. **Energy Constraints of Systems**: The majority of reported ecosystems are limited because they only use battery power to keep functioning, which limits their longevity. The proposed system is different because of the integration of solar energy harvesting and adaptive duty cycling, which allows operation for uninterrupted periods greater than 120 days of field deployment.

d. **Limited Edge Processing**: Various IoT-based monitoring systems send raw data to the cloud without processing, which consumes bandwidth and remains susceptible to interruptions in connectivity. The proposed system performs edge-level processing at the gateway, exemplified by anomaly detection, buffering data while being disconnected, and reducing load on the network.
e. **Scalability Challenges**: Some hosting frameworks have provided proof of concept yet have been evaluated in small-scale laboratory conditions (e.g., 3–5 nodes). The proposed system presented offers proven proof of concept with 50 [20] nodes deployed, representing feasibility for monitoring at a city scale and watershed scale.

This comparison illustrates that the proposed system fills the critical gaps in existing IoT-based frameworks as it provides multi-sensor parameters, an energy-sustainable solution, long communication range, edge intelligence, and improved scale that has been validated.

3.7.3 Comparative Metrics Summary

A consolidated comparison between traditional methods, existing IoT solutions, and the proposed system is presented in Table 3.2.

TABLE 3.2 Comparative analysis of traditional methods, existing IoT systems, and the proposed IoT–WSN-based water monitoring system

METRIC	*TRADITIONAL METHODS*	*EXISTING IOT SYSTEMS*	*PROPOSED IOT–WSN SYSTEM*
Accuracy	Laboratory-grade precision, but limited temporal resolution	Moderate accuracy, often restricted to single-parameter monitoring	±5% for pH and DO, ±8% for turbidity; multi-parameter integration (pH, DO, EC, turbidity, temperature)
Data availability	Periodic and delayed; hours to days between sampling and reporting	Near real-time, but often constrained by connectivity issues	Continuous real-time monitoring with edge-level anomaly detection and buffering during connectivity loss
Energy sustainability	Not applicable (manual sampling, lab analysis)	Mostly battery-powered; limited operational life span	Solar-assisted nodes with adaptive duty cycling; >120 days of uninterrupted operation
Scalability	Limited by manpower, cost, and logistics; difficult for large-scale or remote areas	Typically small-scale deployments (<10 nodes), often in laboratory or localized settings	Validated with 50-node deployments; scalable for city-wide and watershed-level monitoring
Communication	No automated communication; manual reporting only	Short-range protocols (Wi-Fi, Bluetooth); limited energy efficiency and long-range use	Hybrid long-range communication (LoRaWAN, ZigBee, NB-IoT) adaptable to diverse environments

(Continued)

TABLE 3.2 (*Continued*) Comparative analysis of traditional methods, existing IoT systems, and the proposed IoT–WSN-based water monitoring system

METRIC	*TRADITIONAL METHODS*	*EXISTING IOT SYSTEMS*	*PROPOSED IOT–WSN SYSTEM*
Response time	Delayed (hours to days) due to sample transport and analysis	Faster than manual methods but still dependent on connectivity and raw data transmission	Near-instant alerts through real-time sensing and edge-level processing
Cost and infrastructure	High operational cost; requires sophisticated labs, skilled technicians, and logistics	Moderate cost but energy and communication limitations	Cost-effective with sustainable energy harvesting and a scalable, adaptable framework
Application suitability	Best for precision lab testing and regulatory compliance in controlled environments	Suitable for localized, single-parameter, or small-scale monitoring	Comprehensive, scalable, and real-time monitoring in urban, rural, and resource-constrained settings

The IoT–WSN system proposed in this chapter represents an advancement from other conventional methods and previous IoT solutions, as it provides real-time, multi-parameter, and scalable monitoring capabilities with high energy efficiency and utilizes edge intelligence. While laboratory methods continue to be unmatched in precision, and some IoT systems are tailored toward local needs, the present framework proposes a balanced, practical approach for very large-scale comprehensive monitoring with speed for application in diverse and resource-constrained environmental monitoring.

3.8 CONCLUSION

This chapter summarized the development and establishment of an IoT–WSN-enabled real-time water monitoring system. The combination of robust low-power sensors, hybrid communication, energy harvesting through renewable resources, and cloud analytics has addressed the limitations of manually operated and earlier IoT-tested monitoring systems, as well as their protocols and methodologies. Experiments demonstrated that the system can provide sufficient measurement accuracy, sustainable and stable energy continuity, and scalable performance (e.g., urban and rural circumstances). In comparison to traditional laboratory-based monitoring systems, the proposed water monitoring system offers real-time monitoring, rapid response where conditions demand, and can provide monitoring services over broad areas. The proposed system provides capabilities denied to existing systems, namely, monitoring multiple parameters, an adaptive energy-efficient duty cycle, and anomaly detection at the edge. A 50-node deployment verified its scalability and resilience to failure. The IoT–WSN platform provides a smart, efficient alternative to traditional forms of water monitoring over large scales. It demonstrated that IoT–WSN-enabled monitoring solutions can play a significant role in sustainable water management. There are still challenges to overcome, such as potential sensor fouling or maintenance; however, the documented results demonstrate resilience, dependability, and reliability. Overall, the system has demonstrated the potential to improve water quality management and water quality decisions.

REFERENCES

1. L. Parra, A. Ahmad, S. Sendra, J. Lloret, and P. Lorenz, "Combination of machine learning and RGB sensors to quantify and classify water turbidity," *Chemosensors*, vol. 12, no. 3, p. 34, 2024, doi: 10.3390/chemosensors12030034.
2. M. McGarrity and F. Zhao, "Graphene-based chemiresistor sensors for drinking water quality monitoring," *Sensors*, vol. 23, no. 24, p. 9828, 2023, doi: 10.3390/s23249828.
3. H. Sharma, J. Singh, G. Singh, S. K. Chandel, and L. Sharma, "Introduction to water suitability," *Water Suitability Analysis: Advanced Research Approaches for Sustainable and Resilient Resource Management*, pp. 1–22, 2026. doi: 10.1201/9781003683544-1.
4. T. P. Lambrou, C. C. Anastasiou, C. G. Panayiotou, and M. M. Polycarpou, "A low-cost sensor network for real-time monitoring and contamination detection in drinking water distribution systems," *IEEE Sensors Journal*, vol. 14, no. 8, pp. 2765–2772, 2014, doi: 10.1109/JSEN.2014.2316414.
5. T. Bhatnagar et al., "Unlocking the strength of nanocomposites: mitigating the impact of nanoparticle agglomeration on tensile performance," *E3S Web of Conferences*, 2024, doi: 10.1051/e3sconf/202458801007.
6. S. S. Kumar, B. V. S. Rao, and J. R. Prasad, "Design and development of a water quality monitoring system by using IoT," *International Journal of Emerging Trends in Engineering Research*, vol. 8, no. 9, pp. 6441–6447, 2020, doi: 10.30534/ijeter/2020/245892020.
7. S. Popli, R. K. Jha, and S. Jain, "A survey on energy efficient narrowband internet of things (NBIoT): architecture, application and challenges," *IEEE Access*, vol. 7, pp. 16739–16776, 2019, doi: 10.1109/ACCESS.2018.2881533.
8. R. K. Sandal, S. Sharma, A. Chauhan, and L. Sharma, "Experimental methods in water quality assessment," *Water Suitability Analysis: Advanced Research Approaches for Sustainable and Resilient Resource Management*, 2026, pp. 23–44. doi: 10.1201/9781003683544-2.
9. M. Zhu et al., "A review of the application of machine learning in water quality evaluation," *Eco-Environment & Health*, vol. 1, pp. 107–116, 2022, doi: 10.1016/j.eehl.2022.06.001.
10. L. Sharma, S. Singh, A. Anand, and A. Kumar, *Water Suitability Analysis: Advanced Research Approaches for Sustainable and Resilient Resource Management*. 2026. doi: 10.1201/9781003683544.
11. M. Hong, K. Kim, and Y. Hwang, "Arduino and IoT-based direct filter observation method monitoring the color change of water filter for safe drinking water," *Journal of Water Process Engineering*, vol. 49, p. 103158, 2022, doi: 10.1016/j.jwpe.2022.103158.
12. N. Nasir et al., "Water quality classification using machine learning algorithms," *Journal of Water Process Engineering*, vol. 48, p. 102920, 2022, doi: 10.1016/j.jwpe.2022.102920.
13. J. Huan, H. Li, F. Wu, and W. Cao, "Design of water quality monitoring system for aquaculture ponds based on NB-IoT," *Aquacultural Engineering*, vol. 90, p. 102088, 2020, doi: 10.1016/j.aquaeng.2020.102088.
14. O. O. Ajayi, A. B. Bagula, H. C. Maluleke, Z. Gaffoor, N. Jovanovic, and K. C. Pietersen, "WaterNet: a network for monitoring and assessing water quality for drinking and irrigation purposes," *IEEE Access*, vol. 10, pp. 48318–48337, 2022, doi: 10.1109/ACCESS.2022.3172274.
15. N. Sharma, A. Anand, and L. Sharma, "Water security management: future research directions and emerging tools," *Water Suitability Analysis: Advanced Research Approaches for Sustainable and Resilient Resource Management*, 2026, pp. 181–199. doi: 10.1201/9781003683544-10.
16. N. A. Cloete, R. Malekian, and L. Nair, "Design of smart sensors for real-time water quality monitoring," *IEEE Access*, vol. 4, pp. 3975–3990, 2016, doi: 10.1109/ACCESS.2016.2592958.
17. M. S. U. Chowdury et al., "IoT based real-time river water quality monitoring system," *Procedia Computer Science*, vol. 155, pp. 161–168, 2019, doi: 10.1016/j.procs.2019.08.025.
18. A. T. Demetillo, M. V Japitana, and E. B. Taboada, "A system for monitoring water quality in a large aquatic area using wireless sensor network technology," *Sustainable Environment Research*, vol. 29, p. 12, 2019, doi: 10.1186/s42834-019-0009-4.
19. K. S. Adu-Manu, C. Tapparello, W. B. Heinzelman, F. A. Katsriku, and J.-D. Abdulai, "Water quality monitoring using wireless sensor networks: Current trends and future research directions," *ACM Transactions on Sensor Network*, vol. 13, no. 1, pp. 4:1–4:41, 2017, doi: 10.1145/3005719.
20. S. Geetha and S. Gouthami, "Internet of things enabled real time water quality monitoring system," *Smart Water*, vol. 2, p. 1, 2017, doi: 10.1186/s40713-017-0005-y.

21. E. E.-D. Hemdan, Y. M. Essa, M. Shouman, A. El-Sayed, and A. N. Moustafa, "An efficient IoT based smart water quality monitoring system," *Multimedia Tools and Applications*, vol. 82, no. 19, pp. 28827–28851, 2023, doi: 10.1007/s11042-023-14504-z.
22. M. Miller, A. Kisiel, D. Cembrowska-Lech, I. Durlik, and T. Miller, "IoT in water quality monitoring—are we really here?," *Sensors*, vol. 23, no. 2, p. 960, 2023, doi: 10.3390/s23020960.
23. V. Lakshmikantha, A. Hiriyannagowda, A. Manjunath, A. Patted, J. Basavaiah, and A. A. Anthony, "IoT based smart water quality monitoring system," *Global Transitions Proceedings*, vol. 2, no. 2, pp. 181–186, 2021, doi: 10.1016/j.gltp.2021.08.062.
24. P. Sharma, A. Thakur, P. Thakur, S. Chaudhary, and L. Sharma, "Impact of Mn2+ doping on morphological, optical and electrical properties of Co2W-Type barium ferrite," in *AIP Conference Proceedings*, 2026. doi: 10.1063/5.0317860.
25. S. Pasika and S. T. Gandla, "Smart water quality monitoring system with cost-effective using IoT," *Heliyon*, vol. 6, no. 7, p. e04096, 2020, doi: 10.1016/j.heliyon.2020.e04096.
26. R. K. Nishan, S. Akter, R. I. Sony, Md. M. Hoque, M. J. Anee, and A. Hossain, "Development of an IoT-based multi-level system for real-time water quality monitoring in industrial wastewater," *Discover Water*, 2024, doi: 10.1007/s43832-024-00092-y.
27. F. Jan, N. Min-Allah, and D. Düştegör, "IoT Based Smart Water Quality Monitoring: Recent Techniques, Trends and Challenges for Domestic Applications," *Water (Basel).*, vol. 13, no. 13, p. 1729, 2021, doi: 10.3390/w13131729.
28. C. Axiotidis, E. Konstantopoulou, and N. Sklavos, "A wireless sensor network IoT platform for consumption and quality monitoring of drinking water," *Discover Applied Sciences*, vol. 7, p. 15, 2025, doi: 10.1007/s42452-024-06384-1.
29. K. Lal, S. Menon, F. Noble, and K. M. Arif, "Low-cost IoT based system for lake water quality monitoring," *PLoS One*, vol. 19, no. 3, p. e0299089, 2024, doi: 10.1371/journal.pone.0299089.
30. M. Attri and L. Sharma, "GIS and Remote Sensing Applications," *Water Suitability Analysis: Advanced Research Approaches for Sustainable and Resilient Resource Management*, 2026, pp. 63–84. doi: 10.1201/9781003683544-4.
31. R. Wiryasaputra, C.-Y. Huang, Y.-J. Lin, and C.-T. Yang, "An IoT real-time potable water quality monitoring and prediction model based on cloud computing architecture," *Sensors*, vol. 24, no. 4, p. 1180, 2024, doi: 10.3390/s24041180.
32. W. A. Jabbar et al., "Development of LoRaWAN-based IoT system for water quality monitoring in rural areas," *Expert Systems with Applications*, vol. 242, p. 122862, 2024, doi: 10.1016/j.eswa.2023.122862.
33. H. M. Forhad et al., "IoT based real-time water quality monitoring system in water treatment plants (WTPs)," *Heliyon*, vol. 10, no. 23, p. e40746, 2024, doi: 10.1016/j.heliyon.2024.e40746.
34. Y. Singh and T. Walingo, "Smart water quality monitoring with IoT wireless sensor networks," *Sensors*, vol. 24, no. 9, p. 2871, 2024, doi: 10.3390/s24092871.
35. L. Sharma, H. Sharma, J. Kalia, R. K. Sandal, A. Chauhan, and P. Sharma, "Eco-concrete: maximizing strength and minimizing implications of crushed over-burnt on the environment brick substitution," *AIP Conference Proceedings*, 2026. doi: 10.1063/5.0317502.
36. L. Sharma, "Comparative study of composite column behaviour under compressive load," in *AIP Conference Proceedings*, 2026. doi: 10.1063/5.0317503.

Smart Technologies for Water Quality Prediction with AI and ML

4

Gopal Usha

4.1 INTRODUCTION

Urbanization, industrial waste, farm runoff, and global change are all worsening water quality, increasing the need for better monitoring systems. Chemical and biological analyses are still the norm, but they are, by definition, slow. Computer vision and machine learning (ML) enable quick assessments of water quality without physical contact (Doshi et al., 2021; Kumar & Sharma, 2022). This introduction aims to explain the reasons, technological feasibility, and broader social implications of using AI-based water-monitoring systems (Gao et al., 2020; WHO & UNICEF, 2023). Water is one of the most important natural resources, and water quality plays a major role in human health, ecosystems, and economic growth. Due to the growth of industries and cities, farming techniques have become more intensive. Weather conditions are also changing with rapidly growing industries, leading to significant changes in water bodies, which have intensified issues related to heavy metals, pesticides, plastics, organic pollutants, and microbes.

These pollutants harm aquatic ecosystems and affect drinking water sources, food availability, and human health (United States Environmental Protection Agency, 2022). Because of this, there is a growing need for new, scalable, and real-time water quality monitoring systems that go beyond traditional lab testing. Chemical tests, chromatography, and microbiological culture are the main methods used to check water quality. These methods are very accurate, but they take a lot of time and effort, and they need skilled workers and special facilities. These kinds of tools could make environmental monitoring more accessible to everyone, reduce the unfairness of access to clean water, and help policymakers make decisions about water governance based on facts. This is how the convergence of environmental

DOI: 10.1201/9781003714217-4

science, AI, and public health represents a significant shift in how individuals can safeguard water supplies. Not only will we require improved technology, but we will also require ethical, transparent, and inclusive AI systems to ensure that water monitoring is equitable, trustworthy, and sustainable at the local, regional, and international levels. Next, we will discuss computer vision and its principles for water monitoring in detail.

4.2 COMPUTER VISION PRINCIPLES IN WATER MONITORING

Figure 4.1 explains the computer vision principles used for water monitoring. The following section explains the steps in detail.

4.2.1 Image Acquisition and Sensing Platforms

The precision of computer vision–based water monitoring relies profoundly on the quality and nature of images captured. Camera-based systems are the most straightforward and common method (Gao et al., 2020), generally mounted along riverbanks, reservoirs, or water treatment plants to capture ongoing visual information. These systems give useful data on turbidity levels, floating material, and color changes in the water. In deeper or hazardous environments, underwater remotely operated vehicles are critical. They are outfitted with cameras and occasionally hyperspectral sensors for close observation of suspended sediments, benthic communities, and potential sources of pollutants. Drones, also known as unmanned aerial vehicles (Zhang & Qiu, 2022), are used for wide-ranging and large-scale observations. Scientists can identify locations of eutrophication, oil spills, or unusual discharges by equipping drones with RGB, thermal, or multispectral cameras. Fixed riverbank monitoring stations are also important for making observations. The combination of cameras and IoT connectivity enables real-time monitoring and immediate alerts if something unexpected happens. These acquisition systems provide both detailed views and broader landscape views that are beneficial for various water observations.

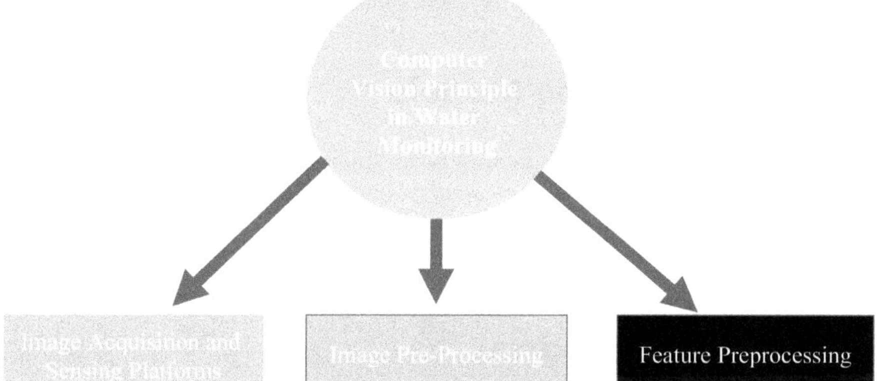

FIGURE 4.1 Computer vision principal in water monitoring.

4.2.2 Image Pre-processing

In image pre-processing, raw image data from natural water bodies is often distorted by noise, changes in light, and water glints, which makes it hard to see useful information. Therefore, pre-processing steps are needed to ensure that the data is of high quality. Denoising methods, such as median filtering and/or wavelet techniques (Islam et al., 2020), help to get rid of noise that comes from the way sensors work, water turbulence, or water scattering. Water environments are very sensitive to changes in light caused by factors such as sunlight, shadows, and the seasons. Illumination adjustment is also used to ensure that light and contrast changes are uniform. Techniques such as histogram equalization help with this.

4.2.3 Feature Extraction

After image pre-processing, we examine the images to extract features that will indicate water quality. Texture features are particularly useful for detecting turbidity (Islam et al., 2020) or foam because they report on changes in pixel intensities that correspond with smooth, rough, or grainy water surfaces. Algorithms that achieve higher levels such as the gray-level co-occurrence matrix and local binary patterns can assist in measuring these subtle changes. Color histograms are also a very basic feature set (Li et al., 2021) and record the color distribution of an image. Since the color and context of water change when there is algal growth, a chemical runoff event, or sediment presence, color histograms offer quantifiable signatures for classification. In addition to visible color, spectral signatures based on multispectral and hyperspectral images are reliable diagnostic features. Each contaminant or naturally occurring substance has an identified reflectance pattern, making it possible to measure chlorophyll, dissolved organic matter, or even microplastics. Shape features are very important, especially when identifying drifting contaminants such as plastics or oil sheens. Contour detection and morphological analysis can yield measures of size, boundaries, and geometric properties of these detectable products.

4.3 ML APPROACHES

Varying modeling paradigms, from conventional supervised learning to state-of-the-art deep learning and mixed strategies, are employed based on data type, task complexity, and available annotated training sets. Figure 4.2 explains ML approaches for water quality prediction in detail.

4.3.1 Supervised Learning Models

Supervised learning models are very common in water quality monitoring because they can categorize samples into known categories such as clean or polluted. Decision trees offer a robust but simple framework where the data is partitioned according to rules learned from attributes such as the color histogram or turbidity.

4.3.2 Deep Learning Models

Convolutional neural networks (CNNs) specifically dominate in learning spatial hierarchies of features automatically without manual feature engineering. CNNs are commonly employed for detecting

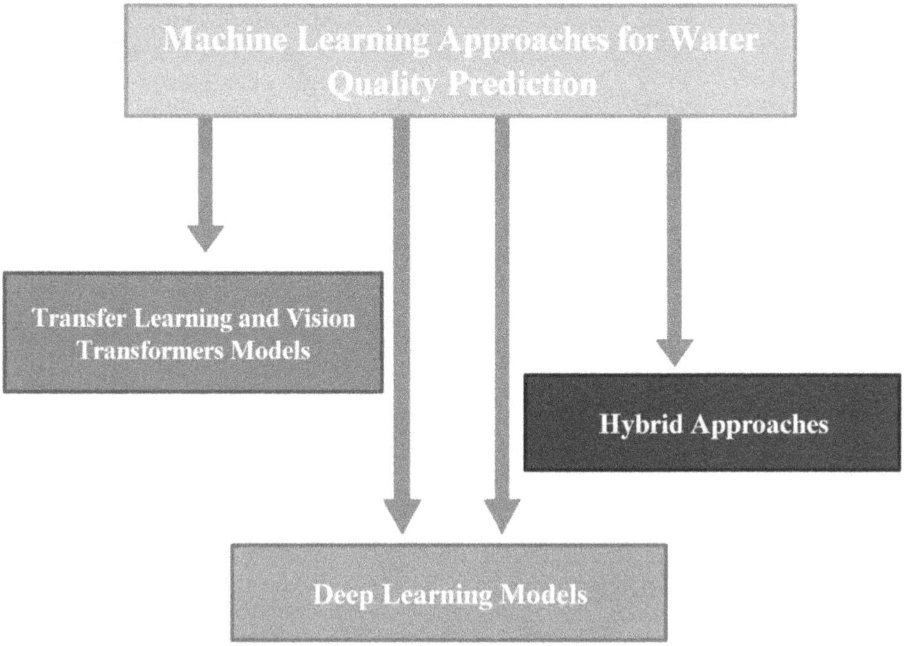

FIGURE 4.2 Machine learning approaches for water quality prediction.

floating debris, algae bloom identification, and oil slicks on the surface of water. For example, CNN models may be trained on aerial photography by drones to separate clear water, clumps of phytoplankton, or plastic debris. Their hierarchical structure enables the model to learn low-level edges and textures in earlier layers, progressing to more abstract concepts such as forms and patterns in later layers. Furthermore, CNN-based segmentation models can delineate polluted areas, offering classification as well as spatial localization of pollution. Deep learning models need quantities of labeled data, but when available, they always outperform traditional approaches in accuracy, robustness, and scalability. Next, we will discuss techniques involving transfer learning and vision transformers (ViTs).

4.3.3 Transfer Learning and ViTs

One of the most important challenges in water quality monitoring is the lack of annotated datasets, given that labeling is domain-specific and resource-intensive. Transfer learning is an approach that circumvents the limitations in training sets by leveraging pretrained models, such as ResNet, VGG, or EfficientNet, because they are pretrained on massive datasets such as ImageNet. Transfer learning techniques are adopted to use pretrained networks. The feature representation makes transfer learning possible on smaller water datasets with only a few annotations. This prevents further high training time and computational costs without sacrificing high performance. Besides CNN-based transfer learning, the rise of ViTs presents opportunities here as well. ViTs break images down into patches and then feed them through self-attention, which can capture long-range dependencies and contextual relationships throughout the water scene. For example, ViTs could differentiate between global oil slicks and local algae blooms based on what they see overall in the image. In summary, transfer learning and ViTs are state-of-the-art methods to combat the insufficiency of datasets and improve generalization across aquatic ecosystems (Banjara et al., 2025).

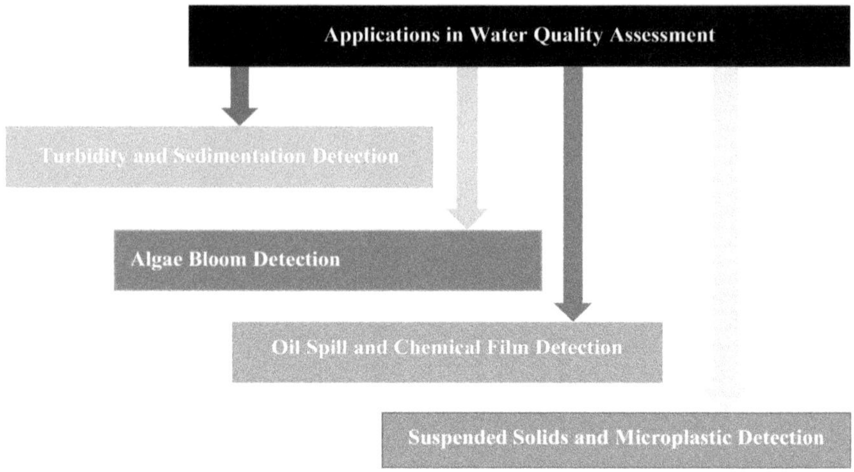

FIGURE 4.3 Applications in water quality assessment.

4.3.4 Hybrid Approaches

In other frameworks, professional-specified water quality regulations—e.g., thresholds in turbidity levels or spectral indices—are combined with AI predictions to yield human-interpretable explanations. Hybrid models, therefore, bring together the expert knowledge of environmental science with the predictive ability of ML, resulting in systems that are both transparent and accurate. This renders them particularly useful in policy-informed settings where accountability and explainability are necessary. Figure 4.3 explains applications in water quality assessment in detail.

4.4 APPLICATIONS IN WATER QUALITY ASSESSMENT

Applications of water quality assessment in various frameworks are explained below in detail.

4.4.1 Turbidity and Sedimentation Detection

Automatic turbidity estimation is based on the effect of suspended particles on water-leaving light by changing it via absorption and scattering. In RGB images, turbid water usually presents high green–red intensity and lower local contrast; in multispectral imagery, backscattering particles enhance reflectance in the visible wavelengths (particularly green) and level off spectral shape. Computer vision systems are trained to map color ratios and texture measures (e.g., local variance and entropy) into turbidity units (NTU) via regression or supervised models learned against field measurements. Temporal differencing and optical-flow features aid in distinguishing transient sediment plumes from constant background color for dynamic rivers. Geometry-sensitive pre-processing—sun-glint removal, atmospheric/underwater correction, and region-of-interest masking close to banks—minimizes bias. Austere deployments combine fixed cameras or UAV (Unmanned Ariel Vehicle) overflights with sporadic grab-sample calibration whereby the model learns to respond to site-specific particle size distributions and natural dyes (e.g., tannins), generating near–real-time turbidity maps and alarms during runoff or dredging activities.

4.4.2 Algae Bloom Detection

Harmful algal blooms exhibit unique spectral behavior due to pigments such as chlorophyll-a and phycocyanin. In this, the spatial morphology of the homogeneous patches with blurry, filamentous margins—and temporal growth patterns (accelerated development over hours to days, wind-driven accumulation on leeward shores)—are used to further differentiate. To minimize false alarms, systems combine temperature (thermal imagery) and motion signals (bloom mats respond differently from oil or foam) and add hydrometeorological context (recent nutrient loads, sunny tranquil spells). Outputs comprise bloom probability maps, estimated pigment concentration, and early-warning triggers when thresholds linked to public health advice are surpassed (Zhang & Qiu, 2022).

4.4.3 Oil Spill and Chemical Film Detection

Oil and surfactant films (Chakraborty & Mitra, 2021) alter surface roughness and create thin-film interference manifesting as iridescent sheens or flat, low-texture slicks. Computer vision takes advantage of three mutually supportive signals: (1) spectral—weak hue changes and diminished blue reflectance; (2) texture—attenuated capillary waves create smoother, lower-frequency patterns measurable through gray-level co-occurrence matrix or wavelet energy; and (3) geometry—coherent, elongated forms aligned with wind/current directions. Edge detectors and active contours define slick edges; polarization and sun-glint suppression restore detail around specular highlights. Classification phases differentiate petroleum slicks from imitators such as biogenic films, algal scums, or peaceful "wind-shadow" areas using a blend of color invariants, micro-texture, and contextual priors (nearness to outfalls, ship tracks). In thick spills, segmentation models estimate area and relative thickness class (sheen, rainbow, metallic) to aid response planning. Permanent riverbank cameras can trigger instant alerts, and UAV missions chart extent and drift for containment.

4.4.4 Suspended Solids and Microplastic Detection

Total suspended solids (TSS) and microplastics (Akhtar et al., 2022) are assessed on two complementary scales. At the scene scale, vision models are capable of predicting TSS from spectral-textural information in a similar way to turbidity (Liang et al., 2023) and site-specific calibrations to mg/L of TSS, as well as plume tracking, calculating load and settling zone boundaries, etc. At the micro-scale, imaging devices—either stereo microscopes, flow-imaging cells, or holographic devices—image particles using transmitted or dark-field illumination. Pre-processing steps, such as background subtraction and deconvolution, separate out particulates, and segmentation provides morphometrics (area, aspect ratio, circularity, perimeter fractal dimension) and surface texture features of microplastics. In terms of microplastics, the use of fluorescence imaging, dyes such as Nile Red, implemented in lab protocols where feasible, and material-aware classifiers (e.g., the use of CNNs trained on representative polymer examples, e.g., PLA, TPU, and PETE) improves separation from organic detritus and minerals. Once we arrive at classifications to note post-classification quality control (size-bin histograms, confidence levels), we can verify a representative subset using spectroscopy (e.g., Raman/FTIR) to keep vision models accountable should there be anomalies. All of these end products (i.e., particle counts, size distribution, and polymer-likelihood scores/concentration) formulate dashboards that provide assessment tools for monitoring hotspots, assessing mitigating factors (e.g., booms/filters), and dimensions for resampling/creation when anomalies are identified. We will move to the discussion of data integration/convergence and the Internet of Things, moving awareness of best practices.

4.5 DATA INTEGRATION AND IOT CONVERGENCE

4.5.1 Sensor Fusion

Sensor fusion (Kumar & Sharma, 2022) in aquatic monitoring involves integrating camera-collected visual data with physicochemical data (pH, dissolved oxygen (DO), conductivity, temperature, turbidity sensors, etc.) so the entire collection is more informative than its constituent parts. Phenomena that are visually detected—unusual color shifts, surface glitter, or spatial anomalies—may not possess the complete context to adequately assess water quality by themselves. A decrease in DO reported by an electrochemical sensor, coupled with an increase in green reflectance and a thermal anomaly in imagery, for instance, is considerably greater evidence of a toxic algal bloom than either indicator alone. Fusion thus enhances detection sensitivity, suppresses false positives due to look-alikes (e.g., tannin-colored water vs. sediment plumes), and provides orthogonal confirmation essential for regulatory or public health response (Figure 4.4).

Technically, fusion can be implemented at several levels. Data-level fusion consumes raw signals (pixel intensities + sensor voltages) and performs joint pre-processing and normalization; feature-level fusion extracts modality-specific descriptors (CNN embeddings from images, statistical summaries, or spectral indices from sensors) and concatenates or carefully weights them into a single feature vector consumed by downstream models; decision-level fusion trains separate classifiers per modality and aggregates their outputs with voting, weighted averaging, or Bayesian model averaging. Temporal fusion algorithms (Kalman/particle filters, state-space models, or recurrent neural networks) smooth and synchronize asynchronous streams such that brief sensor spikes or camera frames don't set off false alarms. State-of-the-art deep learning (Abdelaziz et al., 2023) methods typically employ multimodal networks with attention modules that learn what modality to believe in various environmental situations (e.g., believe DO more at night and imagery more during the day).

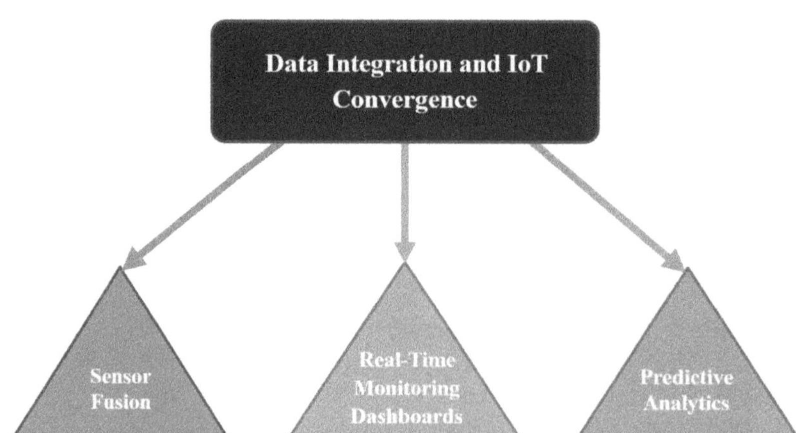

FIGURE 4.4 Data integration and IoT convergence.

4.5.2 Real-Time Monitoring Dashboards

Operational concerns include scalability (sharding time-series data, caching map tiles), latency (edge filtering versus cloud compute trade-offs), data provenance and retention policies (who can change thresholds or labels), and security (TLS, device authentication, logging). Offline and low-bandwidth modes—local edge dashboards or sync-once connectivity—are important for remote field deployments. Finally, dashboards should expose machine-readable APIs so third parties (emergency services, researchers) can integrate the monitoring (Zhou et al., 2024) data into their workflows.

4.5.3 Predictive Analytics

Predictive analytics converts past and real-time combined data into prospective predictions and timely warning signals—foreseeing spikes in turbidity following heavy rainfall, predicting the onset of algal blooms, or forecasting sensor breakdowns. A modeling toolbox is employed based on problem size and data richness: statistical time-series techniques (ARIMA, exponential smoothing) for short-term baselines; tree-based ensemble algorithms (random forest, XGBoost) for rich feature tabular forecasting; and deep learning techniques long short-term memory (LSTM, Temporal Convolutional Networks, or Transformer-based time-series models) for intricate, non-linear spatiotemporal patterns.

4.6 CHALLENGES AND LIMITATIONS

Data availability: lack of sufficiently large, labeled datasets under diverse water conditions. Supervised ML with high performance requires diverse, accurately labeled data, yet water-monitoring datasets are fragmented, costly to produce, and frequently biased toward a handful of sites or sensor types. Ground-truthing requires sampling in the field and laboratory tests (e.g., chlorophyll, TSS, and microplastic confirmation), which are labor-intensive and expensive, and professional labeling (identifying algal mats versus vegetation or biogenic films) is skill-specific. The outcome is models that fail to generalize geographically, sensor by sensor, or season by season. Practical solutions are (1) apply transfer learning and self-/semi-supervised learning to minimize label requirements; (2) generate synthetic and augmented data (rendered scenes, spectral mixing) to increase diversity; (3) structure community datasets and benchmarks with standardized metadata (sensor specs, time, ground-truth protocols); (4) employ citizen science and structured crowdsourcing with validation layers; and (5) employ active learning so models query the most informative labels, reducing expert workload. Next, we will discuss proposed system architecture in detail.

4.7 SYSTEM ARCHITECTURE

The proposed system is implemented through a three-tier architecture consisting of the edge layer, the cloud layer, and the application layer. At the edge layer, data is captured through smart cameras, UAVs, and IoT-enabled physicochemical sensors. Lightweight deep learning models such as MobileNet and quantized CNNs are deployed on embedded hardware devices, including NVIDIA Jetson and Google

Coral units, to enable real-time inference directly at the source. An edge computing approach is used to minimize latency and reduce dependence on continuous internet connectivity. Then, at the cloud layer, computationally intensive tasks such as deep learning model training, multimodal data fusion, and predictive analytics are executed, which are used for efficient storage and analytic techniques. The centralized servers are responsible for managing large-scale data storage, re-training of models, and integration of environmental datasets. In the application layer, a real-time dashboard presents the processed results in the form of geospatial visualizations. Additionally, it displays time-series plots and automated alerts, providing actionable intelligence for policymakers, environmental agencies, and communities. Figure 4.5 explains the proposed system architecture in detail. The proposed system was implemented using a three-tier architecture consisting of edge, cloud, and application layers. In the edge layer, various components such as smart cameras, UAVs, and IoT-enabled physicochemical sensors are deployed to collect multimodal data such as images, videos, and environmental parameters (pH, turbidity, air quality, temperature, and humidity). The lightweight deep learning models such as MobileNet and quantized CNNs, are executed on embedded hardware devices such as NVIDIA Jetson and Google Coral, enabling real-time inference at the data source and reducing latency. In the cloud

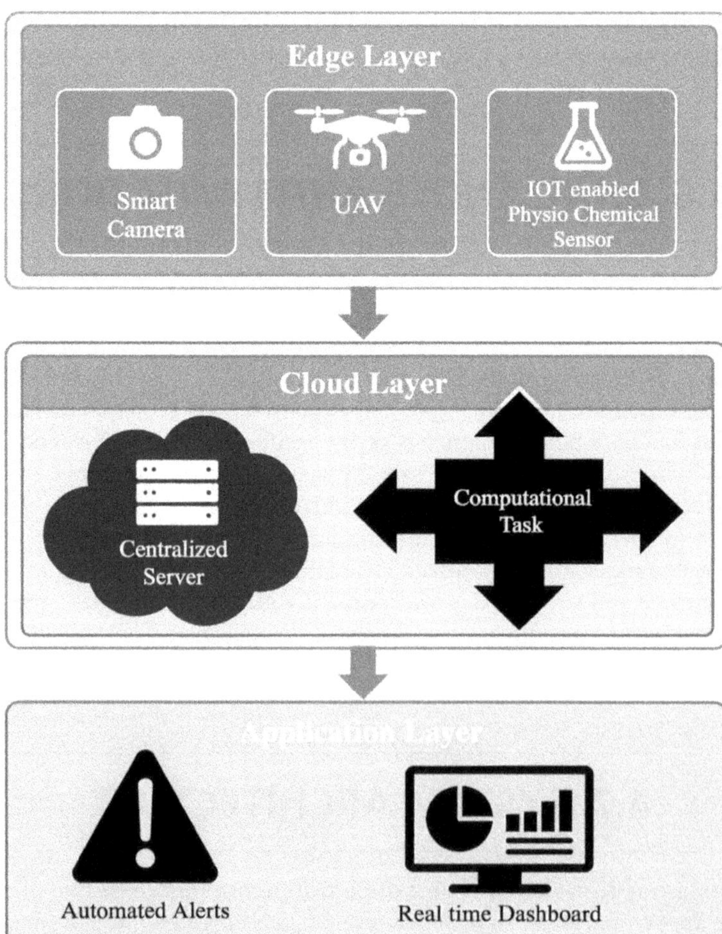

FIGURE 4.5 Layers of proposed system architecture in detail.

layer, high-performance servers manage large-scale data storage, multimodal fusion, and the training and re-training of deep learning models. This layer is integrated with external environmental datasets such as meteorological records and hydrological flows which are used to improve prediction accuracy.

Finally, at the application layer, a web-enabled dashboard is deployed to provide stakeholders with geospatial visualizations, time-series analysis, and real-time alerts. This environment setup ensures both low-latency edge inference and computationally intensive cloud-based learning, thereby providing an efficient and scalable framework for environmental monitoring. Figure 4.5 explains the proposed system architecture in detail with various layers. Next, we will discuss the data pre-processing pipeline in detail.

4.7.1 Data Processing Pipeline

The data processing pipeline starts with obtaining the visual and sensor-based inputs through IoT protocols such as MQTT and HTTP. The data pre-processing is carried out either at the edge or the cloud layer. Then, the feature extraction is conducted using CNNs and spectral descriptors. The inference step applies supervised and deep learning models to classify contamination, which is used to predict pollutant levels and identify anomalies. Next, sensor fusion is performed. The visual indicators such as algal blooms are cross-validated with physicochemical measurements such as DO or turbidity. The outputs are integrated through dashboards, which allow users to monitor trends, visualize spatial pollution patterns, and receive early-warning notifications.

4.7.2 Deployment Scenarios

Various scenarios are used to deploy the proposed system. This deployment facilitates continuous monitoring. Continuous monitoring is achieved through fixed stations installed at rivers and reservoirs to provide uninterrupted surveillance and automatic contamination alerts. Event-driven monitoring is facilitated by UAV deployments during specific incidents such as oil spills or rapid algal bloom proliferation. Industrial discharge points are also monitored with high frequency. The IoT-enabled imaging and sensing ensure regulatory compliance. In addition to real-time monitoring, predictive analytics techniques are integrated using statistical and ML models such as ARIMA, LSTM, and Transformer-based architectures to forecast turbidity (Islam et al., 2020) spikes following rainfall events, predict bloom onset under favorable nutrient conditions, and estimate pollutant propagation along river systems (Sharma et al., 2024).

4.8 RESULTS AND ANALYSIS

The dataset used for this study is a combination of image data and IoT sensor data to create a comprehensive overview of water quality. The images in the dataset are categorized into labeled categories such as Clean, Turbid, Algal Bloom, Oil Spill, and Microplastic, all of which help inform visual classification tasks through CNN, ResNet, and transfer learning models.

The image data came from publicly available water quality image repositories, satellite imagery, or customized field datasets. Besides images, the sensor data elucidates important physicochemical parameters, namely pH, turbidity (NTU), DO (mg/L), and temperature from the IoT water quality sensors. Alongside images, we have the capability of sensor fusion between visual features obtained from images and sensor measurements, improving the accuracy and robustness of the model compared to image classification alone.

The dataset is divided into training, validation, and testing parts to reduce biased evaluation. Pre-processing steps such as image resizing, normalization, data augmentation, and scaling for the sensor data occur before training. Overall, the dataset's multimodality is very useful for constructing reliable water quality predicting and monitoring systems. The implementation consists of ML and deep learning models (decision tree, random forest, SVM, CNN, ResNet, and transfer learning) using a dataset of water quality images (Clean, Turbid, Algal Bloom, Oil Spill, and Microplastic) and IoT-based sensor measurements (pH, turbidity, and DO). The data pre-processing, training, and evaluation are integrated using Python with libraries such as TensorFlow/Keras, PyTorch, Scikit-learn, NumPy, and Pandas, while visualization libraries include Matplotlib and Seaborn. Model evaluations include accuracy, precision, recall, F1-score, receiver operating characteristic (ROC) curves and confusion matrices, with the most effective model using sensor fusion. The system is deployed as a real-time dashboard using Flask/Django integration for the backend and implemented as an IoT-enabled system to support the prediction of water quality and constant water quality monitoring (UNESCO, 2021). Figure 4.6 illustrates the comparison of the accuracy of different ML models.

While checking the accuracy, the decision tree performs the lowest at around 70%, while random forest, SVM, and CNN achieve higher accuracies between 80% and 85%. Apart from everything, transfer learning outperforms with about 90% accuracy, which highlights the best performance of transfer learning. From this, the advanced ensemble and deep learning methods provide more reliable results than simple models such as decision tree.

Figure 4.7 shows the comparison of Precision, Recall, and F1-Score for random forest, CNN, and ResNet models. Random forest performs moderately, with values around 0.8–0.84, while CNN achieves stronger results, with all metrics close to 0.9. ResNet outperforms both, with Precision, Recall, and F1-Score above 0.93, indicating it is the most accurate and reliable model among the three. Figure 4.8 explains that the confusion matrix for the CNN model shows strong classification performance across five water quality categories: Clean, Turbid, Algal Bloom, Oil Spill, and Microplastic. Most samples are correctly classified along the diagonal, with Clean (45), Turbid (40), Algal Bloom (42), Oil Spill (47), and Microplastic (47) achieving high accuracy. Only a few misclassifications occur, such as some Turbid samples predicted as Algal Bloom and minor overlaps between adjacent classes. Overall, the

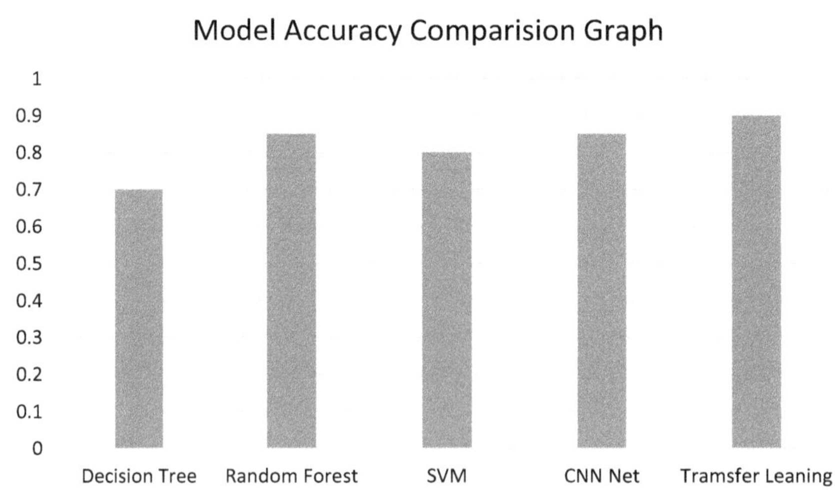

FIGURE 4.6 Model accuracy comparison graph.

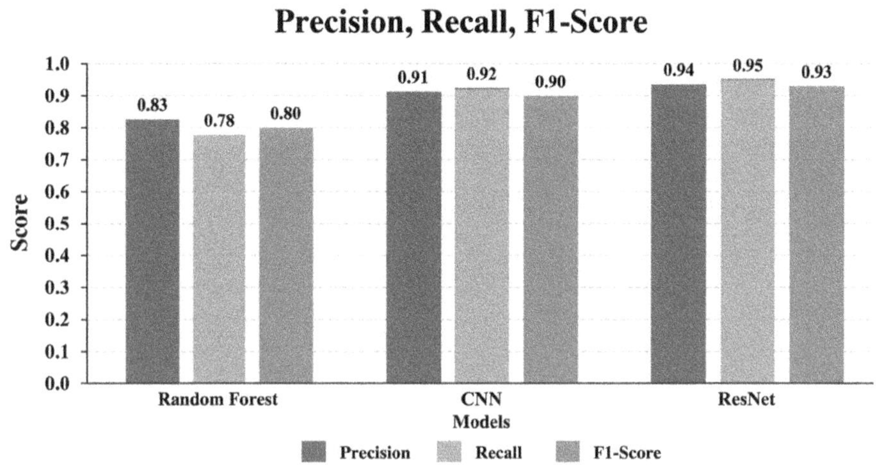

FIGURE 4.7 Precision, recall, and F1 score.

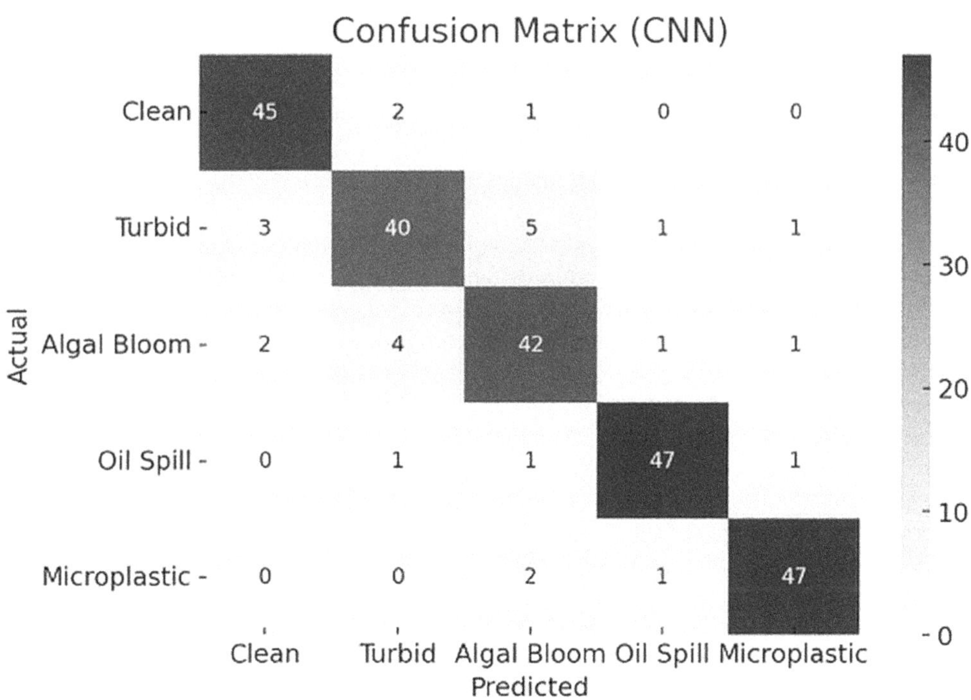

FIGURE 4.8 Confusion matrix.

CNN demonstrates high reliability in distinguishing water quality types, with only small errors indicating strong generalization. Figure 4.9 illustrates the effect of sensor fusion on model accuracy, comparing performance using only image data versus combining image data with IoT sensor inputs. When relying solely on images, the model achieves an accuracy of about 88%, whereas incorporating IoT data significantly boosts accuracy to around 96%. This demonstrates that sensor fusion, by integrating multiple data sources, provides more comprehensive information and leads to more accurate and reliable

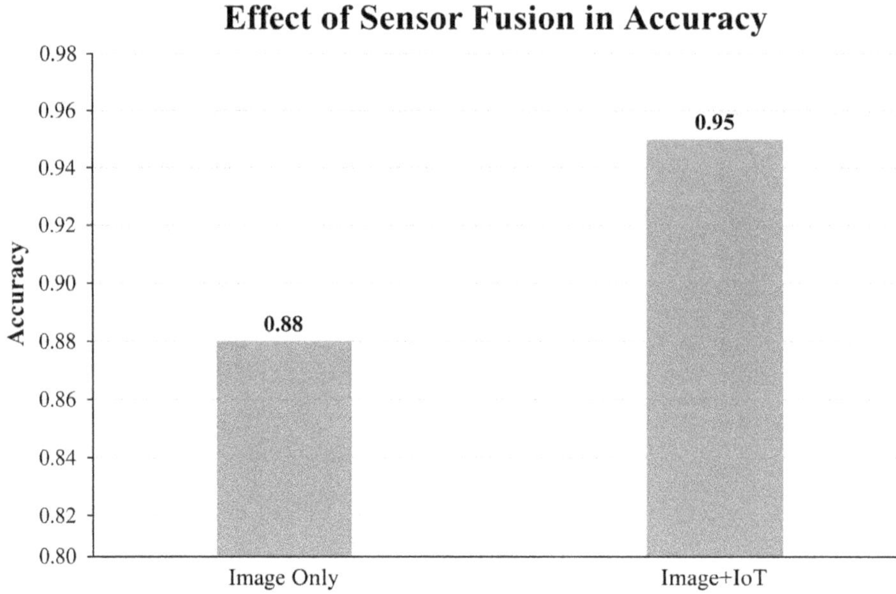

FIGURE 4.9 Effect of sensor fusion on accuracy.

predictions compared to using images alone. The ROC curve in Figure 4.10 illustrates the performance of CNN and random forest models in distinguishing between classes, with the true-positive rate (sensitivity) plotted against the false-positive rate.

The CNN curve (light gray line) lies above the random forest curve (dark gray line), which indicates better classification performance across different thresholds. The area under the curve (AUC) confirms this, with CNN achieving a high AUC of approximately 0.96 compared to random forest's 0.87, meaning that the CNN is more effective at correctly identifying positives, while false positives are minimized.

The diagonal dashed line here represents random guessing (AUC = 0.5), and since both models perform well above it, they are significantly better than chance, with CNN clearly outperforming random forest in predictive accuracy. Figure 4.11 shows a scatter plot of the predicted turbidity values (in NTU) against the ground-truth measurements, with the red dashed line representing the ideal case where predictions perfectly match actual values ($y = x$). The blue data points lie close to this line, indicating that the prediction model is accurately estimating turbidity levels across a wide range from 0 to 100 NTU. There are minor deviations, especially at lower and higher turbidity levels, but the overall alignment suggests strong predictive performance with minimal error. This close clustering around the diagonal demonstrates that the model generalizes well and can reliably estimate turbidity from the input data.

Figure 4.12 represents real-time monitoring of three water quality parameters: pH, turbidity (NTU), and DO (mg/L) for a period of 24 hours. The pH values given by the light gray line remain hardly variable; small oscillations are present around neutral values, which testify to the stable acidity/alkalinity. At the same time, turbidity has shown a reliable growth (the medium-gray line), having jumped from about 10 NTU to more than 60 NTU, which allows one to judge the increased content of suspended particles or contaminants in the water. The line of DO (dark gray) demonstrates a gradual decline from about 8 mg/L to below 5 mg/L, which characterizes possible oxygen depletion as turbidity increases. In general, the output of the dashboard testifies to the deterioration of water quality within the time interval, the increase in turbidity, and the decrease in oxygen being the most disturbing changes.

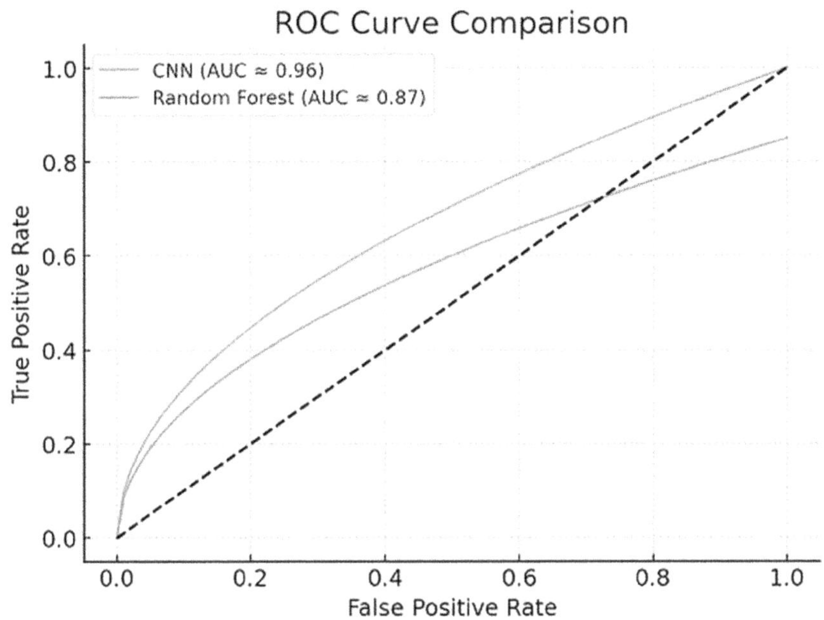

FIGURE 4.10 ROC curve comparison.

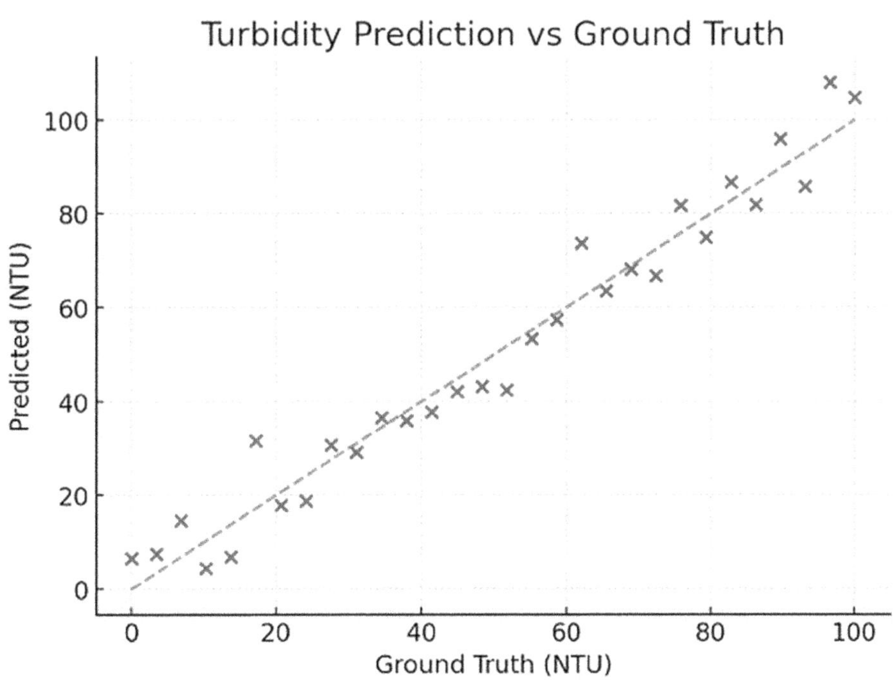

FIGURE 4.11 Turbidity prediction vs. ground truth.

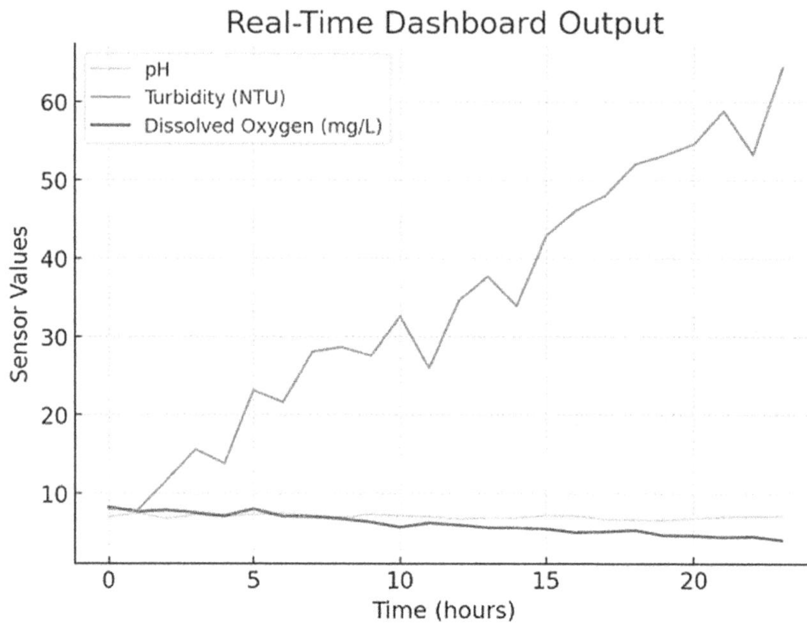

FIGURE 4.12 Real-time dashboard output.

4.9 CONCLUSION AND FUTURE WORK

This chapter highlights that computer vision, ML, and IoT-generated sensor fusion deliver increasingly better water quality monitoring – real-time, automated, and accurate compared to more traditional methods. In particular, experimental results showed deep learning models (CNN, ResNet, and transfer learning) achieved better accuracy, precision, and reliability than more traditional approaches, while sensor fusion further enhances predictive performance. Furthermore, from this chapter, we found that using visual and physicochemical data provided a seamless method for continuous monitoring once implemented with cloud–edge computing and real-time dashboards for observability of sensors, offering a comprehensive early warning of contaminants. All in all, the framework we proposed represents a scalable, intelligent, consistent, and evolving means of environmental management, while also directly contributing to improved protections for public health and support for global SDGs, such as SDG-6 for clean water and sanitation. Future advances in water quality monitoring can focus on engaging ViTs and multimodal deep learning applications to enhance overall accuracy and adaptability across environmental conditions. Large labeled datasets and synthetic data generation will expand the current datasets, which is currently a limitation of existing systems. Scaling will have wide impacts with low-cost IoT sensors, edge AI devices, and satellite-based monitoring. Explainable AI and regulatory-compliance frameworks will allow for transparency and trust-building in future solutions. Additionally, making use of predictive analytics will allow AI to provide early warnings before pollution events occur, centering on proactive water resources management.

REFERENCES

Abdelaziz, A., Elhoseny, M., & Hassanien, A. E. (2023). Edge AI for environmental monitoring using embedded deep learning models. *Future Generation Computer Systems*, 138, 318–330.

Akhtar, N., Prabhu, P., & Shah, M. (2022). Microplastic detection in water using deep learning and microscopy images. *Science of the Total Environment*, 806, 150580. https://doi.org/10.1016/j.scitotenv.2021.150580

Banjara, I. P., Poudel, S., Pariyar, K., Upreti, D., Zafeirakou, A., & Paudel, S. R. (2025). AI for enhanced water quality data imputation: A deep learning perspective. *Environmental Science: Advances*, 4, 817–823. https://doi.org/10.1039/D4VA00367E

Chakraborty, S., & Mitra, P. (2021). Oil spill detection using texture features and deep learning methods. *IEEE Transactions on Geoscience and Remote Sensing*, 59(8), 6781–6794.

Doshi, J., Patel, T., & Bharti, S. (2021). IoT-based smart water quality monitoring systems: A review. *Sensors*, 21(10), 3458. https://doi.org/10.3390/s21103458

Gao, Y., Zhang, Y., & Li, X. (2020). Water quality monitoring using computer vision and machine learning: A review. *Environmental Monitoring and Assessment*, 192(11), 1–18.

Islam, M. S., Rahman, M. M., & Ahmed, T. (2020). Vision-based turbidity estimation using machine learning models. *Journal of Hydrology*, 589, 125169. https://doi.org/10.1016/j.jhydrol.2020.125169

Kumar, A., & Sharma, S. (2022). Sensor fusion techniques for intelligent water quality assessment. *IEEE Sensors Journal*, 22(14), 13890–13902.

Li, S., Wang, Y., & Chen, Q. (2021). Deep learning–based water quality classification using RGB images. *Ecological Informatics*, 61, 101231.

Liang, Z., Sun, Y., & Wang, H. (2023). Transfer learning for water pollution monitoring under limited labeled data. *Expert Systems with Applications*, 213, 118872. https://doi.org/10.1016/j.eswa.2022.118872

Sharma, P., Verma, A., & Singh, R. (2024). Explainable artificial intelligence for environmental decision-making systems. *Artificial Intelligence Review*.

UNESCO. (2021). *Artificial intelligence and water security: Emerging technologies for achieving SDG 6*. UNESCO Publishing. https://unesdoc.unesco.org

United States Environmental Protection Agency. (2022). *Water quality monitoring technologies and standards*. U.S. EPA. https://www.epa.gov

World Health Organization, & UNICEF. (2023). *Guidelines on drinking-water quality and public health*. World Health Organization. https://www.who.int

Zhang, C., & Qiu, Z. (2022). Algal bloom detection using convolutional neural networks and multispectral imagery. *Remote Sensing*, 14(3), 721. https://doi.org/10.3390/rs14030721

Zhou, Y., Li, X., & Zhang, J. (2024). Vision transformers for environmental image analysis and monitoring. *Pattern Recognition*, 146, 109967. https://doi.org/10.1016/j.patcog.2023.109967

5 Blockchain for Secure and Transparent Water Resource Management

Sangeetha Nagamani, Annamalai Selvarajan, Raghini Mohan, John Peter Vincent Paul, Sathiya Priya Rajavelu, and Radhika Rajavelu

5.1 INTRODUCTION

Water crisis is one of the serious issues in the world, and it directly affects the agricultural sector. Contaminants, which include chemicals, fertilisers, and pharmaceuticals, invade the aquatic system and affect human wellness. The major cause of global imbalance and deficit is the inefficient allocation of water resources [1]. Obsolete distribution systems, ineffective governance, a lack of continuous evaluation, and disparity in access are some of the factors that commonly cause loss and overuse in certain sectors [2]. Traditional water management is hampered by a centralised framework, sluggish processes, inaccurate data, and a lack of trust among people. These obstacles limit society's involvement and responsiveness. The innovation underscores the need for more decentralised, transparent, and integrated policies supported by technological development and participatory governance [3]. The rapid adoption of blockchain technology has recently left the research community with many unresolved queries. As a result, there have been a staggering number of research projects in the field of blockchain in recent years [4]. Blockchain technology, originally meant for the foundation of cryptocurrencies, has expanded into a distributed, tamper-proof ledger capable of tracking transactions and events with immutability, transparency, and decentralised trust [5]. In terms of water resource management, blockchain can capture sensor recordings from smart meters and water quality sensors.

DOI: 10.1201/9781003714217-5

5.1.1 Global Water Management Challenges

The key challenges include

1. **Water Scarcity and Quantity Control**:
   The world's population growth contributes to water scarcity, which remains alarming for the sustainability of freshwater resources while meeting the needs of households, industries, and agriculture. Physical and financial accessibility to water resources needs to be considered when ensuring an integrated water scarcity assessment [6].
2. **Pollution and Water Quality**:
   The decrease in water quality globally leads to a threat to human health and the ecosystem. Substantial monitoring and management are important for emerging and traditional pollutants. Inconsistent testing of water quality and data gaps hinder global management performance [7].
3. **Inefficient Allocation and Governance**:
   The traditional centralised water management system is affected by infrastructure and inflexibility, partitioned governance, institutional imperfection, and a lack of teamwork, which obstructs effective water allocation [8].
4. **Climate Change and Human Influence**:
   Climate change intensifies water stress by disrupting ecological cycles, leading to several natural disasters that impact water quality. Human activities impose pressure on water resources, demanding new management systems and technologies [9].

5.2 ROLE OF EMERGING DIGITAL TECHNOLOGIES

In the past few decades, computerised, real-time tracking devices have been developed to assist people in regulating water usage intelligently and measuring it accurately. There are several methods to control water consumption, including the use of smart meters, which leverage the Internet of Things (IoT) and cloud computing.

a. **Internet of Things (IoT)**: The IoT is a broad interconnected network that utilises technologies for sensing, communication, networking, and information processing. It relates to computers with real-world objects; thus, the world is more connected. Using affordable computers, big data, cloud analytics, and mobile technologies for exchanging data with IoT devices makes the services easier. An IoT system uses a constant feedback process to distinguish the services and share the data in particular formats.
b. **Smart Sensors**: The majority of the water infrastructures and smart buildings make use of ultrasonic sensors to track the water levels, Hall-effect flow sensors that provide precise readings, and an accelerometer to detect and identify leaks.
c. **Digital Twins**: Digital twins digitally illustrate the water networks, employing models and real-time data to simulate, observe, and regulate performance. This technique conveys predictive information on maintenance and the best way of utilising the resources. Combinations of models and real-time data display the digital version of specific components of the water system [10].
d. **Remote Sensing and GIS**: Current technologies leverage satellite and aerial data to map groundwater zones, monitor water quality, and forecast the supply and demand sequence. These tools deal with the large shifts in space and time.

e. **Artificial Intelligence and Machine Learning**: Artificial intelligence and machine learning analytics help in identifying the latest developments in how people consume patterns, detecting leaks, and checking water quality to improve the supply chain and resource allocation [11].
f. **Mobile and Web-based Applications**: The applications and dashboards interface for customers and management facilitates the monitoring of usage, gathering feedback on how services are being utilised, charging customers, and refining services in real time [12].

The integration of these technologies is transforming the way we regulate water in an immense way, rendering it smarter, more efficient, and more sustainable. This facilitates stakeholders' compliance with current demands while ensuring future access to this pertinent resource.

5.3 OVERVIEW OF BLOCKCHAIN AND DISTRIBUTED LEDGER TECHNOLOGY

Decentralised systems such as blockchain and distributed ledger technology are developed to maintain records safely and validate transactions with multiple groups by eradicating the necessity for a central authority [13].

5.3.1 What Is Blockchain

Blockchain is an irreversible and decentralised distributed ledger system that effectively processes records and transactions via a dispersed network without centralised authorities, ensuring transparency, security, and trust. Data is exchanged and examined among multiple independent nodes using techniques such as proof of work or proof of stake instead of using a single repository or authority [14]. Data integrity and transparency are assured once data is provided and confirmed in a block; it cannot be altered without updating all the subsequent blocks, which is computationally unfeasible.

5.3.2 Key Blockchain Properties for Water Systems

Table 5.1 shows the interaction between the key blockchain properties and water system management applications. In conclusion, by improving security, transparency, traceability, automation, and operational resilience, blockchain's unique features provide significant advances in the management, monitoring, and governance of contemporary water systems [15].

TABLE 5.1 Blockchain properties for water systems

PROPERTIES	*WATER SYSTEM MANAGEMENT APPLICATIONS*
Decentralisation	Distributed storage and peer-to-peer network
Transparency	Trust for audits, quality control
Immutability	Permanent usage of records and compliance
Security	Robust data storage, anti-tampering measures
Smart contracts	Digitalised billing, water rights management, and maintenance
Traceability	Source-to-destination tracking of the water resource lifecycle

TABLE 5.2 Feature-wise comparison of traditional system and blockchain system

FEATURES	*TRADITIONAL*	*BLOCKCHAIN*
Data transparency	Limited	High (immutable ledger)
Data tampering	Possible	Nearly impossible
Auditability	Manual	Automated & verifiable
Trust level	Medium	Highly decentralised
Efficiency	Moderate	Higher with smart contracts

5.3.3 Blockchain Types (Public, Private, and Consortium)

Blockchain networks are broadly classified into public, private, and consortium types.

a. **Public Blockchain**: The public can sign up, validate transactions, and take part in the consensus-building on the public blockchain. Illustrations of a few public blockchains are Bitcoin and Ethereum. It applies to cryptocurrencies, open-source applications, etc.
b. **Private Blockchain**: A private blockchain is restricted to authorised participants managed by a single entity. It enables strict access restrictions, better privacy, and rapid transaction processing due to minimal hubs. It is applied in all internal business operations in various sectors such as supply chain, healthcare, and finance.
c. **Consortium**: Identical to permissioned blockchain, consortium blockchains are semi-decentralised networks with authority shared among a number of organisations. Transactions are validated by certain companies, ensuring efficiency and strengthening member confidence [16].

5.3.3.1 Comparison of Traditional System vs Blockchain

Table 5.2 depicts the feature-wise comparison of traditional system and blockchain system.

The collaboration of blockchain and IoT ensures secure, effective, transparent water system management by integrating real-time monitoring with tamper-resilient, decentralised data storage and automation [17]. Blockchain, powered by incorporating monitoring, assuring remote control, and further improving accuracy and traceability, promotes smart cities, agricultural water management, and regulatory compliance [18].

5.4 METHODOLOGY FOR BLOCKCHAIN APPLICATION IN WATER MANAGEMENT

Blockchain technology provides a safe and transparent method to deal with the challenges of global water management, such as a shortfall of water, inadequate water supply, and a lack of accountability. Real-time data on water usage, distribution, and quantity are recorded permanently by integrating IoT-enabled smart sensors with blockchain. Smart contracts streamline significant tasks such as accounting and inspection, which lowers the risk of corruption.

5.4.1 Research and Design Framework

The Research and Design Framework for the blockchain water management system is all about illustrating the method for investigating, developing, and implementing blockchain technologies intended for water systems. The first and foremost process is to identify the problem, such as unequal distribution of water, scarcity, corruption, water theft, and inappropriate billing. The next step is to find out the stakeholders, such as government authorities, private suppliers, farmers, and consumers. Define a particular research objective, such as strengthening data quality, automating contracts, or improving stakeholders' confidence [19].

A blockchain-based water management system facilitates decentralised digital ledgers to

1. Keep track of water transactions and sensor data transparently.
2. Provide immutable audit trails to forbid corruption and tampering.
3. Enable decentralised access that leverages communication among suppliers, utilities, and users' trustlessness [20].

5.4.2 Data Acquisition

Data collection includes real-time data from IoT-enabled smart meters and sensors to monitor flow, utilisation, and quality. Integrate the data with the historical dataset by storing the metadata on the blockchain while retaining the raw data in its original storage system.

5.4.3 IoT Sensors

IoT sensors have reshaped several industries in automating real-time data sharing and device connectivity. IoT plays a vital role in increasing the efficiency, sustainability, and reliability of the water management system. IoT devices incorporated with sensors and actuators gather and forward data on water quality, consumption patterns, and infrastructural issues, providing intelligent decision-making [21].

5.4.4 Supervisory Control and Data Acquisition System Systems

The Supervisory Control and Data Acquisition System offers real-time control over large-scale water infrastructure. It integrates with IoT to enable monitoring facilities to operate smartly and detect complications.

5.4.5 Remote Sensing

Collaborate with sensors in the field to leverage large-scale diagnostics and also enable predictive maintenance for a networked system [22].

5.4.6 Smart Contract Frameworks

In blockchain, smart contracts are a fragment of code that is recorded and executed. Without any intermediaries, they are intended to enforce rules automatically, oversee agreements, and trigger actions when certain prerequisites are met. They streamline activities such as device registration, imposing real-time billing, and detecting faults, thus mitigating the demand for agents and enhancing accountability [23].

5.5 CONSENSUS MODEL FOR WATER SENSORS

In a blockchain-based water sensor system, the consensus model serves as the framework that determines the authenticated and recorded data from various sensors. Choosing a consensus model guarantees that the data is secure and trustworthy on the network.

a. **Proof of Authority**:
   Recommended in a confident or collaborative context where validation powers are granted to a limited set of guaranteed nodes, such as water amenities, oversight bodies, or approved agencies. It ensures that real-time water quality and sensor consumption data are published in the ledger only by the accountable stakeholders [24].
b. **Practical Byzantine Fault Tolerance**:
   Various specified nodes collaborate to verify and acknowledge the forthcoming nodes. This model is appropriate for scenarios where several water administrators collaborate to uphold the sensor networks.

5.6 GOVERNMENT AND STAKEHOLDER ENGAGEMENT

Blockchain offers a decentralised governance framework, eliminating power inequality and permitting decision-making by authorities, customers, and government regulators. It facilitates the active engagement of stakeholders in providing transparent and verified interactions and participation in the governance process. Platforms developed on blockchain leverage consortium plans and trustworthy consensus tactics such as proof of authority to collaborate government with customers [25].

5.7 IMPLEMENTATION OF BLOCKCHAIN SOLUTIONS IN WATER RESOURCES MANAGEMENT

5.7.1 Implementation Planning and Roadmap

Figure 5.1 shows the method that involves identifying the use cases for water management where blockchain offers measurable value, such as water quality monitoring (WQM), digital billing, and leak reporting. At the outset of the planning and designing process, water services providers, authorities, and technical staff should be engaged. The stages, such as requirements study, proof-of-concept idea, interfacing with existing assets, pilot run, evaluation, flexibility, and full deployment should be included in the timeframe. The funding sources for IoT devices, blockchain infrastructure, training, services, and sustainability audits are set apart. The mitigation effects for transition periods are being established to safeguard the current system.

5.7.2 Technology Stack and System Architecture

Figure 5.2 shows the integration of blockchain and IoT devices manage with an application layer. A detailed explanation of each component is discussed below.

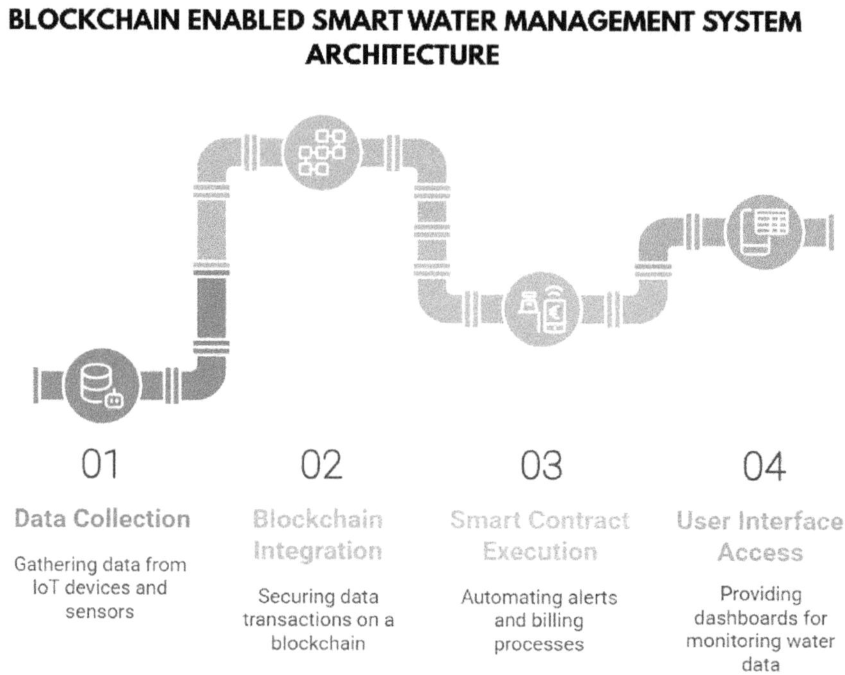

FIGURE 5.1 Blockchain-enabled smart water management system roadmap.

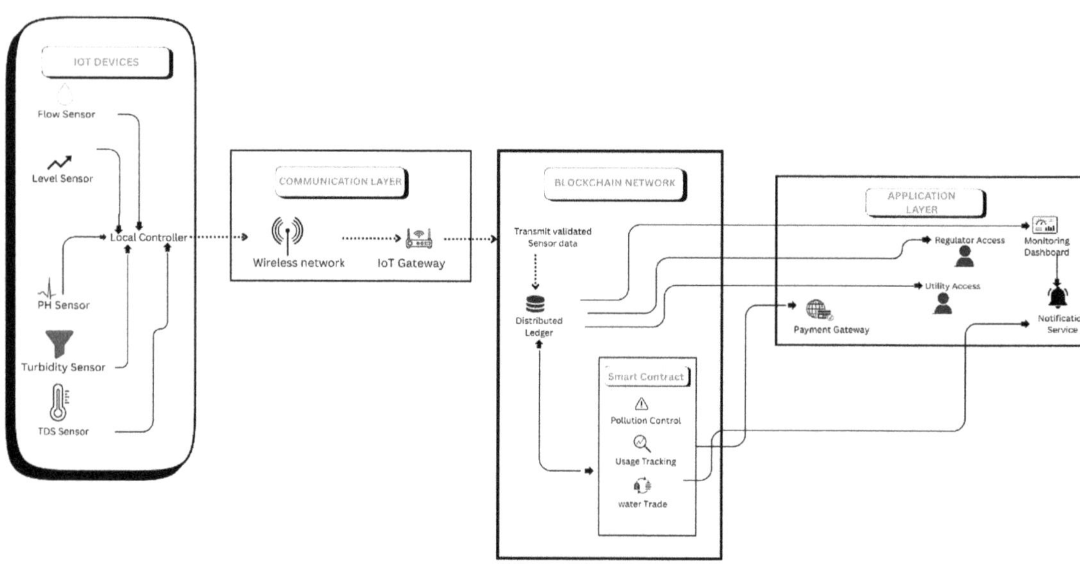

FIGURE 5.2 Technology stack and system architecture.

5.7.3 IoT Devices

a. **Flow Sensors**:

These tools check how much water is moving through pipes and systems. They help track how much water is being used and can spot leaks.

b. **Level Sensors**:
These sensors monitor the amount of water in tanks, storage areas, or natural water sources. They help maintain the right balance between water supply and demand, and prevent water from overflowing.

c. **pH Sensors**:
This sensor checks whether the water is acidic or alkaline. This is necessary to ensure that the water is safe, meets the quality standards, and follows the rules.

d. **Turbidity Sensors**:
These measure the amount of dirt particles suspended in water. If the particles are cloudier, then the water is polluted.

e. **TDS Sensors**:
TDS sensors measure the amount of substances dissolved in water, such as salts and minerals. This helps determine the cleanliness of the water.

5.7.4 Communication Layer

a. **Wireless Network**
It provides the facility for the wireless exchange of sensor data across cellular networks, Wi-Fi, and LoRaWAN.

b. **IoT Gateway**
IoT gateways implement and interpret the data from different sensors or controllers, and then send the acknowledged sensor data.

5.7.5 Blockchain Network

a. **Distributed Ledger**
A database that shares and maintains sensor data information and ensures that the data is stored securely for analysis and checking.

b. **Smart Contracts**
- **Pollution Control**: It ensures that the water quality meets standards and directs warnings or takes action if the parameters are not followed.
- **Usage Tracking**: It tracks the usage of water to manage it efficiently and allows for automatic bill generation.
- **Water Trade**: Based on sensor data and established guidelines, it permits the entities to trade or exchange water rights or amounts.

5.7.6 Application Layer

- **Regulator Access**
The government and regulatory bodies monitor the real-time data, perform monitoring to ensure compliance, and maintain water quality standards and legal requirements.
- **Utility Access**
Water utilities and providers use a restricted portal to perform operational evaluations, examine the data usage, and manage payments or settlements.
- **Payment Gateway**
Financial transactions are coordinated by integrating smart contracts.
- **Monitoring Dashboard**
Enables regulators to facilitate decision-making and operational supervision.

- **Notification Service**

  It is a component for notifying appropriate stakeholders via push alerts/event-driven notifications (e.g., pollution breach and unusual consumption) for a timely response.

5.7.6.1 Process Flow

- The sensor data is transferred to the blockchain network through a communication network.
- Based on real-time data and historical data, smart contracts automate enforcement and alert the processes.
- Regulators and utilities access dashboards, review operations/compliance, and receive alerts, ensuring system transparency and accountability.

5.7.6.2 Pseudo Code for Implementing Water Resource Management Using Blockchain

```
Begin
INPUT: Dataset with attributes
(Time, location,pH value, Turbidity,Usage, Blockchain Tx Hash
    Status, TDS value, Customer ID,Unique Device ID)

STEP 1: Load and Preprocess data
        -Read dataset from CSV file
        -Check for missing or invalid values
        -Handle or remove inconsistent records
STEP 2: Water Safety Level Analysis
    -FOR each record in dataset DO
        -IF (pH between 6.5 and 8.5) AND (Turbidity <5) AND (TDS <500) THEN
            Mark record as "Safe"
        ELSE
            Mark record as "Unsafe"
        ENDIF
        ENDFOR
    -Compute percentage of Safe vs Unsafe
    - Plot pie chart or bar chart for distribution
STEP 3: Average Water Consumption per Device
    - Group dataset by Unique DeviceID
    -For each device, compute mean (Usage in Liters)
    -Sort devices by consumption
    - Plot bar chart of average usage per device
STEP 4: Water Trade Analysis (Blockchain)
    - Use Blockchain Tx Hash as unique trade identifier
    - Aggregate Usage per CustomerID or Location
    - Map trades over Timè dimension
    - Plot line chart or bar chart showing trade volumes
```

5.8 RESULT ANALYSIS

The result graphs depicted below show the relationship between the water usage trend in Figure 5.3, the water safety trend in Figure 5.4, and the blockchain-based water trade success rate in Figure 5.5. The result has been derived using the above algorithm, which utilises the dataset of ten attributes such as Time, Location, pH Value, Turbidity (NTU), Usage (litres), Blockchain Tx Hash, Status, TDS Value, Customer ID, and Unique Device ID. A sample dataset of 5,000 records has been used to obtain these results.

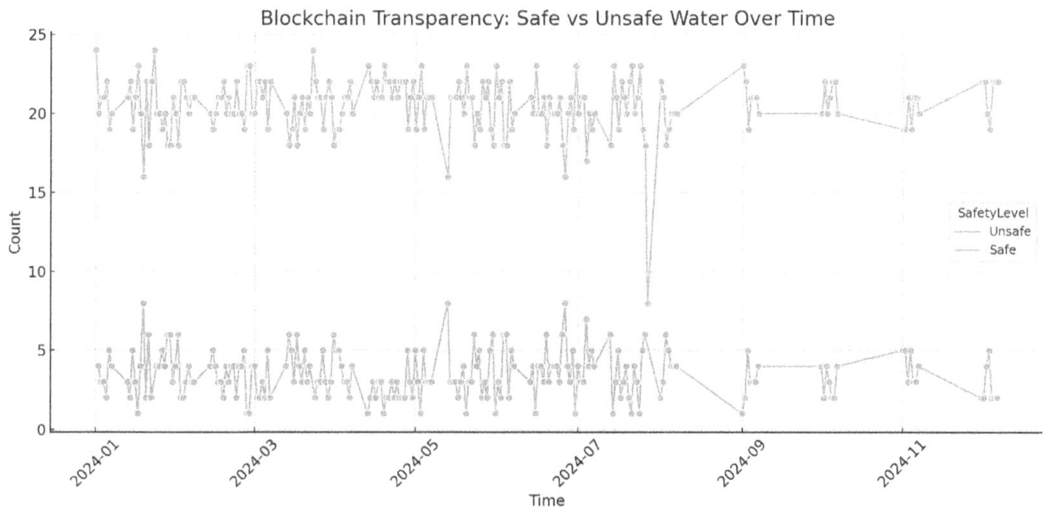

FIGURE 5.3 Blockchain transparency: safe vs unsafe water over time.

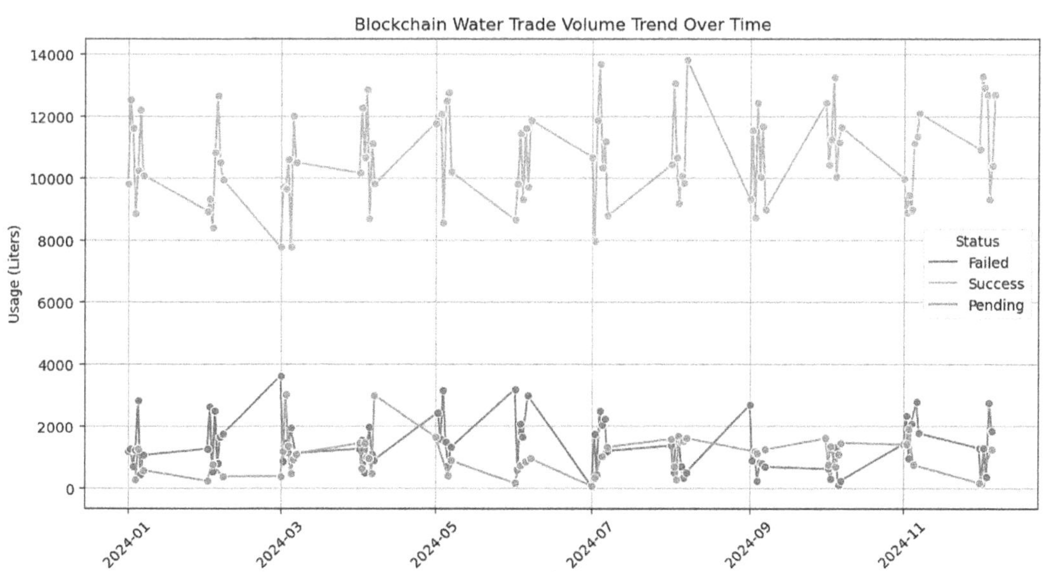

FIGURE 5.4 Blockchain-based water trade volume trend over time.

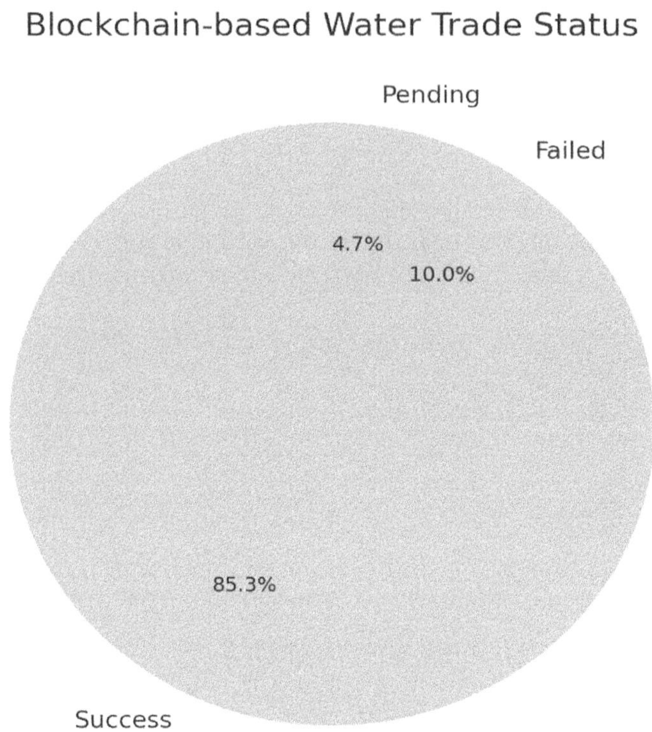

FIGURE 5.5 Blockchain-based water trade status.

TABLE 5.3 Comparison Metrics

METRIC	*WITHOUT BLOCKCHAIN*	*WITH BLOCKCHAIN*
Total records	5000	4264
Valid transactions (%)	85.28%	85.28%
Average usage per device (litres)	522.54	524.73
Safe water (%)	14.26%	25.38%

Table 5.3 represents a data-driven comparison between traditional water resource management without blockchain and blockchain-enabled water management. The metrics are evaluated using the synthetic dataset, which records water quality, consumption, and blockchain transaction details. This depicts how the blockchain enhances transparency, accountability, and reliability.

5.9 BLOCKCHAIN FOR TRACKING USE

5.9.1 IoT and Metering Infrastructure

Field instrumentation is where modern water use tracking begins: pressure level, valve placement sensors throughout the system, ultrasonic or electromagnetic flow meters on utility lines, and soil moisture probes and weather stations on farmland. Edge devices synthesise the data and redistribute it over satellite, cellular, or LPWAN uplinks.

5.9.1.1 *Design Notes*

- **Edge Stack:** sensor → microcontroller/RTU → local buffer → message broker (MQTT/AMQP).
- **Data Model:** timestamped tuples (meter_id, reading, units, quality flags), plus device metadata (firmware hash, calibration cert).
- **Quality Controls at the Edge:** range checks, rate-of-change, and basic leak heuristics; only valid packets are forwarded.

The combination of immutability and provenance (device ID, firmware version) delivers the log of documentary quality. Blockchain has the potential to transfer hydrological telemetry within entities without a single point of trust based on the reviews [26].

5.9.2 Blockchain Data Logging Mechanism

Because water telemetry is high-volume, hybrid architectures are common:

1. **Off-chain storage** (utility data lake, IPFS/object store) holds raw/continuous streams.
2. **On-chain commitments** store cryptographic hashes, anchors, and event summaries.
   (e.g., hourly usage totals and anomaly flags).
3. **Smart contracts** enforce append-only rules, device registration, and signer key rotation.

Peer-reviewed frameworks demonstrate IoT → blockchain pipelines where the ledger authenticates and shares real-time data to dashboards and billing systems; some add leakage detection modules and flow control policies fed by the same stream [27].

5.9.3 Benefits: Transparency and Accountability

- **Single Source of Truth**: collaborative, tamper-evident usage records across utilities, regulators, and users.
- **Operational Efficiency**: automated exception handling (leaks, reverse flow), reduced manual readings and reconciliations.
- **Regulatory Confidence**: evidence of support for tariff disputes and audits.

5.9.3.1 *Practical Challenges and Mitigations*

- **Scalability**: exchange raw streams off-chain; anchor time - buckets, hashes on-chain.
- **Device Trust**: secure boot, signed firmware, HSM/TPM in gateways; revoke keys on compromise.
- **Connectivity Gaps**: use delay-tolerant buffers and satellite relay where cellular is sparse (as in constant amplitude (CA) pilot) [28].

5.10 BLOCKCHAIN FOR POLLUTION CONTROL

5.10.1 Monitoring Parameters and Data Acquisition

Stations monitor pH, turbidity, conductivity, temperature, and—context permitting—nutrients, heavy metals, pathogens, and emerging contaminants. Literature demonstrates that blockchain secures continuous WQM by ensuring the integrity and shareability of sensor data captured via in situ probes and wireless links. Comprehensive reviews of blockchain IoT for environmental monitoring have reported improved data trust, provenance, and inter-system interoperability for pollution management [29].

5.10.2 Smart Contract-Based Pollution Alerts

Thresholds (statutory or permit-specific) are encoded in contracts. When an oracle submits a reading beyond limits (e.g., turbidity > NTU threshold; pH outside 6.5–8.5), the contract:

1. Emits an alert event to subscribed agencies.
2. Logs the breach immutably with timestamp and station ID.
3. Optionally triggers automated responses (e.g., instructing a plant PLC to divert flow to holding).

Scholarly reviews document this pattern—automated regulatory alerts from blockchain-anchored WQM—reducing reporting lag and dispute potential.

5.10.3 Polluter Traceability

Link station networks with asset/permit registries on-chain: upstream dischargers, outfall coordinates, and permit limits are keyed to IDs. Immutable sequences of exceedances plus hydrologic models support trace-back to suspected polluters—strengthening enforcement and remediation.

5.11 SMART CONTRACTS FOR WATER TRADING

5.11.1 Tokenization of Water Assets

Water rights, allocations, and temporary trades can be digitised as tokens that encode quantity, priority class, location, seasonality, and constraints (e.g., return-flow obligations). Token standards (e.g., ERC-style licences) have been applied in Australian pilots to represent permits and entitlements.

Data to Encode: right ID and provenance, basin/zone, extraction works, flow constraints, carry-over rules, metering linkage (meter_id), compliance hooks.

5.11.2 Market Mechanism for Water Trading

A blockchain-based mechanism for water trading integrates ecological and economic rules in every agreement by digitising, transparently managing and promoting water trade between various stakeholders with the use of smart contracts.

Blockchain technologies simplify real-time pricing, smart contracts and peer-to-peer trading of water rights as part of the fundamental market structure. By encouraging direct trade between farmers, businesses, utilities, and users, these networks reduce uncertainty and improve customer trust.

Conservation credits, watershed restoration financing, and restricted trades are examples of ecological contracts and incentives. Fees are paid to the restoration projects managed by decentralized autonomous organization (DAO), and transactions are carried out in compliance with current ecological situations.

5.11.3 Automated Compliance and Settlement

Smart contracts can settle instantly once metered usage and delivery factors are verified by oracles. Integration with IoT meters prevents overspending against allocations and reduces disputes by anchoring evidence (usage, delivery, timestamps) on-chain. California pilots described the same loop—sensors → satellite link → blockchain → transparent trading and accounting.

Dispute resolution flow: freeze tokens involved, run an on-chain challenge period, compare off-chain meter logs to on-chain anchors, and execute refunds/penalties per rules.

5.12 CHALLENGES, RISKS, AND POLICY CONSIDERATIONS

Deployment of blockchain is filled with potential risks and complications of cybersecurity, data security, technological standards, legal frameworks, and ethical issues.

a. **Technological Challenges and Standards**:
   Issues with blockchain technology include managing user databases, large expenditure of energy, low transaction throughput, unambiguous standards and protocols that hinder interoperability, and upgrading is slowed down by software glitches and lack of industry standards [30].
b. **Data Privacy and Cybersecurity**:
   Due to the risk posed by blockchain transparency and immutability, effective encryption, confidentiality, and access control are essential. Consent management is crucial for collaborative stakeholders for data sharing, cybersecurity threats such as social engineering [31].
c. **Regulatory and Legal Frameworks**
   Undefined legal status and enforceability of smart contracts in different countries, limited frameworks, and lack of global uniformity of compliance and anti-money laundering/know your customer standards are the trademarks of cryptocurrency regulation [32].

5.12.1 Ethical Considerations and Risk

Despite disrupting renowned financial institutions and increasing crime and accountability, blockchain technology raises ethical concerns about privacy and transparency [33].

Blockchain's significant promise can be achieved only with strong technological, legal, and ethical governance, as well as continuous improvement of privacy, compliance, and interoperability standards [34].

5.13 CONCLUSION

The groundbreaking move towards accomplishing transparency, efficiency, and sustainability in confronting one of the globe's most critical problems is the inclusion of blockchain technology in water resource management. There is a dire demand for creative proposals to address the world's water shortage, rising pollution, and inadequate distribution systems. Blockchain's special features such as openness, decentralisation, and immutability, provide a robust authentication mechanism among users, especially local communities, businesses, and governments.

Moreover, several issues exist in the implementation of blockchain in water management, such as scalability, energy consumption, regulatory uncertainties, and interoperability with existing infrastructures, which raise certain concerns. Furthermore, financing for technological tools is required to encourage users to utilise the new digital technology.

Blockchain can extensively improve water governance by motivating openness, cooperation, and immutability techniques. While certain legislative hurdles may persist, the positive benefits outweigh all the issues. Hence, it guarantees the sustainable usage of resources for future generations.

REFERENCES

1. A. Biswas, S. Sarkar, S. Das, S. Dutta, M. Roy Choudhury, A. Giri, B. Bera, K. Bag, B. Mukherjee, K. Banerjee, D. Gupta, D. Paul (2025). Water scarcity: A global hindrance to sustainable development and agricultural production – A critical review of the impacts and adaptation strategies. *Cambridge Prisms:Water*, 3, e4, 1–22. https://doi.org/10.1017/wat.2024.16
2. C. Li, J. Wu (2023). How can we achieve more sustainable water allocation in water-deficient areas? A case study from China's Fen River Basin. *Frontiers in Environmental Science*, 11, 1302845. https://doi.org/10.3389/fenvs.2023.1302845.
3. R. Voogd, P.M. Rudberg, J.R. de Vries, R. Beunen, A.A. Espiritu, N. Methner, R.K. Larsen, G.E. Fedreheim, S. Goes, E. Kruger (2022). A systematic review on the role of trust in the water governance literature. *Water Research X*, 16, 100147. ,https://doi.org/10.1016/j.wroa.2022.100147.
4. A.G. Gad, D.T. Mosa, L. Abualigah, A.A. Abohany (2022). Emerging trends in blockchain technology and applications: A review and outlook. *Journal of King Saud University - Computer and Information Sciences*, 34 (9), 6719–6742. https://doi.org/10.1016/j.jksuci.2022.03.007.
5. S. Nakamoto (2008). Bitcoin: A Peer-to-Peer Electronic Cash System. Available online: https://bitcoin.org/bitcoin.pdf (accessed on 21 November 2022).
6. M. Brennan, J. Rondón-Sulbarán, L.P. Sabogal-Paz, P. Fernandez-Ibañez, A. Galdos-Balzategui (2021). Conceptualising global water challenges: A transdisciplinary approach for understanding different discourses in sustainable development. *Journal of Environmental Management*, 298, 113361. https://doi.org/10.1016/j.jenvman.2021.113361.
7. A. du Plessis (2022). Persistent degradation: Global water quality challenges and required actions. *One Earth*, 5 (2), 129–131. https://doi.org/10.1016/j.oneear.2022.01.005.
8. J.P. Adamopoulou, A.A. Frantzana, I.P. Adamopoulos (2023). Addressing water resource management challenges in the context of climate change and human influence. *European Journal of Sustainable Development Research*, 7 (3), em0223. https://doi.org/10.29333/ejosdr/13297.
9. J.C. Algallar, D. Montecastro (2025). Managing urban water resources: A review of challenges, techniques, and sustainability strategies. *Engineering Technological Applications and Science Research*, 15 (3), 23149–23162.
10. P. Gourbesville (2008). Challenges for integrated water resources management. *Physics and Chemistry of the Earth, Parts A/B/C*, 33 (5), 284–289. https://doi.org/10.1016/j.pce.2008.02.002.
11. S.A. Palermo, M. Maiolo, A.C. Brusco, M. Turco, B. Pirouz, E. Greco, G. Spezzano, P. Piro (2022). Smart technologies for water resource management: An overview. *Sensors (Basel)*, 22 (16), 6225. https://doi.org/10.3390/s22166225.
12. C. Parra-López, S.B. Abdallah, G. Garcia-Garcia, A. Hassoun, H. Trollman, S. Jagtap, S. Gupta, A. Aït-Kaddour, S. Makmuang, C. Carmona-Torres (2025). Digital technologies for water use and management in agriculture: Recent applications and future outlook. *Agricultural Water Management*, 309, 109347. https://doi.org/10.1016/j.agwat.2025.109347.
13. C. Walter (2024). Digital technologies for the future of the water sector? Examining the discourse on digital water. *Geoforum*, 148, 103918. https://doi.org/10.1016/j.geoforum.2023.103918.

14. A. Shukla, P. Jirli, A. Mishra, A.K. Singh (2024). An overview of blockchain research and future agenda: Insights from structural topic modelling. *Journal of Innovation & Knowledge*, 9 (4), 100605. https://doi.org/10.1016/j.jik.2024.100605.
15. G. Tripathi, M.A. Ahad, G. Casalino (2023). A comprehensive review of blockchain technology: Underlying principles and historical background with future challenges. *Decision Analytics Journal*, 9, 100344. https://doi.org/10.1016/j.dajour.2023.100344.
16. Y. Cao, H. Li, L. Su (2024). Blockchain-driven incentive mechanism for agricultural water-saving: A tripartite game model. *Journal of Cleaner Production*, 434, 140197. https://doi.org/10.1016/j.jclepro.2023.140197.
17. R. Dhillon, P. Sivabalan (2025). Exploring dimensions of governance for different types of blockchain systems. *The British Accounting Review*, 101588. https://doi.org/10.1016/j.bar.2025.101588.
18. S. Sundaresan, K. Suresh Kumar, T. Ananth Kumar, V. Ashok, E. Golden Julie (2021). Chapter 4 - Blockchain architecture for intelligent water management system in smart cities, In S. Krishnan, V. Emilia Balas, E. Golden Julie, Y. Harold Robinson, R. Kumar (Eds.), *Blockchain for Smart Cities*, Elsevier, pp. 57–80, ISBN 9780128244463. https://doi.org/10.1016/B978-0-12-824446-3.00006-5.
19. R. Chowdhary, S.K. Mahato, A.K. Biswas, S. Datta, K. Tiwari (2024). Blockchain enabled IOT-based water management system for smart cities. *International Journal of Engineering Research & Technology (IJERT)*, 13 (05), 297–302.
20. A.R. Furones, J.I.T. Monzón (2023). Blockchain applicability in the management of urban water supply and sanitation systems. *Spain Journal of Environmental Management*, 344, 118480. https://doi.org/10.1016/j.jenvman.2023.118480.
21. Adarsh Gurudas Vernekar (2020). Blockchain-based water management system. *International Research Journal of Engineering and Technology (IRJET)*, 7 (5), e-ISSN: 2395-0056
22. C. Prabha (2024). IoT oriented approach for rural and urban area based on 'Smart Water Management Systems' through sensors. *International Journal of Intelligent Systems and Applications in Engineering*, 12 (23s), 301. Retrieved from https://ijisae.org/index.php/IJISAE/article/view/6740
23. B. Dwarakanath, P. Kalpana Devi, A. Ranjith Kumar, A.S.M. Metwally, G.A. Ashraf, B.L. Thamineni (2023). Smart IoT-based water treatment with a Supervisory Control and Data Acquisition (SCADA) system process. *Water Reuse*, 13 (3), 411–431. https://doi.org/10.2166/wrd.2023.052
24. M. Homaei et al. (2025). Smart water security with AI and blockchain-enhanced digital twins. *arXiv preprint arXiv:2504.20275.*
25. G.D. Bhatt, A. Emdad (2025). Blockchain's role in environmental sustainability: Bibliometric insights from scopus. *Green Technologies and Sustainability*, 3 (4), 100236. https://doi.org/10.1016/j.grets.2025.100236.
26. N. Testi, R. Marconi, E. Pasher (2025). Exploring the potential of blockchain technology for citizen engagement in smart governance. *Open Research Europe*, 3, 183. https://doi.org/10.12688/openreseurope.16153.4.
27. T.L. Thuy, M.-K. Nguyen, T.D. Bui, H.P.H. Yen, N.T. Hoai, N.V.C. Ngan, A. Pradiprao Khedulkar, D.P. Van, A. Halog, T.-D. Hoang (2025). Blockchain-enabled water quality monitoring: A comprehensive review of digital innovations and challenges. *Water*, 17, 2522. https://doi.org/10.3390/w17172522.
28. M.T. Naqash, T.A. Syed, S.S. Alqahtani, M.S. Siddiqui, A. Alzahrani, M. Nauman (2023). A blockchain based framework for efficient water management and leakage detection in urban areas. *Urban Science*, 7, 99. https://doi.org/10.3390/urbansci7040099.
29. I.U. Ajakwe, S.O. Ajakwe, J.M. Lee, et al. (2025). Internet-of-Things-blockchain integration in environmental pollution monitoring data management: Trends and techniques. *International Journal of Environmental Science and Technology*. https://doi.org/10.1007/s13762-025-06615-x.
30. M. Mehraeen, L. Mahmoudi (2024). Tracing the blockchain challenges in healthcare: A topic modeling and bibliometric analysis. *Blockchain Healthc Today*, 7. https://doi.org/10.30953/bhty.v7.335.
31. F.M. AbdelSalam (2023). Blockchain revolutionizing healthcare industry: A systematic review of blockchain technology benefits and threats. *Perspectives in Health and Information Management*, 20 (3), 1b.
32. M. Hasnain, I. Ghani, D. Smith, A. Daud, S.R. Jeong (2025). Cybersecurity challenges in blockchain-based social media networks: A comprehensive review. *Blockchain: Research and Applications*, 6 (3), 100290. https://doi.org/10.1016/j.bcra.2025.100290.
33. J. Rahman, H. Rahman, N. Islam, T. Tanchangya, M. Ridwan, M. Ali (2025). Regulatory landscape of blockchain assets: Analyzing the drivers of NFT and cryptocurrency regulation. *BenchCouncil Transactions on Benchmarks, Standards and Evaluations*, 5 (1), 100214. https://doi.org/10.1016/j.tbench.2025.100214.
34. J. Szabo, C. Bernard, L. Philip (2024). Legal implications and challenges of blockchain technology and smart contracts. *Computer Life*, 12 (2), 6–10. https://doi.org/10.54097/ztn2w848

Digital Twin Modeling for Water Distribution Network

6

Bhupinder Kaur Srao, Poonam Kukana, Chiman Saini, Sukhjinder Kaur, and Ashima Mockoul

6.1 INTRODUCTION

Water is among the most essential natural resources that promote the survival of humans, ecosystems, and economies. Globally, urbanization, climatic changes, and the rising demand for dependable water services have left untold pressure on water utilities and governments. The proper operation of water distribution networks (WDNs) – the complex system of pipes, pumps, reservoirs, and valves used to supply water to consumers – in an efficient, safe, and sustainable way is one of the main activities of water utilities. Traditionally, scheduled maintenance and simulation and modeling tools such as EPANET have been used in traditional WDN management [1]. Although helpful, these strategies usually do not have real-time flexibility, predictability, and cross-platform (digital) integration.

The concept of the digital twin (DT) has changed the paradigm. A DT may be thought of as a changing virtual model of a physical system that is being fed data collected by sensors, actuator-driven inputs, and external data. In contrast to a frozen model, DTs are alive; they reflect the exact condition and performance of the real things they model in real time or close to it. They allow operators to model various scenarios, foresee failures, and run systems in an efficient manner. DT technologies are being used in critical infrastructures such as water systems, first implemented in aerospace and manufacturing.

The implication of digital twin model (DTM) to WDNs is improved resiliency, performance, and sustainability. DT architectures can be complemented by live data streams from Internet of Things (IoT) devices, analytics, and artificial intelligence (AI) to provide actionable intelligence. For instance, utilities can apply these DTs to identify early leaks, schedule pumping in an optimized energy-wise manner until sewers are drained for repair or renewal, and plan network extensions under population growth and climate change scenarios. Furthermore, the DTs could foster collaboration among engineers, planners, and decision-makers through a common interactive platform [2].

DOI: 10.1201/9781003714217-6

This chapter presents an in-depth review of DT modeling for WDNs. It is divided into the following sections. Section 6.2 reviews the background literature and discusses the evolution of DT technologies in the context of water infrastructure. The central ideas and models for DT shape modeling are presented in Section 6.3, which includes the definition and frameworks of DT architecture, key components, and enabling technologies. Selected case studies and applications of using DTs are shared from both industrial practice and academic research to show the application of these DTs in Section 6.4. Section 6.5 provides an analysis of challenges and limitations, whereas Section 6.6 delves into future directions and opportunities. Key insights and thoughts are summarized in Section 6.7.

By combining theoretical insights, practical applications, and critical evaluations, this chapter aims to contribute to the ongoing dialogue on digital water technologies. This book is intended for those who work or research in the area of water supply and wastewater management, including water utilities, and will appeal to both non-experts interested in smart water systems and experts concerned about urban service sustainability.

6.2 BACKGROUND AND LITERATURE REVIEW

6.2.1 Evolution of DT Technology

The term DT was first introduced by Michael Grieves in 2002 within the product lifecycle management discipline, specifically to support aerospace engineering. DTs were originally conceived as the digital counterparts of real-world systems to enhance product design, manufacturing performance and maintenance. NASA saw the potential in this and used it to simulate systems in spaceships and for predictive maintenance of critical systems. The strategy has since expanded, beyond aerospace and manufacturing, to find applications in urban infrastructure, healthcare, and energy systems [3].

The adoption of DTs in the water sector has been slow but is becoming increasingly important. The first generation of water infrastructure modeling used hydraulic and hydrological models such as EPANET, which had the ability to simulate water flow and pressure in a WDN. Despite the high effectiveness of such a model, it is not dynamic and must be calibrated manually. The trend toward real-time measurement and dynamic modeling, made possible by sensors, along with Supervisory Control and Data Acquisition (SCADA) and Geographic Information Systems (GIS), provided the foundation for DTs when applied to water management.

6.2.2 Digital Transformation of Water Utilities

The ecosystem of smart technologies, including IoT, cloud computing, machine learning (ML), and big data analytics, has entirely changed the manner in which water utilities handle networks. The new paragraph would read, "Smart water meters, pressure sensors and leak detection devices can generate massive amounts of data that can be added to digital twin platforms." For example, predictive maintenance models can identify pipe bursts before they occur, reducing non-revenue water losses and improving system reliability [4].

In addition, urban water systems are being afflicted by the impacts of climate change and population growth [5]. Simulations of drought can be further used to identify scenarios for which the implications of changes in demand or contaminant concentrations are calculated using DTs for planning purposes. This allows the utilities to develop resilient systems that are able to resist short-term shocks as well as long-term perturbations [6].

6.2.3 State-of-the-Art and Case Studies

A number of previous works have shown the applicability and advantages of DT to water systems:

- Ahmed and Khan (2021) [2] proposed a digital network framework for leak detection in WDNs, using pressure sensor information combined with hydraulic models to enhance early fault detection.
- Booysen et al. (2019) suggested how DTs can be used in urban drainage systems (UDS) for real-time control that can aid in better stormwater management [7].
- Khadir et al. (2022) [8] developed a DT platform for the water environment based on integrated monitoring and modeling, showing its importance in ensuring drinking water security.
- Several industrial deployments by Thames Water (UK) and Singapore Public Utilities Board (PUB) have been the front-runners in adopting DT at a utility scale – employing them for network optimization, demand prediction, and predictive asset maintenance.

These works together underscore that although DTs bring transformative opportunities, their deployment is emergent and frequently obstructed by concerns regarding data integration, interoperability, and computational overhead.

6.2.4 Gaps in the Literature

However, despite increasing interest, there are a few gaps. First, there are few studies that have been carried out to study the standard guidelines for developing and managing DTs in water supply system (WSS). Current works are purely pilot or ad hoc based and lack a reference framework for scalability. Second, there is insufficient insight into the cybersecurity and data aspects of DTs, despite the undeniable fact that critical infrastructure is increasingly exposed to cyber threats [1]. Third, while hydraulic performance is widely researched, few studies consider other sustainability aspects (i.e., energy consumption, carbon emissions, or equity).

6.3 CORE CONCEPTS AND FRAMEWORKS

The development of DTs in WDNs mandates a well-organized vision concerning their constituents, architecture, and technological enablers. DTs are not static models similar to traditional models, but rather objects that combine up-to-date information and computational models/decision-support systems in real time using feedback control as a guide for physical and cyber activities. This section provides the core ideas and principles that make DTs effective for WDN applications.

6.3.1 Defining the DT in Water Systems

A DT in the context of water distribution refers to a virtual representation of a physical system that is continuously adjusted with information collected through sensors, control systems, and user interactions [9]. It is not just a model for simulation; it continuously synchronizes with empirical data.

A DT technically consists of three elements (Tao et al., 2019) [10]:

- The material: the pipes, valves, pumps, etc., up to the reservoirs.
- Digital model Hydraulics and Quality Model – The digital, virtual, or numerical model that is composed of computer-based hydraulic and quality models.
- The relationships they have – data streams and the algorithms that ensure synchronization.

In practice, this means that when a pressure sensor somewhere in the physical grid senses an anomaly, it is mirrored digitally – allowing operators to test fixes.

6.3.2 Architecture of a DT for WDNs

The architecture of a DT can be conceptualized in four interconnected layers:

a. **Physical Layer**
   - Consists of the physical assets: water pipelines, pumping stations, storage tanks, sensors, and actuators.
   - Equipped with IoT devices that capture operational data such as flow, pressure, and water quality parameters.

b. **Data Layer**
   - Data collection, data storage, and data management.
   - Includes data acquisition systems (SCADA), IoT, and cloud-based storage.
   - Must process non-homogeneous streams of data in real time.

c. **Model Layer**
   - Water quality, hydraulic, and hydrological models.
   - Encompasses AI/ML algorithms to guide predictions (e.g., leak detection and demand forecasting) [8].
   - Uses calibration systems to ensure alignment with the real world.

d. **Application Layer**
   - Offers operator, engineer, and planner user interfaces and decision-support systems.
   - Allows visualization, scenario testing, and optimization (e.g., energy-efficient pump scheduling).

Modularity, scalability, and interoperability are assured by this layered approach, as illustrated in Figure 6.1.

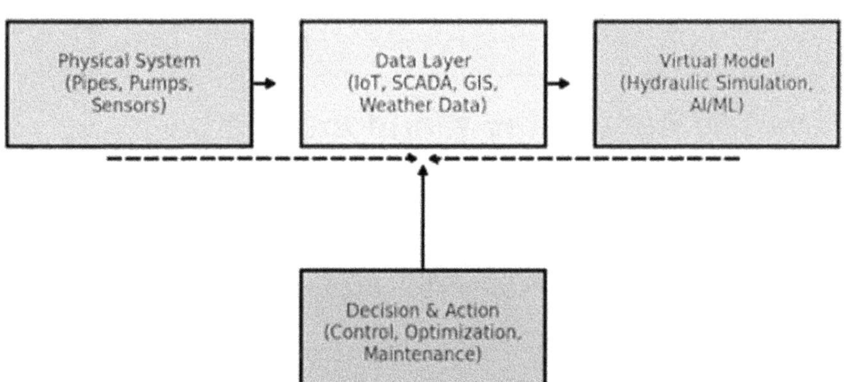

FIGURE 6.1 Digital twin conceptual framework diagram of a water distribution networks.

6.4 ENABLING TECHNOLOGIES

Determining the technological enablers to support DT development within the framework of WDNs includes

- **IoT**: Provides the ability to monitor continuously in real time with the help of smart sensors, flowmeters, and other connectable devices.
- **Big Data Analytics**: Handles large volumes of data from distributed sensors (high frequency rates).
- **AI and ML**: Can perform predictive modeling, anomaly detection, and optimization. For instance, neural networks can predict water demand trends (Zhang et al., 2018) [11].
- **Cloud Computing**: Provides scalable storage and processing capabilities for simulation and analytics [12].
- **GIS**: Used to give spatially explicit mapping of network assets and integrate topographic and demographic information.
- **Cyber-Physical Systems**: Facilitates real-time communication between physical objects and digital models with the help of control and automation.

6.4.1 Types of DTs

DTs may be classified into three types depending on the scope and functionality:

a. **Component Twin**: Models individual equipment, e.g., pumps or valves, that can be used for predictive maintenance.
b. **System Twins**: Whole subsystems, e.g., a district metered area (DMA) of a city.
c. **Process Twin**: Concentrates on working processes, e.g., pumping processes, treatment, and distribution.

In the case of water utilities, a hybrid between system and process twins is usually the most pertinent approach that will guarantee physical asset management and operational optimization.

6.5 CASE STUDIES

6.5.1 Thames Water (United Kingdom)

Thames Water, the largest water utility in the UK, has been a trendsetter in embracing the DT. Nevertheless, there are still several key issues that the utility struggles to address, including old pipes and huge amounts of non-revenue water. To fight this, Thames Water collaborated with technology companies to develop a DT that integrates IoT sensors, hydro-models, and AI algorithms [13].

- **Application**: Leak inspection and predictive maintenance.
- **Procedure**: DMAs were fitted with pressure and flow transducers. Information was continuously fed into a DT that was adjusted with hydraulic models.

- **Impact**: The DT assisted the utility in reducing the number of leaks by observing abnormalities in hours rather than days, ultimately saving on operation costs and improving customer service.

This demonstrates the economic prospects of DTs based on the reduction of water losses and asset performance.

6.5.2 Singapore Public Utilities Board (Singapore)

Singapore's PUB is a world leader in intelligent water management. Water-scarce Singapore looks to technology for water security. As a country with no natural water resources, Singapore depends on technological innovation for its own drinkable supply. PUB DTs its whole water supply and sanitation network at the national level [14].

- **Application**: Integrated water resource management.
- **Method**: The DT integrates SCADA systems, GIS, weather forecasts, and AI predictive demand models.
- **Impact**: Operators can adjust the pump schedule, manage stormwater discharge, and predict water demand. The DT also contributes to flood risk control under torrential rains by simulating drainage.

The PUB case shows how DTs enable integrated supply-side and demand-side management in a holistic manner.

6.6 ACADEMIC PILOT: UNIVERSITY OF EXETER (UNITED KINGDOM)

A research-focused DT for the resilience of WDN's was also adapted from [6] at the University of Exeter to test resilience under climate change scenarios.

- **Application**: Resilience assessment [15].
- **Method**: DT coupled EPANET-based hydraulic models with stochastic weather generators and demand models.
- **Impact**: The results demonstrated how changing temperatures and rainfall patterns would affect water availability, allowing utilities to design adaptation strategies.

This work demonstrates the potential that DTs have for long-term planning and climate resilience research.

6.6.1 Industry Pilot: SUEZ (France)

The multi-utility company SUEZ trialed a DT for operating their services in some French cities.

- Application: Low-energy pumping and in situ monitoring of water quality.
- **Method**: IoT-enabled sensors for flow and water chemistry values; optimization algorithms were regulating the eligible pump operations [15].
- **Results**: Up to 15% energy could be saved with improved adjustments to water quality.

TABLE 6.1 Comparison of digital twin applications across case studies

CASE STUDY	*UTILITY/ INSTITUTION*	*APPLICATION AREA*	*TECHNOLOGIES USED*	*REPORTED BENEFITS*
Thames Water (UK)	Utility	Leak detection, maintenance	IoT, AI, hydraulic models	Faster leak detection, reduced costs
PUB Singapore	National agency	Integrated water management	SCADA, AI, GIS, weather models	Demand forecasting, flood management
University of Exeter (UK)	Academic	Resilience under climate change	EPANET, stochastic models	Long-term planning
SUEZ (France)	Utility	Energy-efficient pumping, water quality	IoT + optimization algorithms	15% less energy use

This case illustrates that DTs may be suitable systems to fulfill sustainability goals and not only operational reliability.

6.6.2 Emerging Applications

In addition to these highly publicized cases, DTs are increasingly being developed for the following:

- **Management of Contamination**: Fast simulation and response in the networked spread of contaminants [8].
- **Asset Renewal**: Prioritization of aging infrastructure through predictive modeling.
- **Consumer Involvement**: Offering personal water use insights to end consumers using DT-enabled apps (Table 6.1).

6.6.3 Lessons Learned from Case Studies

There are several lessons to be learned from these cases:

- Integration we need to get right – We will live or die on the successful integration of IoT, hydraulic models, and AI in DT.
- Why is this so important? "First of all, it's a scalability issue: Pilot projects work quite well but to scale them up thorough whole cities you need a very strong IT infrastructure."
- **Economic Justification**: The business case is economically justified by cost savings in leakage control, power consumption, and failure prevention [16].
- **Cooperation**: Utilities work with universities, vendors and government to advance the adoption of DTs.

6.7 CHALLENGES AND LIMITATIONS

While DTs are heralding the potential to revolutionize urban WDNs, realizing them is riddled with technical and non-technical barriers. These are the concerns that should be acknowledged for realistic attitudes and adapted practices.

6.7.1 Data Availability and Quality

A challenge for DTs is the requirement of good quality continuous data streams. But a large number of water utilities, particularly in the developing world, "do have enough sensors out there or are working with older SCADA systems." Problems include

- **Lack of Sensor Data**: Fewer samples can lead to poor calibration and prediction of the model.
- **Data Noise**: Inaccurate sensing data can lead to useless simulation results.
- **High Maintenance Needs**: IoT sensors require calibration and constant upkeep, which adds up to OpEx [6].
- A DT model might remain a stagnant model rather than a dynamic online model because of the absence of real data.

6.7.2 Integration and Interoperability

The water infrastructure is characterized by multi-vendor, heterogeneous technologies spanning a period of many years. Integration of hydraulic models, SCADA databases, GIS databases, and AI modules into one single DT platform is not trivial.

- **Compatibility Issues**: Proprietary information and architectures may not be easily connected [6].
- **Standard Paucities**: The absence of standards for DT data exchange is a barrier to implementation.
- **Existing Infrastructure**: Many current systems do not easily integrate with modern sensing and communication.

This renders interoperability a systemic constraint to scaling DTs beyond pilot experiments.

6.7.3 Computational and Cost Barriers

Simulating large urban networks in real time can be power-consuming.

- **Expensive for Small Utilities**: Cloud-based or high-performance computing (HPC) systems may be too costly for small-budget utilities.
- **Complex Model**: Including reverse dynamics, water quality, and AI optimization increases computational demand [10].
- **Scalability Issues**: Pilot DTs may perform well at a district level but poorly as they are applied to large metropolitan areas.

The business case for DTs is unclear to most utilities until they see a demonstrated and definitive return on investment.

6.7.4 Cybersecurity and Data Privacy

Water systems are part of the critical national infrastructure, and cybersecurity is a significant concern.

- **Susceptibility to Attacks**: IoT-based DTs increase the cyber-attack surface.
- **Data and Privacy Breach**: Unauthorized access to operational data could compromise safety and customer privacy [10].

- **Absence of Standards**: Water utility DT deployments lack well-developed cybersecurity standards.

It is therefore necessary to have trustworthy and secure services for widespread adoption.

6.7.5 Organizational and Human Barriers

The challenge of digital transformation is not simply a technical one; it is, in fact, a cultural and organizational issue.

- **Skills Gap**: Many utilities lack staff trained in data science, AI, or advanced modeling.
- **Adoption Barrier**: Operators may be reluctant to use DT systems since they are familiar with traditional SCADA-driven workflows and can learn from skills with conventional processes.
- **Bringing the Disciplines Together**: Without cross-disciplinary cooperation between engineers, IT staff, data scientists, and business decision-makers (which siloed organizations are not exactly keen on), design thinking could be overlooked.

Institutionalized resistance occurs when organizations resist the changes inherent in DT if there is no support mechanism in the form of a trained, quality strategy for implementing these changes.

6.8 LIMITATIONS OF CURRENT RESEARCH

From a research perspective, several limitations remain as follows:

- **Limited Validation**: Many academic prototypes are experimented within ideal scenarios, not in real environments.
- **Short-Term View**: Studies rarely focus on long-term sustainability, such as the carbon footprint or lifecycle costs of DTs.
- **Equity**: Consideration of how the adoption of DTs could further widen the gap between well-resourced and inadequately resourced utilities is largely missing.

These gaps highlight the need for more applied, cross-disciplinary research and policy engagement (Table 6.2).

TABLE 6.2 Key challenges and possible mitigation strategies

CHALLENGE	*DESCRIPTION*	*MITIGATION STRATEGY*
Data quality	Incomplete or noisy sensor data	Use redundancy, AI-based cleaning
Interoperability	Different legacy systems	Open standards, APIs
Cybersecurity	Risk of hacking critical infrastructure	Encryption, secure IoT protocols
High costs	Sensor deployment, cloud infrastructure	Public–private partnerships, phased adoption

6.8.1 Summary of Challenges

In summary, DT implementation for WDNs is constrained by

1. **Data challenges** (scarcity, noise, reliability)
2. **Integration issues** (interoperability, standardization gaps)
3. **Economic barriers** (computational and financial costs)
4. **Cybersecurity risks** (data privacy, system vulnerability)
5. **Organizational hurdles** (skills, resistance, collaboration)
6. **Research limitations** (validation, long-term sustainability)

These barriers underscore that while DTs are **innovative and powerful tools**, their success depends on addressing these limitations through research, funding, and policy interventions.

6.8.2 Future Trends and Opportunities

Despite the fact that the concept of the DT of WDNs is in its infancy, new opportunities are emerging due to the rapid development of computing, sensing, and data science [17]. The next generation of research and work will probably be related to intelligent automation, resilience, the incorporation of sustainability, and the democratization of DTs.

6.8.3 AI and ML Integration

The successor generation of DTs will be based on heavy use of AI and ML to operate beyond descriptive modeling to offer prescriptive intelligence.

- **Self-learning DTs**: Models that can recalibrate both on historical and real-time data.
- **Predictive Analytics**: AI-powered DTs can predict the failure of pipes, optimize pumps, and predict demand in a more accurate way.
- **Automation of Decisions**: Networks connected with reinforcement learning have the potential to make networks semi-autonomous, learning to make the best decisions over time [4].

These developments will transform DTs into decision-making partners as opposed to decision-support tools.

6.8.4 Edge Computing and Real-time Control

Conventional DTs rely on cloud computing; however, the increasing data volumes and latency issues are compelling the use of edge computing.

- **On-site Analytics**: This processing occurs near the source and avoids delays and bandwidth consumption.
- **Resilience Control**: Edge-enabled DTs can sustain operations when they lose connectivity to central servers.

- **Quick Response**: In case of leak detection or contamination events, milliseconds mark the difference; edge processing offers real-time response.

This will be of special importance to large-scale urban WDNs in which centralized cloud-only solutions could become ineffective.

6.8.5 Blockchain in Data Protection and Openness

Since WDN DTs deal with operational and customer data of a sensitive nature, blockchain technology provides opportunities for data management that is secure, transparent, and tamper-proof.

- **Data Provenance**: Guarantees the credibility and integrity of working data.
- **Smart Contracts**: Automation of water trading, billing, or inter-utility deals [4].
- **Collaborative DTs**: Multiple stakeholders (utilities, regulators, customers) can interact with common DT platforms without undermining trust due to blockchain.

This would improve cybersecurity and governance.

6.9 CLIMATE RESILIENCE AND SUSTAINABILITY INTEGRATION

Climate resilience and sustainability will be integrated into the project in ways that ensure the site and its functionality are maintained in extreme weather events. Future DTs will be instrumental in modifying WDNs to meet climate change and promote sustainability objectives.

- **Integration of Climate Modeling**: DTs are used in conjunction with climate scenarios to forecast the effects of water demand, drought, or flood in the long term [6].
- **Energy-Water Nexus Optimization**: DTs will be able to optimize the distribution of both water and energy, which will help minimize carbon footprints.
- **Sustainable Design**: Help plan resilient infrastructure improvements that strike a balance among social, economic, and environmental goals.

This makes DTs a key to sustainable urban water futures.

6.9.1 DT Democratization

At present, DTs tend to consume heavy resources; thus, they can only be implemented by large utility companies. One shift ahead? Diffusing these tools to small towns and local providers. Tools that are free of charge may allow individuals to create simple digital models in any location. Such systems may propagate easily since every person can adjust them. Other teams are already testing more basic hardware-compatible versions. The exchange of improvements would assist in the gradual upgrading of everyone. Basic instruments assist in collecting data about the quality of water. Low-cost devices monitor variations in the field. People start working together to share what they observe around them. Such measures do not cost much. Learning develops when teams practice new tools together. Laborers receive live hands-on training with models. Skills are developed gradually with the help of real tasks. It is easier to remember what is applied every day. Improvement is observed in the improved workflow and the reduction of errors. Cheap equipment can bring parity in the way wealthy and underprivileged districts deal with water data.

6.9.2 Integrated Urban Digital Twins

In addition to WDNs, city-scale DTs that combine numerous infrastructures are becoming increasingly popular: water, energy, transport, and waste.

- **Cross-Sector Optimization**: The use of renewable energy to generate water pumping activities to reduce costs and emissions will be achieved through coordination between water pumping and renewable energy generation.
- **Urban Planning Support:** Modeling water demand, housing development, and transportation development.
- **Disaster Management:** Coordinating responses to floods, droughts, and pandemics with integrated DTs.

These combined strategies emphasize the role of water DTs in smart cities.

6.9.3 Research and Policy Opportunities

Finally, several opportunities exist for researchers and policymakers:

- **Standardization Efforts:** Development of shared protocols and ontologies for DT data exchange.
- **Public–Private Partnerships:** Collaborations to fund and scale DT projects.
- **Regulatory Frameworks:** Guidelines for cybersecurity, privacy, and ethical AI use in DTs.
- **Cross-Disciplinary Research:** Integrating water engineering with computer science, climate science, and social sciences.

These initiatives will help move DTs from experimental pilots to **mainstream infrastructure management tools**.

6.10 CONCLUSION

DT modeling is revolutionizing the design, operation, and management of WDNs. (2) A DT is an online virtual image of real-world infrastructure and acts as a bridge between in silico and in vivo. They help utilities transition from being reactive to predictive, adaptive, and increasingly data-driven. This section has presented the background of DTs for WDNs, starting from their definition and significance [18]. The background discussion and the review of literature showed that the development of hydrodynamic models, sensor networks, and computational intelligence have indeed created a basis for deploying DT tools in water systems. We also investigated the core DT development aspects, including data integration, model calibration, and online synchronization. Futuristically speaking, DTs might weave into everyday urban networks, linking transport, energy, and utilities more smoothly [11]. Still, making that happen won't come from tech alone; it'll take researchers, planners, and lawmakers moving together. Without shared rules, accessible tools, and teamwork between fields, growth could stall or favor only certain areas. One step at a time, DTs could reshape how cities manage water flow – bringing smarter control into everyday operations. Over time, they might do more than track performance; their real role may lie in shaping long-term plans for healthier urban environment.

REFERENCES

1. Brilakis, I., Pan, Y., & Borrmann, A. (2020). Built environment digital twins: State of the art. *Automation in Construction, 116,* 103225. https://doi.org/10.1016/j.autcon.2020.103225
2. Ahmed, F., & Khan, M. S. (2021). Digital twin technology for smart water systems: A comprehensive review. *Journal of Hydroinformatics, 23*(5), 987–1004.
3. Alvisi, S., & Franchini, M. (2020). Near real-time leak detection in water distribution systems: A digital twin approach. *Water, 12*(5), 1379. https://doi.org/10.3390/w12051379
4. Wu, Z. Y., Walski, T., Yang, S., & Bowdler, D. (2021). Advances in smart water system modeling with digital twins. *Journal of Water Supply: Research and Technology – AQUA, 70*(7), 923–938. https://doi.org/10.2166/aqua.2021.042
5. Batty, M. (2018). Digital twins. *Environment and Planning B: Urban Analytics and City Science, 45*(5), 817–820. https://doi.org/10.1177/2399808318796416
6. Sun, S., Zheng, F., & Kapelan, Z. (2020). Digital twin-driven resilience assessment of water distribution systems. *Water Research, 183,* 116070. https://doi.org/10.1016/j.watres.2020.116070
7. Booysen, M. J., Visser, M., & Burger, R. (2019). Temporal case study of household leak detection using smart metering. *Water SA, 45*(1), 80–90. https://doi.org/10.4314/wsa.v45i1.10
8. Khadir, A., Massou, S., & Chibani, A. (2022). Digital twin for contamination event management in water distribution networks. *Journal of Water Resources Planning and Management, 148*(5), 04022014. https://doi.org/10.1061/(ASCE)WR.1943-5452.0001507
9. Grieves, M., & Vickers, J. (2017). Digital twin: Mitigating unpredictable, undesirable emergent behavior in complex systems. In F.-J. Kahlen, S. Flumerfelt, & A. Alves (Eds.), *Transdisciplinary perspectives on complex systems* (pp. 85–113). Springer. https://doi.org/10.1007/978-3-319-38756-7_4
10. Tao, F., Zhang, H., Liu, A., & Nee, A. Y. C. (2019). Digital twin in industry: State-of-the-art. *IEEE Transactions on Industrial Informatics, 15*(4), 2405–2415. https://doi.org/10.1109/TII.2018.2873186
11. Zhang, C., Han, D., & Butler, D. (2018). Linking data-driven models and water distribution system simulations for online anomaly detection. *Environmental Modelling & Software, 108,* 1–14. https://doi.org/10.1016/j.envsoft.2018.07.006
12. Liu, Y., Yang, C., & Liu, X. (2021). Digital twin-based smart water management for cities. *Water Science and Technology, 83*(7), 1556–1568. https://doi.org/10.2166/wst.2021.109
13. Neupauer, R. M., & Lin, C. (2021). Using digital twins to improve resilience in urban water distribution systems. *Journal of Water Resources Planning and Management, 147*(9), 04021054. https://doi.org/10.1061/(ASCE)WR.1943-5452.0001389
14. Puust, R., Kapelan, Z., Savic, D. A., & Koppel, T. (2010). A review of methods for leakage management in pipe networks. *Urban Water Journal, 7*(1), 25–45. https://doi.org/10.1080/15730621003610878
15. Qi, Q., & Tao, F. (2018). Digital twin and big data towards smart manufacturing and industry 4.0: 360-degree comparison. *IEEE Access, 6,* 3585–3593. https://doi.org/10.1109/ACCESS.2018.2793265
16. Savic, D. A., & Kapelan, Z. (2018). Smart water systems and the digital twin concept. *Proceedings of the Institution of Civil Engineers – Water Management, 171*(5), 229–232.
17. van der Kleij, R., van Egmond-de Wilde de Ligny, E., & Hidayat, R. (2021). Cybersecurity risks in digital twins of water systems: A systematic review. *Water, 13*(22), 3184. https://doi.org/10.3390/w13223184
18. Yang, T., Li, H., & He, S. (2022). Digital twin for urban infrastructure: A framework for smart water-energy management. *Sustainable Cities and Society, 82,* 103880. https://doi.org/10.1016/j.scs.2022.103880

Flood Prediction Using Machine Learning Techniques

7

Aakanksha Sankhyan, Deelip Kumar Sarangi, Laishram Bijialuxmi Devi, Nirav Rameshbhai Amin, Gaurav Mehta, and Shivam Kumar

7.1 INTRODUCTION

7.1.1 Importance of Flood Prediction in Disaster Management

Floods are one of the natural calamities. They cause deaths, destroy homes, businesses, and infrastructure. They have a huge impact on the economy. According to the World Meteorological Organization, from 2000 to 2019, floods caused 44% of all natural disasters, disrupting the lives of around 2 billion people globally [1]. It is essential to have a system that predicts floods early and precisely. Such a predictive system can help reduce the number of deaths due to floods and can also assist in the development of better policies, communities, and infrastructure [2].

7.1.2 Limitations of Traditional Flood Prediction Methods

Flood predictions are currently done using conventional methods. These methods depend on water-based models. The changing climatic conditions, rising development, and the accessibility of vital resources are limitations of conventional methods. There is a rising need for advanced technical methods. An artificial intelligence (AI)-backed system can be developed that predicts floods accurately [3].

Hydrological and statistical models have primarily been used for flood prediction. The complexity and dynamics of hydrological conditions are increasing day by day. Conventional methods are already failing to make accurate predictions, and with changing conditions, their performance is further deteriorating.

DOI: 10.1201/9781003714217-7

a. **Model Simplification**: Conventional hydrological methods take floodplain characteristics between cross-sections in a river system. This limits the reality of the river system while simulating the natural river and floodplain variability. The model's accuracy is further reduced because of deficiencies in guidelines for selection.
b. **Poor Timeliness**: The conventional hydrologic method needs heavy computation, taking a lot of time to process the data, which causes delays in forecasting and prevents timely warnings [4].
c. **Linear Approach**: The conventional methods lack the ability to identify the nonlinear relationships among the attributes while predicting, which is another reason for the poor performance of these methods [5].
d. **Uncertainty and Quantification Issues**: If there is uncertainty in the dataset and model structure, it brings uncertainty into the output, which is flood predictions. Conventional methods lack strong processes to address this uncertainty. Hence, there is a lack of trust in these methods [6].
e. **Availability and Accessibility of Data**: The majority of conventional methods require accurate information on soil conditions, rainfall, river dynamics, etc. Such information is often not available. In such cases, conventional flood prediction methods are not applicable [7].
f. **Generalizability Issues**: Traditional approaches are designed for a specific region and geography, and when such models are applied to another region or geography, their performance is not as good as it was previously.
g. **Computational Burden and Scalability**: For large basins or high-spatial resolution data, simulations become complex and computationally intensive. The applicability of conventional models reduces in such cases as they take time to produce predictions, failing to provide real-time forecasting or generate inundation maps quickly [8].
h. **Integration Challenges**: Combining hydrological models with hydraulic or meteorological models remains challenging because of complexity and compatibility issues. Flood forecasting workflows become complicated if such integration is attempted [9].

7.2 OVERVIEW OF FLOOD PREDICTION TECHNIQUES

Flood prediction techniques can be divided into three major categories: hydrological, hydraulic, and statistical approaches. Each approach has different principles, data requirements, and applications.

7.2.1 Hydrological Models

- They manipulate rainfall–runoff processes.
- They simulate rainfall–runoff processes in a watershed to analyze the discharge from rivers and flood peaks.
- The models use rainfall input, soil characteristics, land use, and river network data.
- A wide range of models is available, varying from simple pragmatic copies to complex physically based distributed models.
- The hydrological models aim to capture the complex processes of infiltration, surface runoff, evapotranspiration, and baseflow.
- They are highly dependent on high-quality input databases and face difficulties.
- They face difficulties in representing spatial heterogeneity.

7.2.2 Hydraulic Models

- These models initiate the flow of water through river channels and floodplains and analyze flood extent and depth.
- They use output from hydrological models, such as discharge, and detailed topography (digital elevation model (DEM)) for the simulation of water flow.
- They can generate flood hazard maps, which are critical for flood risk supervision, town development, and alternative response.
- They sometimes require further topography and land cover data for accuracy.

7.2.3 Statistical Models

- It uses historical data on floods and rainfall to forecast flood severity.
- It uses regression analysis, time span analysis, and probability distribution models.
- It requires less effort and is computer efficient.
- It is limited in understanding multifaceted nonlinear hydrological processes, which may not be addressed under varying climate conditions.

All the approaches mentioned above help in different ways to calculate floods, presenting key learnings by focusing on applying approaches and combining all three models with AI-based algorithms for precision and accuracy.

7.3 ROLE OF BIG DATA AND REMOTE SENSING IN PREDICTION OF DISASTER

A huge amount of data and instruments play a crucial role in the remote detection of floods and other natural calamities:

- **Multiple Data**: To accurately predict disasters and build alert systems for a particular region, it is necessary to gather databases from satellite images along with sensor outputs from meteorological systems. It also needs historical data to train the model and public dissemination systems [10].
- **Flood Surveillance**: By using advanced technologies powered by AI-based learning models, real-time satellite-based information can be used to track variations in climatic conditions and water basin levels. Big data analytics ensure accuracy and consistently detect early signs of disasters, i.e., rising water levels and storm formation [11].
- **Improved Predictive Accuracy**: A machine learning (ML) method combined with big data analytics can process complex datasets and information more conveniently. It can handle nonlinear relations among the attributes of the datasets and can predict floods with high accuracy, which conventional models fail to do [12].
- **Timely Damage Assessment**: Once a disaster strikes, a remote sensing system combined with big data analytics can assess damages quickly and precisely. It also assists in providing emergency relief, supplying needed resources, and supervising overall emergency response [13].

- **Facilitating Proactive Disaster Management**: A remote sensing system combined with big data analytics can help the government and disaster management agencies improve readiness and ensure a quick and efficient response.

7.4 SUMMARY OF RECENT STUDIES: ML IN FLOOD FORECASTING

Several studies have been conducted, of which the recent studies from 2023 to 2025 are referenced here. This research highlights significant improvements and the usefulness of ML in enhancing flood prediction accuracy and improving emergency responses:

- A study was conducted on the African Ouémé basin to predict discharge from the river and simulate flood-prone areas. It used high-resolution data obtained from satellites as input and built a model with the help of advanced recurrent neural networks, including LSTM and ConvLSTM. The use of this model achieved a high coefficient of determination, explaining significant runoff simulation and flood risk mapping, even though data is scarce [14].
- Using ML algorithms, a study was conducted on the forecasting of floods in Makassar City, finding random forest (RF) to be the most effective model for calculating multiple datasets from urban areas, supporting accurate and reliable warning systems for metropolitan areas [15].
- A research work where Transformer architecture (AGRS-LSTM-Transformer) combined with a hybrid deep learning model was developed and tested, achieving high accuracy in predicting approximately six times more compared to target models; it also apprehends nonlinear processing dynamics for floods and reduces peak discharge underestimation common in other models [16].
- A study conducted in 2024 focuses on increasing the importance of ML in flood forecasting and classification, also assessing its capacity to learn from large spatiotemporal and Internet of Things information, excluding multiple physical combinations [17].

Various research focuses on incorporating EXAI along with big data analysis, promoting ways to comprehend the mechanisms responsible for generating floods whereas preserving computational efficiency compared to conventional or operational flood forecasting.

7.5 DATA SOURCES FOR FLOOD PREDICTION

- *Hydrological data: rainfall, river discharge, soil moisture, and water levels.*

  For predicting floods, vital data related to water include rainfall intensity, discharge through water, soil moisture, and water level. Such data are important for forecasting water-related events through the use of algorithms. Streamflow, rainwater, soil moisture, and water level identifications are core as they are used in both conventional and modern ML/DL predicting systems because they capture precipitation input, catchment storage state, and river response (peaks and timing).

- *Meteorological data: temperature, humidity, and precipitation forecasts.*

  Meteorological-based information is required to forecast floods, mostly the temperature, humidity, and precipitation. These data will also aid in identifying atmospheric conditions affecting the patterns of rainfall, evapotranspiration, and moisture present in the soil, which are needed for flood risk identification and water-related patterns.

  Such variables are usually obtained from the integration of stations situated at ground levels and innovative numerical prediction algorithms to promote a range of medium predictions.
- *Remote sensing and satellite data: land use, vegetation, and topography.*

  Satellite and remote-based sensing data are very important in providing in-depth information on land use, vegetation, and topography, which influence flood aspects. Some important data may include
  - **Land Use and Vegetation**: Some satellite sensors, such as Sentinel-2 and Landsat-8, can capture different aspects of images, helping in identifying land area coverage and vegetation indicators such as normalized difference vegetation index (NDVI), which further aids in alerting flood prediction areas with low vegetation and higher potential [18].
  - **Topography**: DEMs created via satellite biometry synthetic aperture radar (SAR) images help provide important and crucial angles and high data needed for shaping water flow paths and identifying potential flood-prone areas [19].
- *Social and IoT data: Sensors, drones, crowdsourced flood reports.*

  Flood monitoring and prediction use social and IoT datasets by aiding real-time data collected from different sources such as
  - **IoT Sensors**: Installed in flood-prone areas to detect water levels, flow, rainfall runoff, temperature, and humidity indicators, transferring actual data wirelessly for early warning and risk calculation [20].
  - **Drones**: Cameras and sensors attached to the drones provide aerial images and mapping of flood distributions in inaccessible areas, facilitating post-flood interventions and situational awareness.
  - **Crowdsourced Flood Reports**: Collects data and reports from the community with the help of platforms and mobile apps, supplementing official data and helping to monitor the actual impacts of floods more reliably.

These technologies allow information from multiple sources, which helps in monitoring floods, improving alert systems, and enhancing community resilience.

7.6 ML TECHNIQUES AND FLOOD PREDICTION

Hybrid and ensemble methods combine hydrological models with ML techniques to improve flood prediction accuracy by leveraging the strengths of both approaches. Ensemble methods integrate multiple forecast scenarios from diverse models and post-process precipitation data to reduce uncertainties and increase reliability. This ensemble precipitation data is used as input for hydrological models to simulate flood events with confidence intervals, enhancing the prediction of flood magnitudes and timings. ML models can address many limitations such as

- **Learning from Data, Predicting Nonlinear Relationships**: When used, ML models can assess and process a wide range of data and reliably forecast the nonlinear correlations in flood forecasts compared to historical and remote sensing data.

- **Managing Multi-Source Data**: To provide more accurate and superior prediction rates while working with multi-source data, ML models can incorporate multidimensional data, such as satellite photos, remote sensing, and geodatabases.
- **Accurate and Reliable Predictions**: When ML models are trained and used, they require less computing time than physical simulations, reducing the time needed to conceptualize the results. It has been demonstrated that ML models have a shorter simulation time for mapping flood risk compared to conventional hydrological models.
- **Better Performance in Data-Scarce Regions**: ML models analyze data from satellite imaging, remote sensing, and climate change when hydrological data for a certain area is not fully available.
- **Flexibility**: The flexibility of ML models lies in their ability to be updated and retrained on new datasets while retaining accuracy and precision.
- **Hybrid Approaches**: ML models can be combined with the conventional physics-based models because the hybrid prediction model provides physical consistency while removing physically irrational predictions.

All of the above points reflect the need for the integration of AI in predicting and forecasting floods.

7.7 FRAMEWORK/METHODOLOGY

7.7.1 Data Preprocessing (Cleaning, Normalization, Handling Missing Values)

Data preprocessing is a crucial step for flood prediction using ML, involving the following key tasks (Figure 7.1):

- **Handling Missing Values**: Missing values and data can cause bias in prediction and may cause errors too. Small gaps may be filled by calculating mean, median, and mode values, whereas large gaps might require exclusion of records to maintain data uniformity.
- **Cleaning**: It is also needed to remove outliers, duplicate data, and noise to ensure the quality and reliability of the dataset for building an algorithm.
- **Normalization/Standardization**: Arranging features (e.g., rainfall and river discharge) in a uniform range facilitates ML models to increase their performance as well as integration, particularly for networks and algorithms sensitive to feature magnitude.
- **Feature Selection**: Picking the most relevant variables (e.g., rainfall and soil moisture) by removing superfluous and irrelevant parameters reduces the complexity of models and increases their accuracy.
- **Splitting Data**: Division of data into training and testing sets, most commonly in an 80/20 ratio, increases the chances of unbiased evaluation of model performance on unknown data.
- **Feature Engineering**: Developing a new innovative structure from existing ones enhances model predictability.

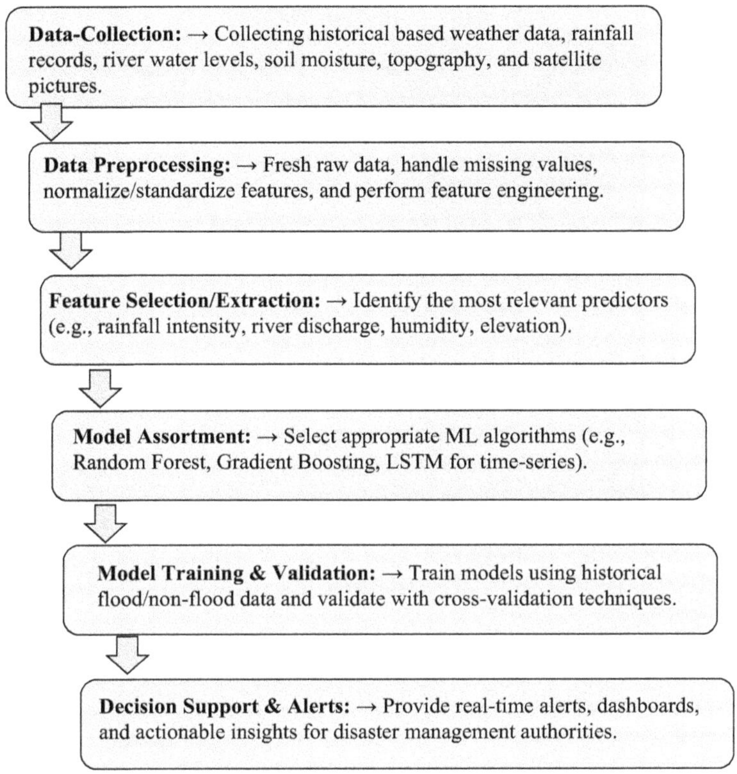

FIGURE 7.1 Flowchart on flood prediction using machine learning techniques

7.7.2 Implementation of LightGBM Model

The LightGBM model has been implemented to predict floods in advance. The results, in the form of a confusion matrix, receiver operating characteristic (ROC) curve, and classification report, are shown below:

The confusion matrix in Figure 7.2 shows that the LightGBM model performs well in identifying events. It can also be observed from the confusion matrix that the model reliably avoids unnecessary warnings; however, the model's flood detection accuracy needs to be improved.

The ROC curve in Figure 7.3 demonstrates high classification performance, with an area under the curve (AUC) of 0.87, indicating that it has maintained a good balance between sensitivity and specificity. It can be observed that LightGBM is effective for predicting floods.

Figure 7.4 shows that the model positively and consistently detects actual flood cases, further confirming moderate overall performance and stressing the need for improvement in capturing flood events while maintaining reliability.

7.7.3 Implementation of XGBoost Model

The XGBoost model has been implemented to predict floods in advance. The results, in the form of a confusion matrix, ROC curve, and classification report, are shown below:

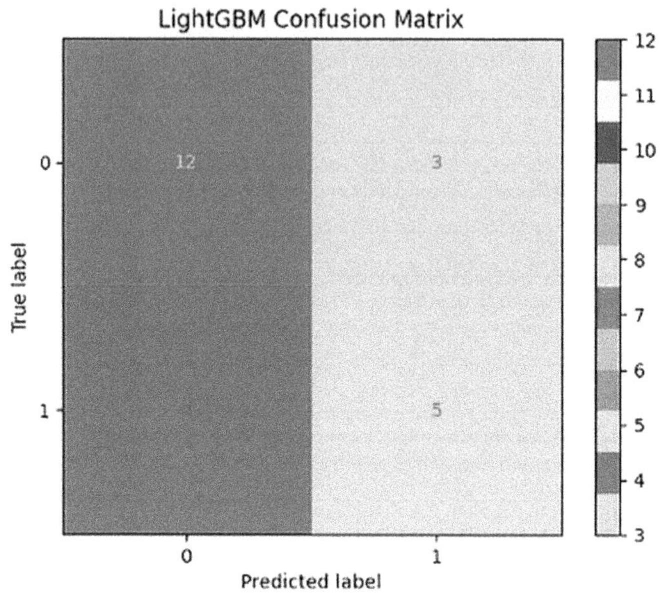

FIGURE 7.2 Confusion matrix: LightGBM confusion matrix.

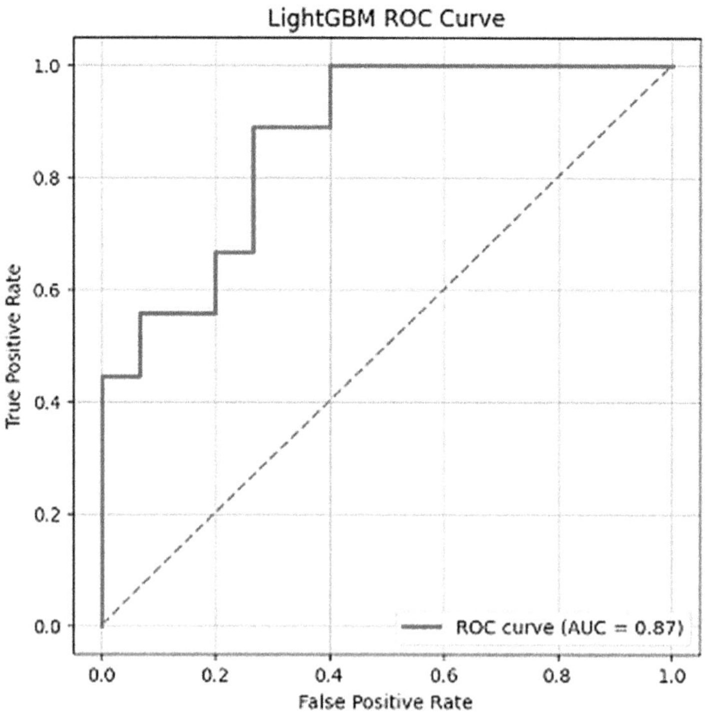

FIGURE 7.3 ROC curve: LightGBM.

Classification Report:

	precision	recall	f1-score	support
0	0.80	0.80	0.80	15
1	0.67	0.67	0.67	9
accuracy			0.75	24
macro avg	0.73	0.73	0.73	24
weighted avg	0.75	0.75	0.75	24

FIGURE 7.4 Classification report: LightGBM.

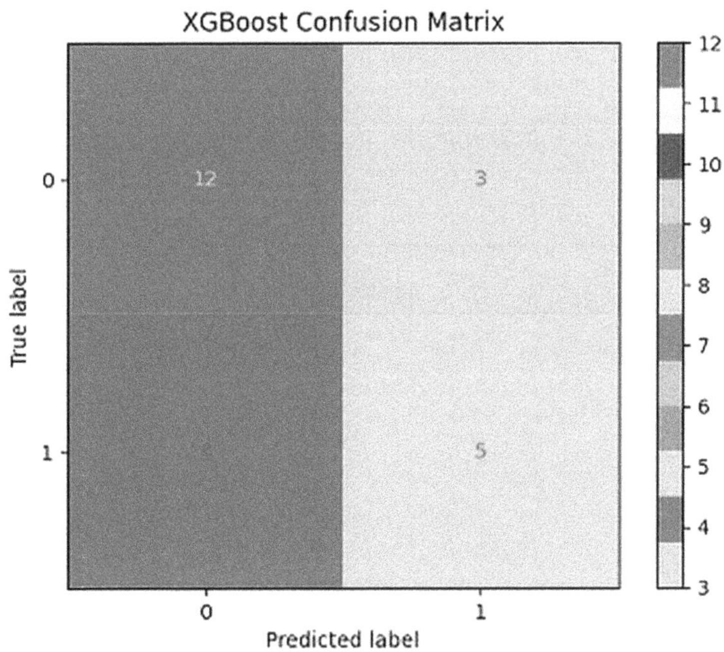

FIGURE 7.5 Confusion matrix: XGBoost.

Figure 7.5 Illustrates that the model shows better performance in predicting class 0, while class 1 has a higher misclassification rate.

The ROC curve (Figure 7.6) for XGBoost illustrates robust ability, with an AUC of 0.82, indicating that it has a good balance and can differentiate between the samples, performing well in predicting the outcomes.

XGBoost attains an overall accuracy of 0.71 (Figure 7.7), indicating moderate overall performance.

7.7.4 Implementation of CatBoost Model

The CatBoost model has been implemented to predict floods in advance. The results, in the form of a confusion matrix, ROC curve, and classification report, are shown below:

Figure 7.8 shows that the method correctly classified 12 instances of class 0 and 7 instances of class 1, indicating that the model has slightly better performance.

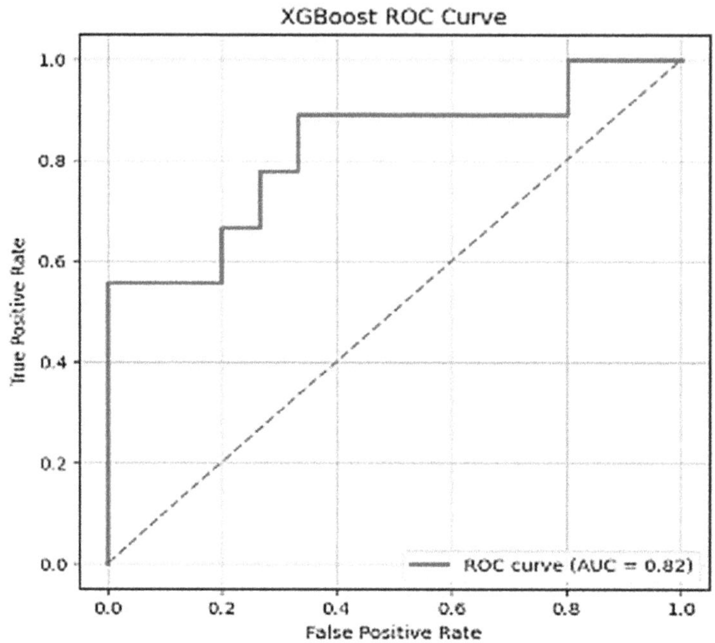

FIGURE 7.6 ROC curve: XGBoost.

Classification Report:

	precision	recall	f1-score	support
0	0.75	0.80	0.77	15
1	0.62	0.56	0.59	9
accuracy			0.71	24
macro avg	0.69	0.68	0.68	24
weighted avg	0.70	0.71	0.70	24

FIGURE 7.7 Classification report: XGBoost.

The ROC curve shown in Figure 7.9 depicts robust performance with an AUC of 0.90, which further indicates a high capability of the model to discriminate between the two classes.

The model achieved 79% accuracy in Figure 7.10, performing better on class 0 (F1 = 0.83) than on class 1 (F1 = 0.74), indicating slightly reduced consistency in predicting class 1.

7.7.5 Implementation of Gradient Boosting Model

The gradient boosting model has been implemented to predict floods in advance. The results, in the form of a confusion matrix, ROC curve, and classification report, are shown below:

Figure 7.11 shows that the model indicates a balanced performance with good predictive ability but still makes some errors in distinguishing between the two classes.

The ROC curve (Figure 7.12) for the Gradient Boosting model shows strong distinguishability with an AUC of 0.87, indicating that the model effectively distinguishes between the two classes, which confirms the model's robustness and reliability in classification.

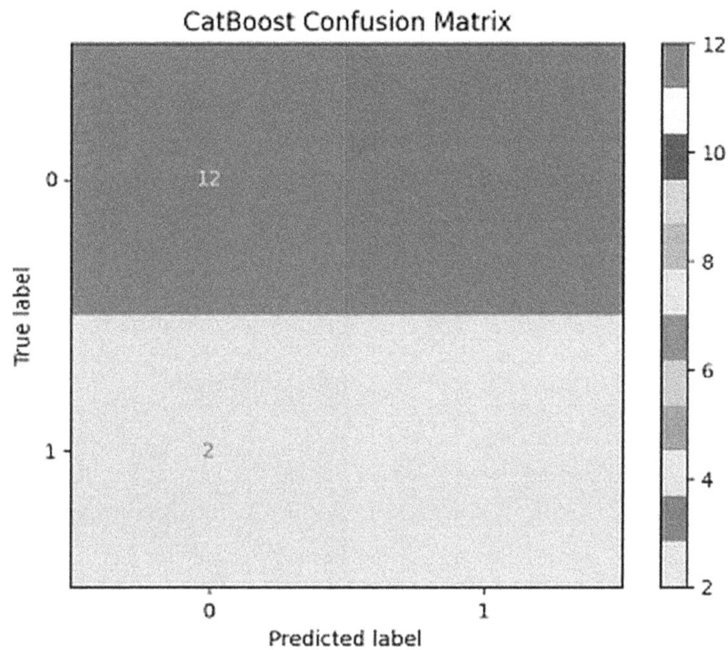

FIGURE 7.8 Confusion matrix: CatBoost.

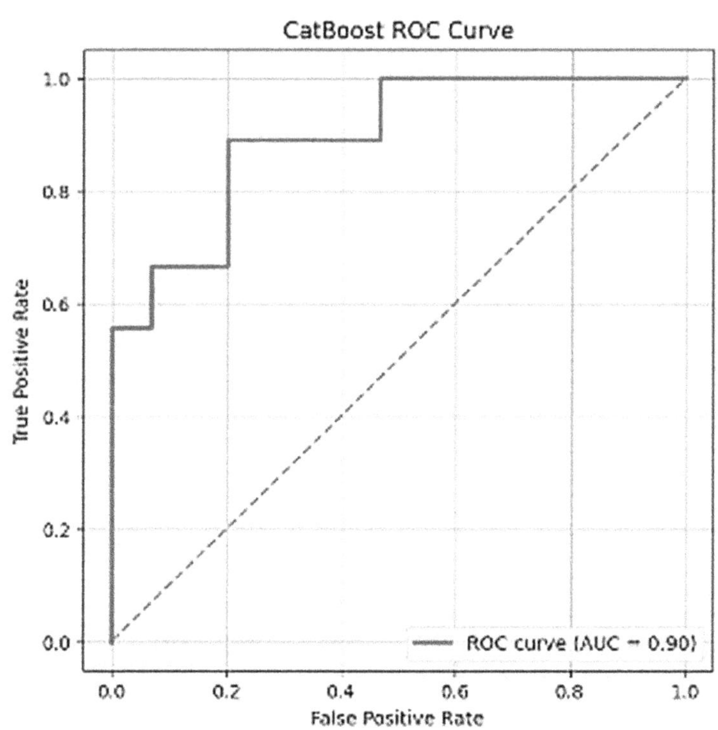

FIGURE 7.9 ROC curve: CatBoost.

Classification Report:

	precision	recall	f1-score	support
0	0.86	0.80	0.83	15
1	0.70	0.78	0.74	9
accuracy			0.79	24
macro avg	0.78	0.79	0.78	24
weighted avg	0.80	0.79	0.79	24

FIGURE 7.10 Classification Report: CatBoost.

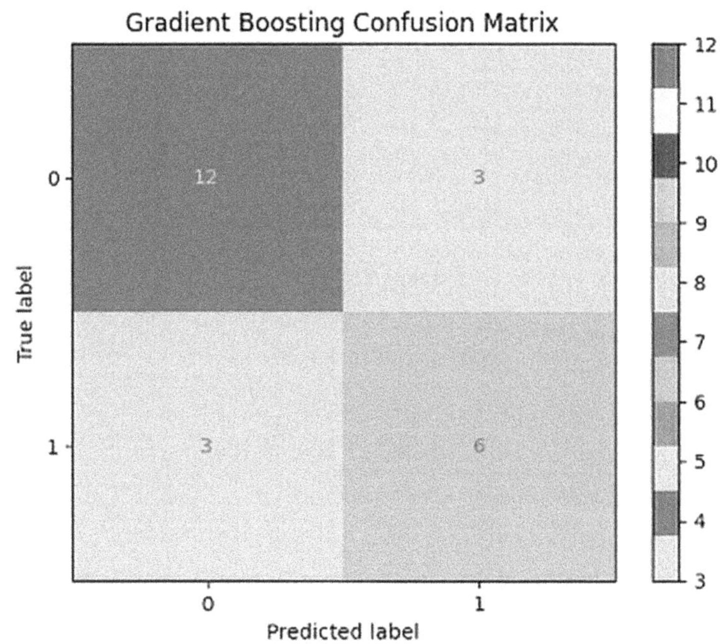

FIGURE 7.11 Confusion Matrix: Gradient boosting.

The classification report shows that the model achieved an overall accuracy of 75%, indicating more misclassifications and highlighting the need for improvement in detecting minority class instances (Figure 7.13).

7.8 CASE STUDIES AND APPLICATIONS

7.8.1 Examples of Flood Prediction Systems Deployed

- **Case 1: Indus Basin, Pakistan**: Asia's one of the most vulnerable river systems prone to flooding is the Indus Basin, where the region is often exposed to severe floods that affect the region's vegetation and animals. By implementing multifaceted data, counting the parameters

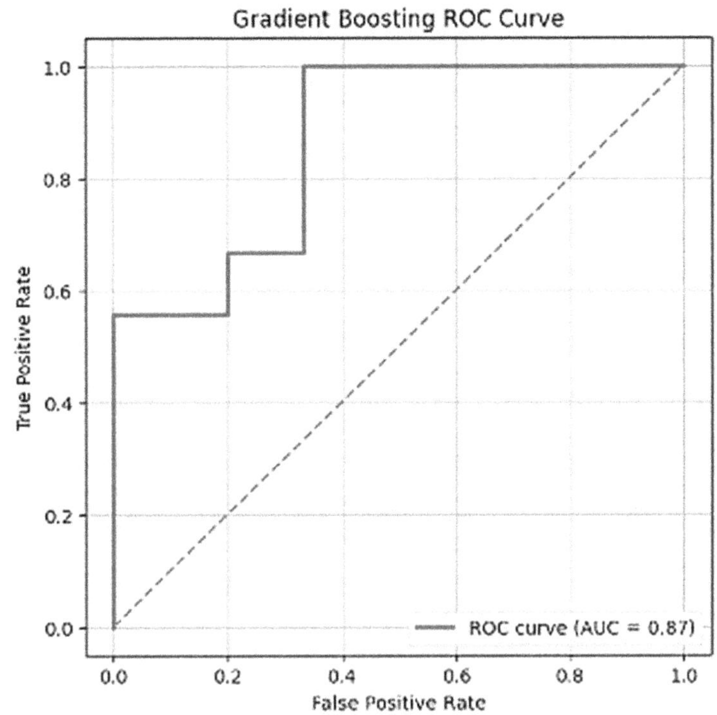

FIGURE 7.12 ROC curve: Gradient boosting.

Classification Report:

	precision	recall	f1-score	support
0	0.75	0.80	0.77	15
1	0.62	0.56	0.59	9
accuracy			0.71	24
macro avg	0.69	0.68	0.68	24
weighted avg	0.70	0.71	0.70	24

FIGURE 7.13 Classification report: gradient boosting.

such as soil condition, water levels, meteorological data, and climatic conditions, different algorithms were integrated in the Indus Basin to foresee the warning signs of floods. The RF algorithm outperformed better than others due to its capability to detect nonlinear interactions. Moreover, it was also found that ML algorithms surpass conventional methods in the context of early warning signs and precise flood detection. Hence, this reflects the potential capabilities of ML compared to older methods for disaster alertness as well as reducing the aftermath of floods [21].

- **Case 2: Mahanadi River Basin, India**: This has been mostly vulnerable to flooding, particularly during days of high rainfall that lead to a rise in water levels. In these instances, certain models are being used to forecast floods. ML algorithms used data on previous rainfall histories, ultimately concluding that incorporating ML algorithms produces high-quality and accurate flood predictions, supports risk relief practices in prone areas, and reduces the environmental impact by providing more precise results [22].

7.9 PERFORMANCE COMPARISON BETWEEN CONVENTIONAL AND ML-BASED METHODS

7.9.1 Role of Early Warning Systems and Real-Time Monitoring

Early informing and projecting structures play an important role in alleviating the outcomes of floods by creating actual and precise predictions (Table 7.1). Some important roles are

1. **Responsiveness**: early warning system (EWS) support the exposed population by informing them in advance of any form of disaster, which may, in turn, help the authorities with adequate time and resources to take action [23].
2. **Mitigation of Damage**: When these models are employed, they aid in mitigating the damages caused by floods to buildings, land, and human lives, ultimately working toward a better ecological balance [24].

TABLE 7.1 Performance comparison between conventional and ML-based methods

ASPECT	*CONVENTIONAL METHODS*	*ML METHODS*	*FINDINGS*
Approach	Physical hydrodynamic & statistical models	Data-driven ML models (random forest, LSTM, SVM)	Traditional methods use physics; ML relies on various datasets
Accuracy	Good for known conditions; struggles with extremes	Often higher accuracy, particularly with multifaceted data	RF and LSTM show high NSE (>0.9); SVM is weaker in extreme events
Computational speed	Slow, computationally intensive	Fast inference and updates	ML models are suitable for real-time prediction
Interpretability	High (physics-based)	Variable; RF is more explainable	Physical insights are well denoted
Handling nonlinearity	Moderate	Good at capturing nonlinear interactions	ML is better at complex temporal/spatial dependencies
Data requirement	Moderate to high historical and spatial data	Requires large, high-quality training data	Hybrid models reduce dependence on very large datasets
Extreme event handling	Limited; model is often simplified	Mixed; stacking ensembles improve extreme event prediction	Best performance achieved by combining ML models
Model updating	Difficult, costly	Easy, flexible with new data	ML is adjustable to shifting conditions
Examples	Hydrodynamic (Saint-Venant), SCS-CN method	Random forest, LSTM (deep learning), SVM	Hybrid stacking models reduce RMSE by >6-13%, improve NSE

3. **Health Safety**: As the system provides early warnings and alerts, it helps individuals take protective actions in advance, directly decreasing the chances of mass casualties and health-related concerns [25].
4. **Surveillance**: The ML-based models enhance the ability to track the river basin and make it possible to track details in real time, e.g., the movement of the flood, its nature, and initiation point [26].
5. **Advisories and Notifications**: The combination of ML with flood prediction helps produce advisories and notifications for regions prone to floods and helps in coordinating relief interventions [27].
6. **Decision-Making Guidance**: Early warning systems helps the authoritative members regarding the steps and interventions to be taken prior to the initiation of any disaster.
7. **Uncertainty Management**: The hybrid models currently in use enable the issuance of warnings well in advance of the actual onset of the problem and also assist in managing uncertainties [28].

7.10 CHALLENGES AND LIMITATIONS

7.10.1 Challenges and Considerations in Utilizing ML Models for Flood Prediction

Even though ML models provide a number of benefits over conventional flood prediction and estimation techniques, several issues still need to be addressed:

1. **Data Quality, Availability, and Preprocessing**: The predictive accuracy of some models can be compromised by gaps such as missing values, limited historical records, unreliable sensor inputs, or irregular measurement intervals. To enhance both the applicability and performance of these models, it is essential to apply rigorous feature engineering, thorough data preprocessing, and robust techniques for handling missing values [29].
2. **Generalizability and Transferability**: Once the model is used for a certain area, it can forecast with high accuracy and reliability; however, when used in a variety of settings, it may not offer the same levels of accuracy and prediction, creating a problem for generalization and for developing a model that can operate in several places and yield accurate results [30].
3. **Interpretability and Trust**: Many ML models are criticized for their incomprehensible procedures since it is important for government officials and scholars to understand the algorithms that the models use to predict floods. To improve these models' interpretability and trustworthiness in terms of prediction generation, numerous sophisticated technological and hybrid approaches are being used [31].
4. **Computational Resources and Infrastructure**: Even though integrating ML models results in more accurate and superior predictions, training these models and processing high-resolution data necessitate significant computational resources as well as infrastructure investment and assistance [29].
5. **Uncertainty Quantification**: While the prediction of floods innately includes doubt due to inconsistencies in weather forecasts, structural prototyping, and experimental errors, integrating possibility in forecasting and ambiguous predictions is crucial for real-world applications [32].

6. **Temporal and Spatial Resolution**: The actual duration for predicting floods and the three-dimensional resolution needs strongly developed algorithms as well as their performance. ML algorithms may not always be accessible [33].

7.10.2 Model Overfitting and Lack of Generalization to New Regions

Model overfitting and lack of generalization to new regions are significant tasks in ML-based flood prediction:

- **Overfitting**: ML models might perform well with the data they are trained on but can perform poorly with uncertain datasets due to their learning of specific patterns, which can lead to poor predictions when the same model is applied to different locations or settings [34].
- **Lack of Generalization**: As the models are created with specific data and attributes of one region, they may lead perform poorly of same model when applied to a different region because of the novel characteristics of the new region [35].

7.10.2.1 Solutions and Mitigation Techniques

- **Cross-validation and Regularization**: Technical advancements such as K-fold cross-validation, when applied, may aid in decreasing the issue of overfitting. Furthermore, using multiple models helps improve the generalization of the models.
- **Data Augmentation and Transfer Learning**: Considering previously trained systems or synthetic data from related areas increases the model's flexibility. Additionally, using multiple features and attributes to train the model enhances its performance. Continuous retraining of the model with new data and features aids in maintaining optimal performance [36].

7.10.3 Computational Complexity and Real-Time Constraints

Traditional hydrodynamic models, while physically accurate, are processor-demanding and frequently too slow for real-time forecasting, limiting their practical use in rapid decision-making for flood management. While ML algorithms provide faster analysis and accurate predictions, making them suitable for application in real-time flood prediction. However, there are some model that are very much complex, e.g., long short-term memory networks, and need systems with higher configuration and careful hyperparameter regulation. Such algorithms may not find application in real-time flood prediction due to their time-sensitive nature.

ML models such as random forest and others, offer good prediction accuracy, interpretability, and system efficiency, creating favorable applications in active sites, specifically for tidal and complex flood scenarios. Further advancements in hardware, optimization in algorithms, and quantum ML may further reduce processing time and improve real-time flood prediction capacity.

7.11 FUTURE PERSPECTIVE IN AI-BASED FLOOD PREDICTION

7.11.1 Combining Deep Learning with Climate Models

There are different emerging technologies and trends in forecasting floods. A few of them are discussed below [31].

- **Reasonable AI**: The use of explainable AI in building AI models brings trustworthiness among users, i.e., Shapley Additive explanations can be implemented with the predictable AI model.
- **Hybrid Models**: Joining physics-based hydrological replicas with AI ensures physical uniformity along with model-based precision.
- **Allocation Learning and Meta-Learning**: Transfer learning allows models trained in one place with a good amount of data to be used in another place where there is not enough data with little revision to the trained model.
- **Integration of IoT and Edge Computing**: Future systems, with the help of edge AI, can help in the functions of IoT sensors and in processing information in the vicinity.
- **Climate Change Adaptation**: AI should be able to address the unpredictability caused by climate change. Adaptive models that adjust to the changes in the environment while making predictions can be useful here.

7.11.2 Community-Based Flood Prediction Using AI-Powered Mobile Apps

Active community engagement is one of the important and necessary steps in mitigating the risk of disasters. With the increasing prevalence of AI-backed mobile applications, flood prediction and alert systems are becoming more convenient, localized, and participatory. Mobile applications can combine real-time hydrological data with AI and ML models to provide more localized and precise flood forecasts [37].

7.12 CONCLUSION AND SUMMARIZATION

- *Role of ML in improving flood forecasting.*

  ML enhances flood forecasting by analyzing large datasets, i.e., rainfall, soil moisture, and river discharge, with greater expertise. ML models can recognize patterns and analyze time–space aspects, which strengthens correctness and decreases prediction time compared to conventional methods. Furthermore, data based on satellite images, IoT, and sensor data can help AI and ML models make accurate and precise predictions.

- *Emphasis on ML as a complement, not a replacement, for hydrological expertise.*
  Even with the strong predictive capacity of ML, it cannot completely replace the existing hydrological systems due to its dependence on training data and inadequate interpretability. By providing in-depth understanding of watershed systems, data related to river basins, and physical mechanisms, hydrologists aid in generating prediction results based on contextual accuracy.
- *Prospective research approach.*
  Future studies should emphasize the integration of ML with existing conventional methodologies, as this would make the whole model more robust and produce better results. Furthermore, researchers should also work on making the AI process interpretable, as it would enhance the trustworthiness of these processes.

REFERENCES

1. World Meteorological Organization (WMO). *Atlas of mortality and economic losses from weather, climate and water extremes (1970–2019).* Geneva: WMO; 2021.
2. Jongman B. Effective adaptation to rising flood risk. *Nat Commun.* 2018;9(1):1986.
3. Zhou J, et al. The comparative study of machine learning agent models and traditional hydrodynamic models in flood prediction. *Sci Rep.* 2025;15(1):4633.
4. Sharma S, Prasad V. Innovations in hydrological modeling for predicting flood risk. *Hydrol Curr Res.* 2024;15(3):112–119.
5. Liu T, et al. Flood simulation using hydrological and hydrodynamic models: data, complexity, and limitations. *J Water Clim Change.* 2024;14(10):3496–3505.
6. Singh R, Kumar A, Singh A. Flood prediction using hydrologic and ML-based modeling. *Int J Adv Comput Sci Appl.* 2023;14(11):55–67.
7. Vassar Labs. Machine Learning for Flood Forecasting. 2025 Jul 20 [cited 2025 Sep 16]. Available from: https://vassarlabs.com/machine-learning-for-flood-forecasting/
8. Tang Y, et al. Flood forecasting based on machine learning pattern recognition: advances and challenges. *J Hydroinform.* 2023;25(4):679–694.
9. Propulsion Tech Journal. A Comparative Study of Machine Learning Algorithms for Flood Prediction. 2025 [cited 2025 Sep 16]. Available from: https://www.propulsiontechjournal.com/index.php/journal/article/download/7641/4893/13040
10. Kaler S, Sharma A. Big data predicting natural disasters. *J Emerg Technol Innov Res.* 2023;10(9):231–239.
11. Jain H, Dhupper R, Shrivastava A, Kumar D, Kumari M. Leveraging machine learning algorithms for improved disaster preparedness and response through accurate weather pattern and natural disaster prediction. *Front Environ Sci.* 2023; 11:1194918.
12. Jones B, Lee C. Applications of big data and AI in disaster damage assessment. *Nat Hazards Rev.* 2021;22(4):04021019.
13. Sarker MNI. Disaster resilience through big data: Way to environmental sustainability. *Sustain Cities Soc.* 2020;61:102223.
14. Obada E, Biao EI, Zohou PJ, Yarou H, Hounnondaho FZ, Alamou EA. Using machine learning and satellite data to improve flood forecasting: The case of the Ouémé basin at the Bétérou outlet. *Hydrol Res* [Internet]. 2025 [cited 2025 Sep 19];56(2):153. doi:10.2166/nh.2025.133.
15. Hariadi M, Saputra Y, Fadillah M, et al. Enhancing Flood Prediction in Urban Areas: A Machine Learning Approach for Makassar City. ETASR. 2025; Received: 25 Oct 2024. Available from: https://etasr.com/index.php/ETASR/article/view/9402
16. Tang Y, Sun Y, Han Z, et al. An interpretable hybrid deep learning model for flood forecasting based on LSTM and transformer. *J Hydroinform.* 2025; Available from: https://www.sciencedirect.com/science/article/pii/S2214581824002210
17. Refadah SS. Machine learning applications in flood forecasting and classification: A comprehensive review. *Environ Sci.* 2024; Available from: https://www.aimspress.com/article/doi/10.3934/environsci.2025004?viewType=HTML

18. Awasthi Y, Chinzvende J. Deep learning-based flood forecasting using satellite imagery and IoT sensor fusion. *Int J Comput Appl.* 2025;187(21):37–52. Available from: https://ijcaonline.org/archives/volume187/number21/awasthi-2025-ijca-925261.pdf
19. Remote sensing and climatic data for flood impact assessment. *Sci Rep.* 2025; (Article). Available from: https://www.nature.com/articles/s41598-025-12234-w
20. Boltiot. IoT Based Flood Monitoring and Alerting System [Internet]. 2021 Feb 22 [cited 2025 Sep 16]. Available from: https://www.boltiot.com/projects/iot-based-flood-monitoring-and-alerting-system
21. Abbas S, Yaseen ZM, Salih SQ, Shahid S, Ismail T, Zounemat-Kermani M, et al. Developing a machine learning-based flood risk prediction model for the Indus Basin in Pakistan. *Water Pract Technol.* 2024;19(6):2213–2225.
22. Kumari S, Sharma S. Comparison of machine learning models for flood forecasting in the Mahanadi River Basin, India. *J Water Clim Change.* 2024;15(4):1629–1652.
23. VassarLabs. Flood Early Warning System for Disaster Management - aquaFLOOD. VassarLabs [Internet]. [cited 2025 Aug 15]. Available from: https://vassarlabs.com/flood-early-warning-system-for-disaster-management/
24. European Commission, Joint Research Centre. *The benefit of continental flood early warning systems to reduce the impact of flood disasters.* Publications Office of the European Union; 2025.
25. United Nations Development Programme (UNDP). What Are Early Warning Systems and Why Do They Matter for Climate Action? Climate Promise, UNDP [Internet]. 2025 Feb 28 [cited 2025 Aug 15]; Available from: https://climatepromise.undp.org/news-and-stories/what-are-early-warning-systems-and-why-do-they-matter-climate-action
26. European Commission, Joint Research Centre. Floods - European Commission - EU Science Hub (preparedness, monitoring, flood risk via in situ + forecasts). European Commission; [cited 2025 Aug 15]. Available from: https://joint-research-centre.ec.europa.eu/projects-and-activities/natural-and-man-made-hazards/floods_en
27. UNDP. Climate Information and Early Warning Systems. Adaptation, UNDP [Internet]. [cited 2025 Aug 15]. Available from: https://www.adaptation-undp.org/climate-information-and-early-warning-systems
28. European Commission, Joint Research Centre. New EFAS Partnership Boosts Early Flood Warnings in Europe. Innovations-Report; 2025.
29. Intelligent flood forecasting and warning: a survey. Intelligence & Robotics. 2023;3(2):190–212. Available from: https://www.oaepublish.com/articles/ir.2023.12
30. Sidrane C, Fitzpatrick DJ, Annex A, O'Donoghue D, Gal Y, Biliński P. Machine learning for generalizable prediction of flood susceptibility. *arXiv* [Preprint]. 2019 Oct 15 [cited 2025 Sep 14]. Available from: https://arxiv.org/abs/1910.06521
31. Mosavi A, Ozturk P, Chau K-W. Flood prediction using machine learning models: Literature review. *Water.* 2023;15(3):566.
32. Guzman SM, Paz JO, Tagert ML, Mercer AE, Bingner RL. Spatially explicit prediction of streamflow and uncertainty analysis using machine learning. *Environments.* 2021;12(8):259.
33. Li W, Kiaghadi A, Dawson CN. High temporal resolution rainfall–runoff modelling using long-short-term memory (LSTM) networks. *arXiv* [Preprint]. 2020 Feb 7 [cited 2025 Sep 14]. Available from: https://arxiv.org/abs/2002.02568
34. Shrestha S, Tuladhar A, Aryal N. Flood forecasting with machine learning models in an Alpine catchment: Challenges and opportunities. *Hydrol Earth Syst Sci.* 2022;26(8):4013–4027. Available from: https://hess.copernicus.org/articles/26/4013/2022/hess-26-4013-2022.html
35. Kumar G, Venkata M. Flood prediction using classical and quantum machine learning models. *Bioresscientia.* 2024;10(4):1–20. Available from: https://bioresscientia.com/uploads/articles/1731751295Galley_Proof-Flood_Prediction_Using_Classical_and_Quantum_Machine_Learning_Models.pdf
36. Singh R, Kumar D. Using machine learning and satellite data to improve flood forecasting in diverse regions. *Hydrol Res.* 2025;56(2):153–172. Available from: https://iwaponline.com/hr/article/56/2/153/106992/Using-machine-learning-and-satellite-data-toiwaponline
37. Annis A, Nurdin N, Fitriani A. Community-based flood early warning system using mobile applications: Case study in Makassar, Indonesia. *IOP Conf Ser Earth Environ Sci.* 2022; 1063:012012. doi:10.1088/1755-1315/1063/1/012012

8 Smart Irrigation and Precision Agriculture for Water Resource Management

Gurwinder Singh

8.1 INTRODUCTION

In agricultural systems, water represents a vital as well as decisive element for global food security. Hence, globally, societies are confronting widespread challenges due to population growth, climate change and rapid urbanization. Thus, under these pressures, the long-term sustainability of freshwater resources has become a primary concern for policymakers and practitioners. In addition, agriculture has accounted for the largest share of global freshwater withdrawals over recent decades. Therefore, irrigation efficiency has emerged as a decisive factor in balancing food production and water availability.

In response to these challenges, the agricultural sector is undergoing a transition from resource-intensive to knowledge-intensive practices. In this sequence, smart irrigation and precision agriculture have emerged as transformative approaches. Thus, these approaches have leveraged advancements in sensing technologies and Internet of Things (IoT) infrastructures. Taken together, there is also automation of systems to optimize water usage at spatial and temporal scales. In addition, these approaches have recognized and exploited the variability (within fields) in soil properties as well as crop growth stages. In specific terms, the smart irrigation systems have focused enhancing water-use efficiency while maintaining or improving agricultural productivity. Therefore, it has aligned with the irrigation decisions of actual crop water demand.

Due to the rapid development of low-cost sensors and wireless communication networks, there is a rise in the practical deployment of data-driven irrigation systems. Therefore, it's very important to consider the soil moisture sensors as well as plant-based stress indicators. Thus, they can be utilized in weather monitoring stations to provide continuous observations of the soil–plant–atmosphere. On the other hand, these data streams can be integrated with physically based models and machine learning

DOI: 10.1201/9781003714217-8

(ML)-based predictive models. This leads to a robust foundation for intelligent irrigation scheduling. In addition, the translation of complex datasets into actionable recommendations (for farmers and water managers) has been facilitated. Therefore, this can help manage variable rate irrigation (VRI) technologies as well as decision support systems (DSS).

In potential terms, the chapter offers an analysis of smart irrigation and precision agriculture practices. In this analysis, the integrated decision-oriented frameworks for water resource management have been reported. In addition, the contemporary irrigation innovations are also situated within a wide sustainability domain. Thus, while addressing these challenges, this chapter also examines key technologies used for agile irrigation systems. Likewise, the underlying motives of these approaches are critically discussed. Similarly, the chapter brings together different perspectives. These aspects include agronomy, computer science and environmental engineering. Taken together, this discussion emphasizes both conceptual understanding and practical considerations.

8.2 LITERATURE REVIEW

Over the years, progress in smart irrigation and precision agriculture has evolved through the convergence of multiple research fields. These domains have explored plant water usage/loss, sensor-based monitoring, etc. In practice, some fundamental studies have reported on crop water demand/usage. These have been very influential, providing the scientific foundation for contemporary irrigation approaches. In this discourse, the seminal FAO-56 guidelines developed by Allen et al. [1] are very effective. In fact, these have standardized the Penman–Monteith method for estimating reference evapotranspiration (ET_o). Therefore, this has formed the backbone of water balance modelling and irrigation scheduling practices worldwide [1]. Complementary to this work, a comprehensive physiological understanding of plant–microclimate interactions was provided by Jones, emphasizing the role of plant water stress indicators in guiding irrigation decisions [2]. Collectively, these early studies have established the theoretical foundation for knowledge-intensive water management practices. Subsequently, they have motivated later innovations in sensor-based and automated irrigation systems.

In the sequence, the automation in irrigation was highly adopted in the soil moisture-based control systems. Abdullah and Rahman [3] have demonstrated the integration of soil moisture sensors with automated drip irrigation systems, which leads to significant reductions in water losses and improvements in application uniformity [3]. While in response to the demand for scalable and real-time solutions within the agricultural sector, the wireless sensor networks (WSNs) were subsequently enacted as a transformative technology. One of the most comprehensive early surveys on the application of WSNs in agriculture was presented by Ojha et al. [4], wherein the system architectures for communication protocols and the key design challenges of IoT-enabled irrigation systems were systematically examined. In parallel, advances in satellite-based monitoring of evapotranspiration (ET) were achieved by Melton et al. [5]. Thereby, it has been reviewed to improve the generation of spatially distributed irrigation estimates.

As part of developments, the transition toward precision agriculture was accelerated through the integration of IoT platforms with remote sensing technologies. Therefore, the critical role of sensor innovations (in soil, plant and atmosphere) to enable the continuous monitoring for precision irrigation was highlighted by Hawkins and Wang [6]. It was further demonstrated by Jim´enez et al. using the coupling of in situ sensor data with meteorological forecasts. It also addressed how a DSS can substantially improve irrigation timing and significantly reduce water usage [7]. During the same period, one of the earliest field-scale evaluations of VRI was reported by Sui and Yan [8], wherein the spatially adaptive irrigation strategies were shown to enhance water productivity [8]. Similarly, an IoT–ML-based automated irrigation system was successfully implemented by Goap et al. [9]. Furthermore, some advancements in plant-based sensing were demonstrated by Bhandari et al. [10], wherein the canopy temperature measurements and infrared thermography were employed to detect early indicators of plant water stress.

The subsequent stage of innovation was marked by the emergence of ML techniques and predictive analytics. It was demonstrated by Ferreira and Jones [11], wherein the ML-based models outperform traditional empirical approaches for forecasting crop water demand under climatic variability conditions. On the other hand, the suitability of long-range communication technologies (LoRaWAN) for distributed agricultural sensor networks was examined by Raj and Patel [12]. In addition, the concept of precision irrigation was revisited by Smith et al. [13]. They supported the integration of field-level variability with crop modelling through sensor-based analytics. In the same sequence, empirical validation of WSN-based soil monitoring systems was provided by Ali and Nguyen [14].

Likewise, the high-resolution monitoring of crop biophysical parameters has been enabled through the adoption of drones and satellite-based remote sensing technologies. It was demonstrated by Zhao et al. that drone-based multispectral imagery can effectively support VRI [15]. On the contrary, the robust integration of IoT platforms with soil moisture sensors was efficiently demonstrated by Maheswari and Swaminathan [16]. On the other hand, some global assessments conducted by the Food and Agriculture Organization (FAO) have underscored the adoption of water-efficient irrigation practices. Thus, it has linked the deployment of smart irrigation technologies with the sustainability objectives [17].

In recent years, there has been a rapid proliferation of artificial intelligence (AI)–driven irrigation systems. This has been recorded in terms of a comprehensive review of deep learning models by Kukkapalli and Roy [18]. In addition, some advancements in automated irrigation control strategies were synthesized by Bwambale et al. [19]. In the same relevance, the effectiveness of thermal imaging-based scheduling was demonstrated by Rana and Katerji [20]. Furthermore, there are some extensions of IoT-based irrigation architectures that were proposed by Montes et al. [21]. Therefore, the emerging concept of digital twins was examined by Singh and Singh [22]. In addition, the performance of sensor-based irrigation scheduling tools was evaluated by Kim et al. [23]. In this regard, the most recent advances have been centred on the integration of AI and deep learning techniques. Likewise, a comprehensive review was published by Sharma et al. [24]. In addition, Wang et al. have evaluated the deep learning models such as the long short-term memory (LSTM) networks and hybrid CNN–LSTM architectures [25]. In the same context, these contemporary studies indicate a transition towards a refined irrigation ecosystem. Therefore, it is very helpful to understand the integration of AI with IoT infrastructures.

8.3 FOUNDATIONAL TECHNOLOGIES FOR SMART IRRIGATION

In smart irrigation systems, a tightly integrated technological stack supports continuous monitoring and informed decision-making. Thus, in practice, these systems perform as cyber-physical infrastructures. Therefore, the field conditions can be sensed, processed and then translated into precise control actions. In addition, the foundational technologies can be classified into three core components: (1) sensing and data acquisition, (2) communication and networking, and (3) data processing and computation for effective control. Taken together, these components are capable of forming the operational backbone of precision irrigation systems.

8.3.1 Sensing and Data Acquisition Technologies

In certain terms, real-time data collection is essential for understanding the dynamic interactions within the soil–plant–atmosphere continuum. Smart irrigation systems can deploy heterogeneous sensors to capture soil conditions. They also capture the plant physiological responses and atmospheric variables. Precisely, they have the capacity to influence crop water demand/usage.

8.3.1.1 Soil-Based Sensors

Soil-based sensors can provide direct information on the availability of water in the root zones. Therefore, they become the primary determinant of irrigation requirements. The commonly used soil sensing technologies include the following points:

- **Time-Domain Reflectometry**: Measures soil volumetric water content. It also analyses the travel time of an electromagnetic pulse from a probe inserted into the soil. Time-domain reflectometry sensors are able to offer high accuracy and stability but are relatively expensive.
- **Capacitance and Frequency-Domain Reflectometry Sensors**: Estimate soil moisture by measuring changes in dielectric permittivity. These sensors are cost-effective and widely used in field-scale deployments. They require careful calibration for soil texture and salinity.
- **Tensiometers and Matric Potential Sensors**: Measure soil water potential rather than moisture content. They provide an indication of the energy required by plants to extract water. They are particularly useful for irrigation scheduling.

Therefore, the soil sensors are typically installed at multiple depths to represent effective root zones. They are capable of capturing the critical inputs for threshold-based irrigation control strategies.

8.3.1.2 Plant-Based Sensors

In smart irrigation applications, the plant-based sensing technologies are used to assess crop water status. This can be done by capturing the integrated response of plants to soil moisture, atmospheric demand and physical conditions. On the other hand, unlike soil-only measurements, these sensors provide direct insight into plant water stress. They can also be adopted for the early detection of irrigation deficits.

Among the commonly employed plant-based sensors, canopy temperature sensors and infrared thermography are widely used. Thus, this is important for measuring leaf or canopy temperature. They can also enable the estimation of stress indicators such as the Crop Water Stress Index . Similarly, dendrometers can monitor small variations in stem diameters, which can be associated with plant water balance during diurnal transpiration cycles. In addition, sap flow sensors quantify the rate of water movement within the plant stem and can provide direct estimates of transpiration dynamics.

8.3.1.3 Atmospheric and Environmental Sensors

Local meteorological conditions can significantly influence crop ET. In addition, smart irrigation systems can integrate on-site or nearby weather stations for

- Air temperature and relative humidity
- Solar radiation
- Wind speed and direction
- Precipitation

These measurements enable the accurate estimation of reference evapotranspiration (ET_o). They can also improve the reliability of irrigation scheduling models for weather forecasts.

8.3.2 Communication and Networking Infrastructure

The effectiveness of smart irrigation systems depends on reliable data transmission between sensors, controllers and computing platforms. In addition, IoT architectures can provide the communication backbone for distributed irrigation systems. Figure 8.1 illustrates the layered communication and networking architecture. This enables data flow from field-level sensors to cloud and edge-based decision-making platforms.

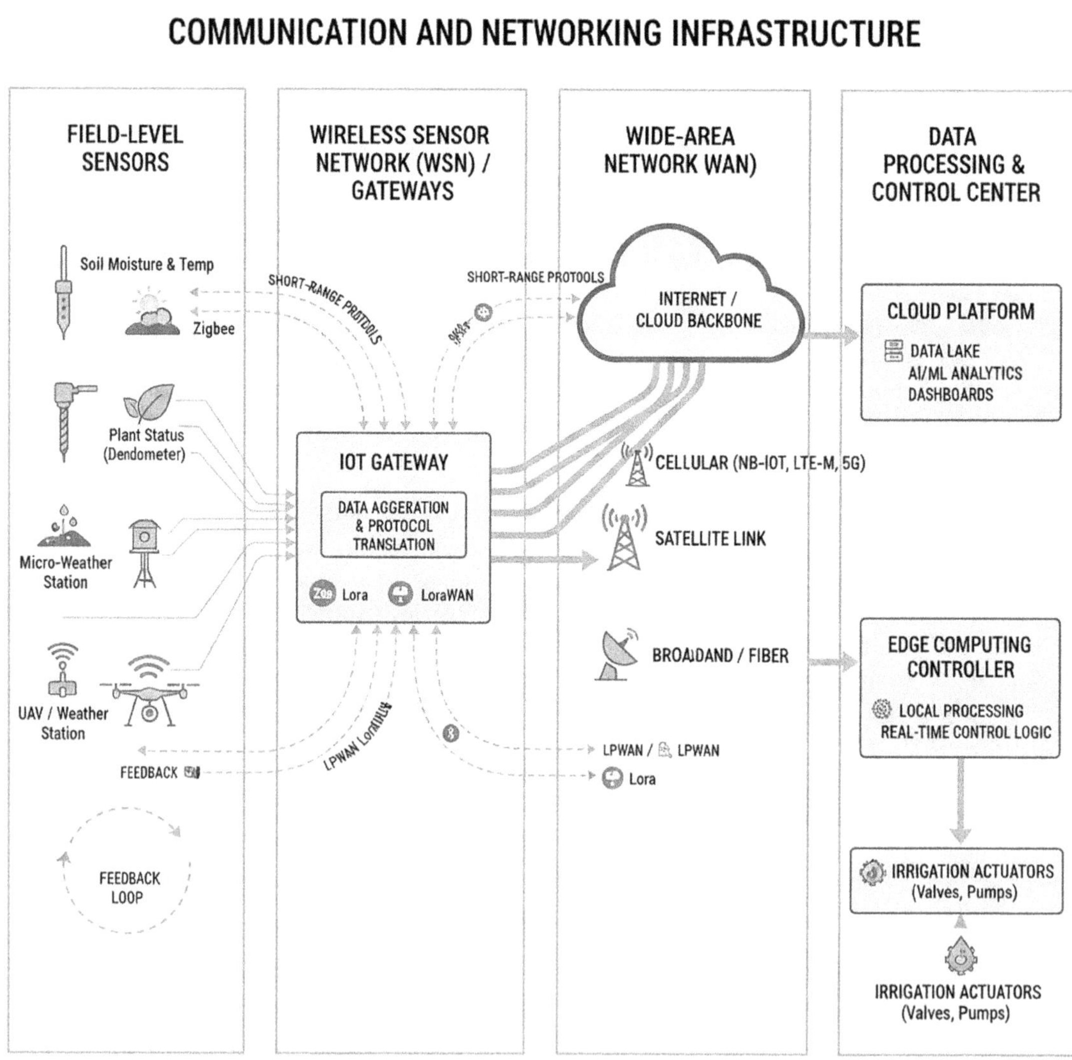

FIGURE 8.1 Communication and networking infrastructure for smart irrigation systems.

8.3.2.1 WSNs and IoT Gateways

The WSNs consist of spatially distributed sensor nodes that collect field data and transmit it to a central gateway. Therefore, the IoT gateways can aggregate data from multiple sensors. Then, it performs the preliminary processing and forwards the information to local servers or cloud platforms. Therefore, these gateways serve as the interface between the field-level devices and the higher-level analytics.

8.3.2.2 Communication Protocols

The choice of communication protocol depends on coverage area, power consumption, data rate and deployment cost.

- **Short-Range Protocols**: Zigbee and Bluetooth Low Energy are suitable for dense sensor clusters and greenhouse environments.
- **Low-Power Wide-Area Networks**: Technologies such as LoRa and LoRaWAN enable long-range and low-power communication, making them well suited for large agricultural fields.
- **Cellular Communication**: The emerging 5G technologies can provide reliable connectivity in areas with existing cellular infrastructure. Also, they can support higher data throughput as and when required.

Furthermore, hybrid communication architectures are often employed to balance performance in terms of energy efficiency and cost.

8.3.3 Control and Actuation Technologies

The final layer of the smart irrigation stack converts digital decisions into the physical water application.

8.3.3.1 Controllers and Automation Units

Likewise, microcontrollers (e.g., Arduino and Raspberry Pi) and industrial programmable logic controllers are able to execute irrigation control logic. These controllers can integrate sensor inputs with scheduling algorithms. Also, it can combine user-defined rules to determine irrigation timing and volume.

8.4 SENSOR-DRIVEN IRRIGATION SCHEDULING AND CONTROL

This section presents the theoretical foundations and practical implementation of sensor-driven irrigation scheduling. The sensors can continuously monitor the soil–atmospheric conditions. Therefore, the smart irrigation systems dynamically determine the temporal and spatial distribution of irrigation. On the contrary, the sensor-driven irrigation scheduling can operate as a closed-loop control system. Thus, it is useful in field heterogeneity and climatic variability. As a result, this adaptive behaviour can contribute to improving water-use efficiency and enhanced crop resilience.

8.4.1 Principles of Automatic Irrigation Scheduling

In automated irrigation systems, the scheduling is designed to keep the soil moisture within an optimal range. In addition, it balances the crop water demand and minimizes losses. On this basis, the sensor-derived data can provide the basis for timely irrigation decisions. It can also support the implementation of both reactive and predictive control strategies.

8.4.1.1 Soil Moisture Dynamics

The temporal evolution of soil moisture can be described using a discrete-time water balance model as

$$\theta_{t+1} = \theta_t + \frac{1}{Z_r}\left(I_t + P_t - ET_{c,t} - D_t - R_t\right) \tag{8.1}$$

where θ_t denotes the volumetric soil moisture at time t and Z_r represents the effective root-zone depth. In addition, I_t is the applied irrigation for P_t precipitation and $ET_{c,t}$ is the crop evapotranspiration. Also, the terms D_t and R_t account for deep percolation and surface runoff, respectively. Therefore, this equation establishes a direct mathematical relationship between sensor-based soil moisture measurements and irrigation control actions.

Threshold-Based Control: The threshold-based control is a reactive scheduling strategy. This strategy for irrigation decisions is triggered when soil moisture crosses predefined limits.

Therefore, the control law can be expressed as follows:

$$I_t = \begin{cases} I_{\max,0}, & \text{if } \theta_t \geq \theta_{\max} \\ 0, & \text{if } \theta_t < \theta_{\max} \end{cases} \tag{8.2}$$

where $\theta_{\min}$ and $\theta_{\max}$ represent lower and upper soil moisture thresholds, respectively, and $I_{\max}$ depicts the maximum irrigation depth applied in a control interval. Threshold values are calibrated according to the soil texture for crop types.

Model-Based Control: The model-based control can adopt a proactive approach by forecasting soil water depletion and crop water demand. Therefore, a commonly used formulation is the soil water deficit (SWD) model:

$$\text{SWD}_t = \text{SWD}_{t-1} + ET_c - I - P_e + R + D \tag{8.3}$$

where P_e denotes effective precipitation. Irrigation is scheduled to keep SWD within an allowable depletion range. This often integrates short-term weather forecasts to avoid unnecessary irrigation prior to rainfall. In addition, the model-based control generally outperforms the threshold-based methods even under variable climatic conditions.

8.4.2 ET-Based Irrigation Requirement

The net irrigation requirement (NIR) over a scheduling interval is computed as

$$\text{NIR} = ET_c - P_e - \Delta S \tag{8.4}$$

where ΔS represents the change in soil water storage estimated from sensor data. Thus, this formulation can measure the precise estimation of irrigation. Also, it can serve as a foundation for automated scheduling and DSS.

8.4.3 Automated Water Delivery Systems

The automated water delivery systems can execute irrigation decisions through some integrated hardware components. In addition, the smart controllers can be implemented using microcontrollers (e.g., Arduino and Raspberry Pi). On the contrary, it can also be employed through industrial programmable logic controllers . It can process the sensor data and control algorithms to generate the actuation signals.

These controllers interface with

- **Solenoid Valves**: For the zone-level irrigation control.
- **Pump Control Units and Variable Frequency Drives**: For regulating discharge pressure and energy consumption.
- **VRI Hardware**: For spatially adaptive water application on the centre-pivot and linear-move systems.

8.4.4 Variable Rate Irrigation

The VRI has the ability to enable site-specific water application by dividing the field(s) into multiple management zones. Therefore, the field can be partitioned into N zones. This can lead to the irrigation depth for zone j being expressed as

$$I_j = ET_{c,j} - P_{e,j} + \alpha(\theta_{j*} - \theta_j) \tag{8.5}$$

where θ_j depicts the measured soil moisture in zone j and θ_{j*} denotes the desired moisture level. Also, α denotes a control gain. Precisely, the prescription maps can be generated from soil and remote sensing data. This can help to guide real-time modulation of sprinkler or nozzle flow rates.

8.5 CROP WATER-USE OPTIMIZATION USING COMPUTER APPLICATIONS

In this section, crop water-use optimization is described with respect to the various variants. It lies at the core of smart irrigation and precision agriculture. It also involves accurately estimating crop water demand and applying computational techniques to allocate water efficiently. In addition, the computer-based optimization can integrate the physical crop–water models with the real-time sensor data.

Thus, it can help predictive analytics as well as decision support algorithms. Apart from the others, it is very important in ensuring the irrigation schedule.

8.5.1 Estimating Crop Water Requirements and Evapotranspiration (ET_c)

Crop evapotranspiration (ET_c) represents the combined effect of soil evaporation and plant transpiration. In addition, it serves as the primary indicator of crop water demand. Moreover, it is estimated through the crop coefficient approach as

$$ET_c = K_c \times ET_o \tag{8.6}$$

where K_c denotes the crop coefficient. It varies with crop type and growth stage. Also, the ET_o denotes the reference evapotranspiration by representing the atmospheric evaporative demand.

The standardized FAO Penman–Monteith equation is widely accepted for computing ET_o:

$$ET_0 = \left(0.408\Delta(R_n - G) + \gamma\left(900/(T+273)\right)u_2(e_s - e_a)\right) \Big/ \left(\Delta + \gamma(1 + 0.34u_2)\right) \tag{8.7}$$

where Δ is the slope of the saturation vapour pressure curve (kPa °C^{-1}). And R_n is net radiation at the crop surface (MJ m^{-2} day^{-1}), whereas the G denotes the soil heat flux density (MJ m^2/day). And the γ denotes the psychrometric constant (kPa/°C). Also T is mean daily air temperature (°C). In addition, u_2 denotes the wind speed at 2 m height (m/s). But $e_s - e_a$ is the vapour pressure deficit (kPa).

There are some computer applications that can automate this estimation process by ingesting real-time weather data from sensors. It is also enabling the continuous and localized estimation of ET_c.

8.5.2 Soil Water Balance and NIR

In the context of soil–water balance level, a model can be employed to translate crop water demand into irrigation decisions. Therefore, the NIR over a scheduling interval can be expressed as

$$\text{NIR} = ET_c - P_e - \Delta S \tag{8.8}$$

where P_e is effective precipitation and ΔS is the change in soil water storage. In addition, the sensor-derived soil moisture measurements are denoted by ΔS. Clearly, it is able for to reduce the uncertainty in irrigation while doing the depth calculations.

8.5.3 Modelling and Predictive Analytics

As far as the computational models are concerned to enhance crop water-use optimization, the forecasting can be best utilized for the future water demand and soil moisture dynamics. Therefore, these models can be enabled for irrigation decisions. Thus, it can shift from reactive responses to predictive and anticipatory strategies.

8.5.3.1 ML for Water Demand Forecasting

As well known, the supervised ML models can be employed for prediction-based models. Thus, these models such as random forests and support vector regression are increasingly and efficiently implemented. Also, they can be used to predict ET_c or optimal irrigation volumes. So, a general regression formulation can be expressed as

$$ET_{c,t} = f\left(ET_{o,t} - 1:t-n, \theta_t - 1:t-n, W_t, S_t\right) \tag{8.9}$$

where θ denotes soil moisture. In addition, W_t represents forecast weather variables and S_t denotes the crop growth stage. Thus, $f(\cdot)$ is a learned non-linear mapping. These models can capture complex interactions among climatic variables. Therefore, they can help improve the forecasting accuracy under variable conditions.

8.5.3.2 Artificial Neural Networks and Deep Learning

Importantly, artificial neural networks and deep learning architectures are well suited for the modelling of the non-linear soil–plant–atmosphere continuum. In particular, recurrent neural networks and LSTM models are effective for time-series prediction of soil moisture

$$\theta_{t+1} = F\left(\theta_t, ET_{c,t}, I_t, P_t\right) \tag{8.10}$$

where $F(\cdot)$ represents a neural network model trained on historical sensor data. These models support predictive irrigation scheduling by estimating future soil moisture states. They also help identify potential stress events before they occur.

8.5.4 Optimization of Irrigation Allocation

In a significant direction, crop water-use optimization can be formulated as an optimization problem. This problem can help balance crop stress as well as water consumption. Therefore, a common objective function is expressed as

$$\min_{I_t} \sum_{t=1}^{T} \left[\alpha\left(\theta_t - \theta^*\right)^2 + \beta I_t\right] \tag{8.11}$$

where $\theta*$ is the desired soil moisture level and I_t is the irrigation applied at time t. In addition, α and β denote the weighting factors. Thus, they reflect the relative importance of crop stress reduction and water conservation. Hence, computer applications can solve these problems by using (meta)heuristic or data-driven optimization techniques.

8.5.5 Water-Use Efficiency Metrics

Likewise, the effectiveness of crop water-use optimization strategies is evaluated using standard efficiency metrics:

- **Water-Use Efficiency (WUE)**: Crop yield per unit of water consumed by evapotranspiration,

$$\text{WUE} = \frac{Y}{ET_c} \tag{8.12}$$

where Y is the crop yield.

- **Irrigation WUE**: Crop yield per unit of irrigation water applied,

$$\text{IWUE} = \frac{Y}{I} \tag{8.13}$$

where I is the total irrigation applied.

Therefore, smart irrigation systems aim to maximize WUE. This can be done by aligning ET_c closely with actual crop demand. They also help improve irrigation WUE by minimizing non-beneficial water losses. These water losses are in terms of runoff and deep percolation conditions.

8.6 DSS FOR WATER MANAGEMENT

Moreover, the DSS play a pivotal role in water management; they are also beneficial in translating large volumes of heterogeneous data into actionable irrigation decisions. In the context of smart irrigation, a DSS acts as an integrative layer as well. This layer can connect the sensing infrastructure and computational models, thus automating the control mechanisms alongside human decision-making. For combining real-time observations with predictive models, there are optimization logics to help DSS tools. Therefore, both operational (day-to-day irrigation scheduling) and strategic (seasonal water planning) water management play an important role in water management.

8.6.1 Architecture of an Irrigation DSS

Likewise, a robust irrigation DSS is typically designed using a modular and multi-tier architecture. This architecture ensures scalability, flexibility and interoperability. Therefore, a commonly adopted structure is the three-tier architecture, as described below:

1. **Data Layer**: This layer is responsible for data collection/acquisition and pre-processing. It also includes relational (SQL) or NoSQL databases. In addition, data lakes can store time-series data from soil moisture sensors. Thereupon, plant stress indicators and remote sensing platforms can work with external weather APIs. However, tasks such as data cleaning/validation and synchronization are also performed here.
2. **Logic and Model Layer**: Likewise, the logic layer constitutes the analytical core of the DSS. Therefore, it can integrate crop growth and water balance models. Thus, business rules and agronomic constraints are encoded to generate irrigation recommendations. Hence, in some advanced systems, this layer may also include adaptive learning mechanisms.
3. **Presentation Layer**: Henceforth, the presentation layer provides an intuitive interface for end users. This interface acts through web dashboards or mobile applications. Additionally, it can visualize the spatial and temporal patterns of soil moisture. Also, it can simulate the crop stress and irrigation events on maps and charts. Therefore, this interface also allows users to receive alerts and/or review recommendations.

Similarly, the modern DSS implementations leverage Big Data technologies to handle high-velocity and high-volume data streams. These include Apache Kafka for real-time ingestion and Apache Spark for large-scale analytics. Thus, Figure 8.2 presents an integrated cloud–edge smart irrigation framework,

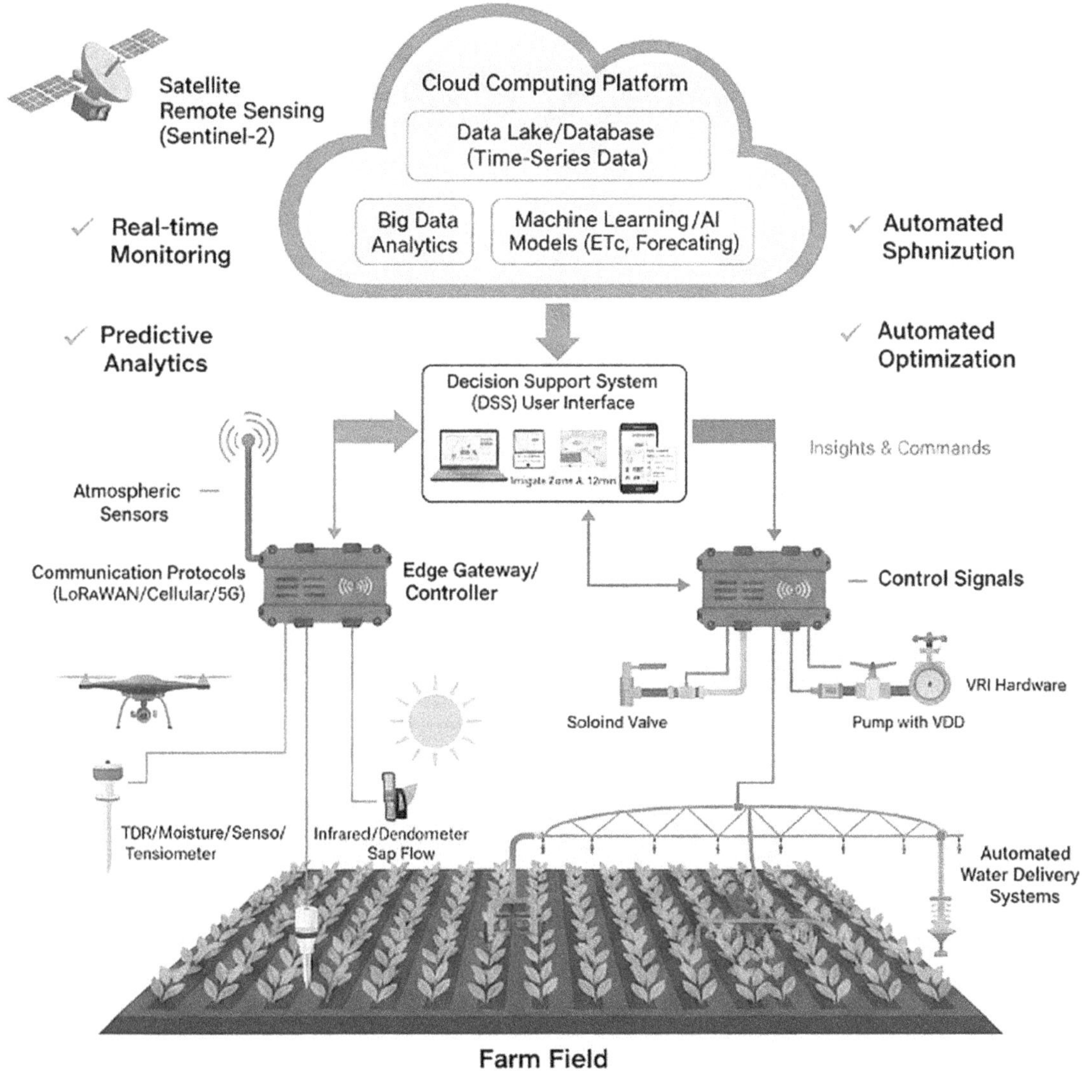

FIGURE 8.2 Cloud–edge–based smart irrigation architecture.

wherein real-time sensing with ML-based analytics has been employed for DSS collectively. Thus, it can enable automated and optimized water delivery.

8.6.2 Functionalities of Smart Irrigation DSS

From an operational perspective, the DSS platforms assist both real-time monitoring and informed irrigation decision-making. In this way, these capabilities will be able to process sensor data and model outputs into useful insights:

- **Automated Monitoring and Alerts in Real Time**: In this case, the DSS platforms can facilitate timely visualization of sensor-derived information.

- **Scenario Analysis and Forecasting**: In this case, the scenario-based simulation tools can compare different irrigation strategies.
- **Prescriptive Recommendations**: In practice, the DSS platforms produce context-based irrigation advisories. In this way, it is able to combine prediction models with real-time feedback.
- **Integration and Reporting**: In this case, farm management information systems tend to be associated with the DSS implementations. Therefore, it can enable the coordinated management of energy resources. Thus, the automated reporting functions facilitate regulation compliance.

8.6.3 Human-in-the-Loop Decision Support

Even in the automation models, good management of irrigation demands human control. Based on this, DSS platforms are usually based on the paradigm of human-in-the-loop. In this way, it is possible to model this system as a transparent and interpretable model. Also, it helps to enable farmers and water managers to incorporate contextual judgement into decision-making. Therefore, this kind of design approach supports user trust and mitigates risks associated with fully autonomous irrigation control.

8.6.4 Case Studies and Implementation Examples

Case Study 1: Precision Irrigation in California Almond Orchards

In a commercial almond orchard setting, a DSS was deployed to integrate the soil moisture sensors with the UAV-based thermal imagery. This is further employed for canopy temperature monitoring and an irrigation scheduling framework. On the other hand, under operational field conditions, the DSS has delivered zone-specific irrigation recommendations. In addition, the evaluation of field performance indicated an approximate 20% reduction in irrigation water. In practice, it was proved without observable yield loss, which has resulted in an improved WUE.

Case Study 2: IoT-Based Drip Irrigation for Smallholders in India

There is another case study, wherein a low-cost DSS was developed using capacitive soil moisture sensors to deliver irrigation advisories via SMS. Thus, this has designed the priorities for the system in terms of operational simplicity. Also, it can support local languages to enhance accessibility. Therefore, there are results which have indicated water savings of approximately 30%–40%. Alongside, there is a 15%–20% increase in vegetable crop yields. In addition, it also underscores the scalability and socio-economic relevance of DSS-based irrigation solutions. As shown in Table 8.1, the numerical findings suggest that DSS-assisted irrigation is associated with a significant increase in water-use efficiency. Therefore, the energy consumption and economic performance become very valuable insights. Hence, the gains are especially high in smallholder systems, where the existing baseline inefficiencies provide more room for improvement.

TABLE 8.1 Numerical performance outcomes of DSS-based smart irrigation implementations

PERFORMANCE METRIC	*CALIFORNIA ALMOND ORCHARDS*	*SMALLHOLDER FARMS (INDIA)*	*QUANTITATIVE INSIGHTS*
Irrigation water applied	~20% reduction compared to baseline	30%–40% reduction compared to baseline	DSS impact is more useful. The pre-DSS irrigation inefficiency is high.
Crop yield change	No significant yield loss (±2%)	15%–20% yield increase	The optimized timing has improved yield stability and productivity.
Water-use efficiency (WUE)	Increase of 18%–22%	Increase of 25%–35%	WUE gains correlate with improved precision.
Irrigation frequency	Reduced by 10%–15%	Reduced by 20%–30%	Lower irrigation frequency has reduced energy/labour requirements.
Energy consumption	15%–20% reduction in pumping energy	10%–15% reduction in pumping energy	Energy savings are proportional to decreased irrigation volume.
Decision accuracy	>90% alignment with model-optimal schedules	~85% alignment with advisory thresholds	High accuracy is achievable.
Return on investment	Payback period: 2–3 seasons	Payback period: 1–2 seasons	Economic feasibility improves.
System reliability	>95% uptime during irrigation season	~90% uptime	Simpler architectures enhance robustness in rural deployments.

8.7 CHALLENGES AND FUTURE DIRECTIONS

Since there has been notable progress in smart irrigation technologies, it is facing some challenges. In addition, their large-scale adoption is facing technical and social constraints. Thus, these challenges need to be addressed to resolve the precision irrigation issues. Thereby, this will deliver scalable and sustainable benefits across diverse agricultural settings.

8.7.1 Technical and Implementation Challenges

- **Sensor Reliability and Maintenance**: In this case, the deployed sensors are subjected to extreme environmental factors on a regular basis. These include extreme temperatures, soil salinity and moisture. Also, the drift in calibration caused by the loss of components can affect the accuracy of data. In this way, this may result in inefficient irrigation choices. Consequently, current studies are focused on the creation of self-calibrating sensors. They also investigate low-maintenance system designs.
- **Interoperability and Standardization**: As is known, there are no agreed-upon communication and data standards, which results in fragmented ecosystems. In this case, therefore, the proprietary hardware and software systems tend to restrict the flexibility and scalability of the system. Practically, there is a higher focus on open standards and interoperable platforms that is necessary to facilitate the process of seamless integration.

- **Data Security and Privacy**: Here, in this context, the smart irrigation platforms generate sensitive farm-level data. Thus, this data is mostly related to water usage and management practices. So, while ensuring data security and maintaining integrity, privacy concerns are critical for user trust.
- **Connectivity Constraints**: As there is a limitation of reliable and affordable internet connectivity, it is not attainable in all agricultural regions. In addition, there is also restricted bandwidth and limited network coverage. Thus, this has affected the effectiveness of cloud-dependent irrigation systems. Therefore, hybrid system architectures can be a practical approach to mitigating these limitations.

8.7.2 Economic and Social Challenges

- **High Capital and Deployment Costs**: Here, the high cost of communication infrastructure can be prohibitive for smallholder farmers.
- **Knowledge and Skill Gaps**: There is still lack of skills required to operate technological equipment in the fields. In addition, limited digital literacy can constrain the adoption of automation.
- **Uncertainty in Return on Investment**: It has been well reported that long-term gains (water savings, energy efficiency and yield stability) will be realized. In this respect, the perceived returns on investment may be unclear to end users. On the other hand, the use of demonstration projects and region-based economic evaluations will be significant in enhancing trust in the adoption of technology.

Taken together, addressing existing challenges while integrating emerging technologies will be fruitful. In addition, it is central to the continued evolution of smart irrigation systems.

8.8 CONCLUSION

This chapter discussed how smart irrigation and precision agriculture could be used in managing water resources in a sustainable manner. This was also concerned with the constraints of climate change and the excessive global food demand. Furthermore, this discussion highlighted how the paradigm shift between conventional irrigation methods and intelligent systems has taken place. It therefore promotes the use of sophisticated sensibility technologies in IoT-enabled structures. Thus, the intersection of ML methods and the model of crop–water interaction is very useful. Hence, predictive analytics are able to increase the accuracy of irrigation schedules. This, in practice, supports adaptive water utilization in changing environmental conditions.

As part of the overall smart irrigation environment, DSS serve as a key operational node. Practically, it is reflected in case-based reviews that demonstrate quantifiable results on water-use efficiency. Despite these developments, issues associated with interoperability, cost and adoption of systems continue to pose a problem. Eventually, the current innovations in digital technologies and favourable policy guidelines open margins for improvements. Thus, smart irrigation technologies can provide a scalable/adaptive solution to enhance agricultural water management and ensure long-term food security.

REFERENCES

1. R. G. Allen, L. S. Pereira, D. Raes, and M. Smith, *Crop Evapotranspiration—Guidelines for Computing Crop Water Requirements,* FAO Irrigation and Drainage Paper No. 56, Rome, 1998.
2. H. G. Jones, *Plants and Microclimate: A Quantitative Approach to Environmental Plant Physiology*, 3rd ed., Cambridge University Press, 2013.
3. M. Abdullah and R. Rahman, "Automation of drip irrigation using soil moisture sensors," *Agricultural Water Management*, vol. 159, pp. 57–65, 2015.
4. T. Ojha, S. Misra, and N. Singh, "Wireless sensor networks for agriculture: A survey," *Computer Standards & Interfaces*, vol. 50, pp. 113–141, 2016.
5. F. Melton et al., "Satellite-based monitoring of crop evapotranspiration," *Remote Sensing*, vol. 8, no. 6, 524, 2016.
6. S. Hawkins and K. Wang, "Precision agriculture and IoT: Review of sensing technologies," *IEEE Sensors Journal*, vol. 17, no. 23, pp. 7628–7640, 2017.
7. M. Jim´enez et al., "Decision support system for irrigation scheduling using weather forecasts," *Agricultural Systems*, vol. 155, pp. 200–212, 2017.
8. R. Sui and H. Yan, "Field study of variable rate irrigation in humid climates," *Irrigation Science*, vol. 36, pp. 39–49, 2018.
9. A. Goap, D. Sharma, A. Shukla, and C. R. Krishna, "An IoT-based smart irrigation management system using machine learning," *Computers and Electronics in Agriculture*, vol. 155, pp. 41–49, 2018.
10. A. Bhandari et al., "Canopy temperature-based irrigation scheduling using infrared thermography," *Agricultural Water Management*, vol. 204, pp. 190–200, 2018.
11. P. Ferreira and D. Jones, "Machine learning for predicting crop water demand," *Computers and Electronics in Agriculture*, vol. 165, 104936, 2019.
12. K. Raj and A. Patel, "LoRaWAN for smart agriculture irrigation systems," *IEEE Communications Magazine*, vol. 57, no. 6, pp. 120–126, 2019.
13. R. J. Smith, J. N. Baillie, A. McCarthy, and N. Hancock, "Defining precision irrigation: A new approach," *Agricultural Water Management*, vol. 240, 106256, 2020.
14. M. Ali and T. H. Nguyen, "Wireless sensor networks for smart irrigation and soil monitoring," *IEEE Sensors Journal*, vol. 20, no. 22, pp. 13500–13510, 2020.
15. Z. Zhao et al., "Drone-based multispectral imaging for variable rate irrigation," *Agricultural Systems*, vol. 190, 103106, 2021.
16. M. Maheswari and S. Swaminathan, "IoT-based precision farming using soil moisture sensors," *IEEE Internet of Things Journal*, vol. 8, no. 14, pp. 11975–11986, 2021.
17. FAO, *The State of Food and Agriculture 2021: Water Challenges in Agriculture.* FAO, 2021.
18. R. K. Kukkapalli and P. Roy, "Deep learning approaches for evapotranspiration estimation: A review," *ISPRS Journal of Photogrammetry and Remote Sensing*, vol. 190, pp. 29–53, 2022.
19. E. Bwambale, F. K. Abagale, and G. K. Anornu, "Smart irrigation monitoring and control strategies for improving water use efficiency," *Agricultural Water Management*, vol. 260, 107324, 2022.
20. G. Rana and N. Katerji, "Remote sensing–based irrigation scheduling using thermal imaging," *Remote Sensing of Environment*, vol. 268, 112773, 2022.
21. V. Montes, T. Alfieri, and S. Kumar, "IoT–edge computing framework for automated irrigation in water-stressed regions," *Computers and Electronics in Agriculture*, vol. 209, 107812, 2023.
22. A. Singh and V. P. Singh, "Application of digital twins in precision irrigation: Opportunities and challenges," *Irrigation Science*, vol. 41, pp. 455–470, 2023.
23. H. Kim, M. Dukes, and R. Daniels, "Assessment of sensor-based irrigation scheduling tools for water conservation," *Agricultural Water Management*, vol. 278, 108164, 2023.
24. P. Sharma, R. R. Rout, and S. P. Mohanty, "AI-driven smart irrigation systems for sustainable agriculture: A comprehensive review," *IEEE Access*, vol. 12, pp. 14567–14592, 2024.
25. J. Wang, Y. Li, and M. H. Rahman, "Deep learning-based soil moisture prediction for precision irrigation," *Agricultural Water Management*, vol. 296, 108742, 2024.

AI-Based Flood Prediction and Early Warning Systems

9

Mamta Sharma, Abhineet Anand, and Kapil Sethi

9.1 INTRODUCTION

Floods are among the most common and damaging natural disasters worldwide, resulting in a staggering number of human, economic, and environmental losses every year. With climate change, rapid urbanization, and an increase in extreme weather, the need for more accurate, timely, and adaptive processes for flood forecasting is more prevalent than ever. In practice, flood prediction and forecasting have mostly relied on various physics-based hydrological and meteorological models, which in many locations work exceptionally well. However, due to their computationally expensive processes, these models have limited spatial and temporal resolutions, limited performance in ungauged or data-scarce areas, and therefore often been improperly suggestive. In response to such limitations, there is growing interest in using artificial intelligence (AI) particularly machine learning (ML) and deep learning (DL) to re-imagine flood forecasting and early warning systems (EWS) [1]. Recent literature provides multiple examples and demonstrations of how models such as Long Short-Term Memory (LSTM), gated recurrent unit (GRU), convolutional neural networks (CNNs), and hybrid methods show that AI can perform better than traditional flood forecasting methods and models because of increased spatial resolution, longer lead times, and improved forecasting accuracy (Figure 9.1). Furthermore, examples like Google's Flood Hub, the diffusion and reservoir-based uncertainty (DRUM) model, and FloodCast highlight the near-future potential of AI to meet operational needs by integrating multi-source data—satellite imagery, Internet of Things (IoT) sensors, rainfall records, and social media—in real-time feeds to provide more tailored, localized, and actionable flood warnings.

Furthermore, AI-driven EWS are now being used across the globe, from low-lying delta communities in South Asia to urban areas in Australia and Europe, to assist emergency services and the community in preparing for, responding to, and recovering from flood events. EWS not only improve prediction performance; they also help situational awareness and decision-making in emergencies through

 DOI: 10.1201/9781003714217-9

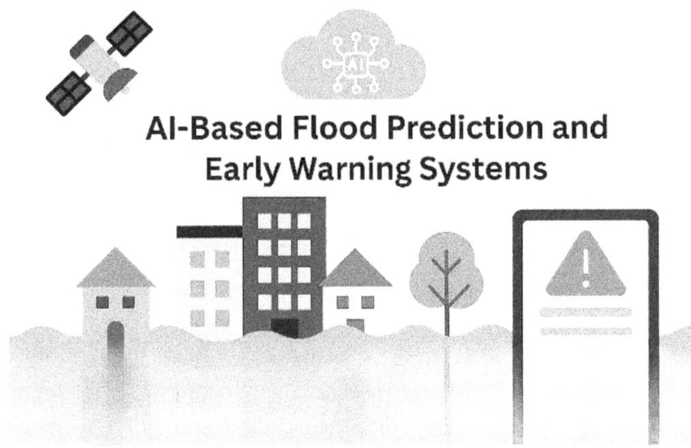

FIGURE 9.1 AI-powered flood forecasting system showing satellite data, urban flooding, and real-time alert delivery.

dynamic visualizations and automated alerts. Nevertheless, challenges remain [2]. The reliability and scalability of AI-enabled systems, especially in disadvantaged areas, can be hampered by data availability, model interpretability, infrastructure limitations, and access to technology equity. However, the fast-growing body of evidence and successful deployments of AI-enabled EWS and flood risk management all over the world indicates a fundamentally innovative role to be played by AI in the future of flood risk management [3]. This chapter has hopefully only opened the door and explored some domain methodology, application, and consequences of AI-driven flood prediction and EWS. By drawing on a wealth of research and examples of flood EWS and experiences, this chapter has attempted to provide a holistic understanding of how intelligent technologies are changing disaster preparedness in a world of uncertainty.

9.2 LITERATURE REVIEW

9.2.1 Evolution of AI in Flood Forecasting

In early literature, there appears to be a certain transition from conventional physics-based hydrological models to data-driven ML and DL methods. Silva and Costa [4] offer a thorough first review in which they highlight robust, accurate, and efficient modeling with hybridization, ensemble modeling, and optimization (Figure 9.2). DL methods include CNNs, recurrent neural networks (RNNs), generative adversarial networks, residual networks, and graph neural networks. There is ongoing research and development applying DL methods in hydrological forecasting, and they have achieved impressive successes. In a review in a similar area, the authors compare DL methods and identify the practical considerations and improvements that can be realized with these methods [5].

Historically, flood forecasting has depended on deterministic hydrological and/or hydraulic models that attempt to simulate the physical processes of rainfall–runoff and flow through a river. For instance, models such as HEC-RAS, MIKE FLOOD, and SWAT have been used to estimate the potential extent of flooding, water levels, and discharge rates across basins. Deterministic hydrological and hydraulic models typically involve significant calibration and are driven by historical hydrological

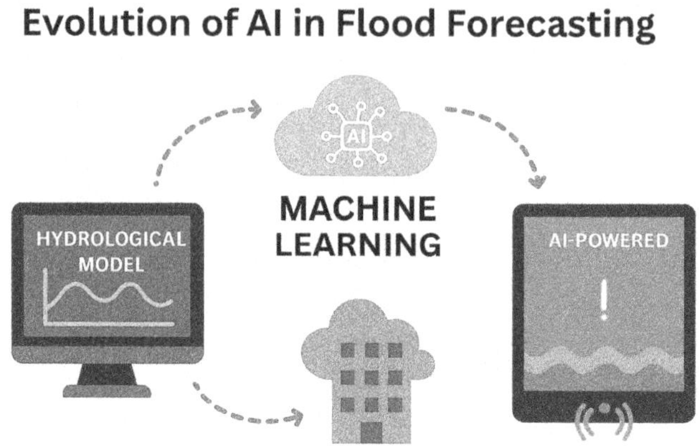

FIGURE 9.2 Timeline showing the shift from hydrological models to machine learning and AI-powered flood forecasting.

data, topography and soil characteristics, and meteorological data/station time-series data [6]. These models often have interpretability and physical realism, but become limited when available data is no longer accessible, leading to computational complexity; outputs produced by model simulations can be become hindrances [7]. A limitation of deterministic models is that they are often ill-equipped to harness changes in environmental conditions and cannot mitigate real-time data streams. Similarly, regions with sparsely distributed hydrometeorological networks and data often produce lower accuracy forecasts when using deterministic-focused models (e.g., hydrological models). Some additional limitations are that these models often cannot change their structure (e.g., they are fixed), often assume static input variables with fixed parameters, and can be considered less adaptive to changing conditions or even extreme weather conditions that can result from climate change. Together, considerations regarding changes to climatic systems become barriers to accurate and reliable forecasts of water systems for hydrological extremes such as floods [8].

An overall enhancement of traditional models has included the incorporation of spatial resolution through Geographic Information Systems (GIS) and remote sensing data. However, spatial resolution still does not allow for modeling transient and dynamic non-linear flooding events, thus shifting the focus to data-driven models that identify patterns in existing and real-time data with no strict physical assumptions. The limitations of the conventional approach have opened the door to AI-based approaches that are more flexible, adaptable, and predictive, transitioning from physics-based modeling to data-driven modeling. This represents a developmental shift in flood forecasting, allowing for responsive and scalable EWS.

9.2.2 Emergence of AI in Hydrological Forecasting

AI has transformed hydrological forecasting by developing models that are able to 'learn' complex connections between input variables and flood outcomes. ML algorithms [9], such as artificial neural networks (ANNs) [10], support vector machines (SVMs) [11], and random forests (RFs) [12], have outperformed traditional models in producing forecasts of river discharge, rainfall intensity, and forecasting flood vulnerability [13]. They require less subject-specific knowledge and can be trained with a variety of datasets, making them useful in data-rich or data-scarce settings [14]. Of all these algorithms, ANNs are particularly suitable for modeling non-linear hydrological processes. Their ability to approximate any continuous function allows them to capture even the most complex interactions in the rainfall–runoff relationship. Multiple studies have shown a marked improvement in the accuracy of flood predictions using ANNs over other traditional models, particularly when both historical and real-time sensor data were used [15]. However, the black-box nature of ANNs raises serious concerns about interpretability and trust in important decision-making situations.

SVMs and decision trees (DTs) provide [16] alternatives that typically provide more interpretability, as these methods have historically been used for classifying sample populations (to create flood-prone zones) or to classify the severity of floods (light vs. moderate vs. heavy flooding) [17]. They also tend to be more robust to noise and are less impacted by the dimensionality of the dataset. Ensemble methods, such as RFs, aggregate several DTs to produce the final prediction. RFs are expected to have improved stability and robustness by aggregating many DTs, improving generalization and reducing overfitting.

The emergence of DL has led to the development of networks such as LSTM networks and CNNs [18], which have grown popular and are increasingly accurate at time-series forecasting and spatial analysis. LSTMs are useful for hydrological data time-series analysis, especially in capturing temporal dependencies (e.g., the temporal variations in flood peaks). CNNs have been used to analyze satellite images of floods to map the extent (e.g., see images using CNNs) . These models further signal advances in AI with a larger role in enabling improved accuracy and timeliness in flood forecasts [19].

9.2.3 Hybrid and Ensemble Models

Hybrid approaches synthesize the utility of both physical and AI-based approaches for improved flood forecasting. They allow researchers to exploit the degree of mechanistic input while also improving model predictions through the use of AI approaches. For example, the rainfall–runoff outputs from SWAT or HEC-HMS used as model inputs in neural networks allow the AI model to learn residual patterns for correcting biased outputs. Hybrid approaches also fill in gaps left by either approach alone. They can be used in areas where models are based on robust, rain-based mechanistic processes and can incorporate information about changing climatic conditions or urban developments. The combination of physics with AI gives better reliability for accurately predicting flood events, especially the uncertainty that can accompany unusual extreme events.

Ensemble techniques clearly improve the accuracy of predictions by combining various methods as ensembles or a multiplicity of full-fledged learning algorithms [20]. Methods such as bagging, boosting, and stacking leverage the derived outputs from a variety of models to minimize both variance and bias in predictions. Ensemble approaches bring together the discriminatory power of the models, resulting in ensemble systems that have resilience against noise in the data and uncertainty in the model predictions, making them especially robust for operational flood forecasting [21].

Recent research has considered multi-model ensembles, which typically contain standard predictors, AI predictors, or both. These new systems can dynamically weigh the contribution of each model, with each model mathematically determining its contribution to the ensemble based on available performance measures to achieve the best possible forecast [22]. The literature has identified hybrid and ensemble models as a favorable area for future flood forecasting systems due to their scalability, robustness, and improvement in the quality of decision support.

9.2.4 Data Sources and Integration

The utility of AI-based flood prediction systems relies on various types of input information from a variety of sources. Remote sensing offers considerable flexibility in providing spatial information on land cover, soil moisture, and flooded areas. Also, to obtain high-resolution satellite data on floods and damage, AI-based cloud computing may be used. Data from Sentinel-1, Landsat, and other satellites can be processed autonomously using AI techniques to identify flood patterns and damages in near real time. Meanwhile, IoT devices and sensor networks can output hydrometeorological data (e.g., rainfall, temperature, humidity, and river water levels) in real time and as time-series data [23]. This data is critical for providing short-term forecasting capabilities for warning systems. Incorporating sensor streams into my AI model allows me to continually provide updates and learn from the data to better cope with and respond to changing flood conditions.

Additionally, crowdsourced data from users via Twitter or other social media can be used as alternative raw data sources for situational awareness [24]. Geotagging posts, images, and user reports can be extracted via natural language processing (NLP) and computer vision techniques to identify floods or community sentiment. Despite reported challenges to privacy and validation of non-traditional data sources, they provide valuable timely information to use with traditional datasets and as a means to support situational awareness. Although data are plentiful, issues of interoperability, latency, and data quality still make the integration of data difficult. Constructing effective AI-based flood prediction systems must involve proper data governance and management of data formats, preprocessing pipelines, and secure data-sharing capabilities to produce viable systems. The literature highlights the importance of complete data governance frameworks to enable effective model deployment.

9.2.5 EWS and AI

EWS are a key component of reducing flood impacts with alerts and actionable information to mitigate risk occurrence or harm. AI utilizes various processes such as predictive analytics, automated alerts, and dynamic risk geography maps to gain information and knowledge. AI can, for example, predict the lead time or arrival and depth of flooding using real-time variables, so that people and organizations can consider appropriate evacuations and mitigation options [25]. Similarly, NLP is used to aggregate or disseminate alerts across many types of applications. AI can convert technical forecasts or information into usable messages and can customize the content based on specific audiences, such as local communities, first responders, and public policymakers, so that each audience can act upon the best available information for their specific needs. Standard alerts can also be prepared or produced in multiple languages. For advocates of accurate risk assessment, alerts can also be sentiment analysis aware, so people can better understand the risk and act upon that understanding.

Geospatial AI uses satellite imagery, GIS datasets, and predictive models to create interactive flood maps and dashboards. The information from analytic models produces even more information for better situational awareness and improved decision-making regarding on responses from maps concerning flood risk zones, evacuation routes, and shelters [26]. There are also mobile applications and portals that share AI tools for real-time alerts and warnings provided by individual users based on their location and risk profile. Research from countries such as India, China, and the Netherlands reveals how effective AI-driven EWS mitigate fatalities and economic losses to a large extent (Table 9.1). However, there are challenges such as equitable access, guaranteeing reliability, and building public buy-in. Research indicates a need for participatory design, allowing end-users to drive system design and use to enhance buy-in and user uptake.

The comparative table (Table 9.1) displays a wide range of studies directed at AI-enhanced flood prediction and EWS that demonstrate different methodologies, from traditional ML models (SVM, RF) to more sophisticated DL models (LSTM and CNN) [27]. The studies themselves are conducted in various geographies and IT/dusty situations and settings including urban India, rural Bangladesh, Turkey, and Brazil, indicating the value of locality data and locality modeling that is respective to context. The various methodological innovations also emphasize some programmatic embedding of emerging technologies (IoT, blockchain, NLP, geospatial AI) with trends toward real-time, decentralized, participatory technologies in the name of flood management processes. The datasets used were again wildly varied, from satellites (Sentinel, MODIS), hydrological simulation tools (SWAT), sensor networks, and even social

TABLE 9.1 Comparative analysis of AI-based flood prediction and early warning system studies

S. NO.	*TITLE*	*AUTHORS AND YEAR*	*METHODOLOGIES*	*TECHNOLOGIES USED*	*DATASET(S) USED*	*KEY CONTRIBUTIONS*	*LIMITATIONS*
1	**Flood Prediction Using Machine Learning Models: Literature Review**	[23, 24]	Comparative review	SVM, RF, ANN, LSTM	Multiple global hydrological datasets	Benchmarked ML models for hydrological forecasting	No regional calibration
2	**IoT-Based Flood Monitoring and Alerting System**	[22]	Real-time alerting	Arduino, sensors, mobile app	Local sensor data (Maharashtra, India)	Community-level EWS	Hardware-dependent
3	**Emerging Technologies in Flash Flood Prediction: A Review**	[19]	Critical review	SVM, RF, IoT, robotics	Literature-based synthesis	Identified trends in flash flood systems	No empirical validation
4	**AI and GIS-Based Flood Intensity and Human Impact Prediction**	[15]	Hybrid AI-GIS	CNN, RNN, satellite imagery	Sentinel-2, Census data (India)	Human impact estimation	Complex integration
5	**Blockchain-Enabled AI Framework for Flood Risk Management**	[22]	Secure integration	ML, blockchain, remote sensing	Landsat, IoT sensor logs	Decentralized monitoring	Scalability issues
6	**Hybrid SWAT-ANN Model for Runoff Prediction**	[28]	Hybrid modeling	ANN + SWAT	SWAT outputs, rainfall data (Vietnam)	Improved runoff accuracy	Dual calibration needed
7	**Flood Susceptibility Mapping Using Ensemble Learning**	[27]	Ensemble ML	RF, gradient boosting	DEM, rainfall, land use (Turkey)	High-accuracy risk mapping	Computationally intensive
8	**Deep Learning for Rainfall–Runoff Forecasting**	[31]	Time-series DL	LSTM, GRU	Yangtze River basin data	Accurate flood forecasting	Black-box limitations
9	**NLP-Based Multilingual Flood Alert System**	[28]	NLP alerting	NLP, SMS, social media	Twitter, SMS logs (India)	Automated multilingual alerts	Misinformation risk

(*Continued*)

TABLE 9.1 (*Continued*) Comparative analysis of AI-based flood prediction and early warning system studies

S. NO.	*TITLE*	*AUTHORS AND YEAR*	*METHODOLOGIES*	*TECHNOLOGIES USED*	*DATASET(S) USED*	*KEY CONTRIBUTIONS*	*LIMITATIONS*
10	**Geospatial AI for Real-Time Flood Mapping**	[26]	Spatial DL	CNN, GIS	Sentinel-1, MODIS	Dynamic flood visualization	Requires high-res imagery
11	**Smart Urban Flood Modeling Using AI**	[25]	Urban modeling	SVM, RF, IoT	Smart city sensor data (Delhi)	Urban flood resilience	Limited rural transferability
12	**Localized ANN-Based Flood Forecasting in Bangladesh**	[24]	Regional AI	ANN	Bangladesh Water Development Board data	Tailored local prediction	Sparse seasonal data
13	**Rapid Flash Flood Detection Using Sensor Fusion and AI**	[24]	Real-time detection	RF, sensor fusion	IoT sensor streams (Brazil)	Fast flash flood alerts	False positives
14	**Crowdsourced Flood Intelligence Using AI and NLP**	[22]	Social media mining	NLP, GIS	Twitter, Facebook, OpenStreetMap	Situational awareness	Data noise, validation issues
15	**AI-Driven Flood Risk Zoning for Mitigation Planning**	[21]	Risk zoning	Decision Trees, GIS	Land-use maps, rainfall data (Gujarat)	Strategic zoning	Static zoning limitations

media platforms to crowdsourced data. As will be important for practitioners, the studies lacked, and even recognized, common criticisms, such as data scarcity, model interpretability, scaling, and ethical implications. These studies appear to offer foundational but still provisional frameworks and methodological approaches. However, their inherent biases suggest the need for a more robust, transparent, and inclusive AI framework, one that is deeply integrated into social justice-oriented disaster resilience planning.

9.2.6 Gaps and Limitations across Studies

Despite the advancements made, there are a number of recurring gaps and limitations that are evident in the review of the studies:

1. **Data Quality and Availability**: Many studies employed regional data (e.g., rainfall, river discharge, and satellite imagery), but this data can often be restricted, dated, or lack key features. For example, Rahman encountered [28] limitations in future predictions in their models due to seasonal bias and static spatial zoning maps.
2. **Model Interpretability**: Models developed with deep neural networks, such as LSTM, CNN, and GRU (e.g., Zhang et al., Li et al., and Nandhini et al.), may help yield high accuracy in their predictions, but they all have inherent black-box limitations in terms of their ability to share their bases/rationale with the general public. In natural disaster flood–response timeframes, this limited model transparency decreases trust in the models and impedes the use of the model for decision-making.
3. **Scalability and Transferability**: Most alert models (e.g., Gulhane et al.'s IoT-based alerting and Kumar et al.'s urban flood model) were designed and implemented around localized contexts and future-specific hardware. This presents a major challenge of scalability and limitations in replicating the study to another region without making extreme alterations, leading to a lack of transferability.
4. **Integration Complexity**: Most hybrid models (e.g., Chau et al. and Nandhini et al.) include both physical simulations and AI, which generally require dual calibration and data fusion. This tends to suggest that a source model with multiple data products could take more resources to put together and require more technical capability to integrate the model effectively.
5. **Real-Time Responsiveness**: Although some studies (e.g., Silva et al. and Ahmed et al.) that use real-time data streams have resulted in valuable quick response and feedback information with latency and false negative risks, there are also systems that utilize NLP-based alerts (Singh et al.) that actually result in an unrealistic situation of misinformation and language-based bias in knowledge transfer and establishment.
6. **Ethics and the Technology**: New technological applications (e.g., blockchain (Adiq et al.) or crowdsourcing data use (Ahmed et al.)) introduce significant topics for data privacy, trust, validation tracking, and ethical use of technology in possibly disaster-prone situations.
7. **Absence of Evaluations That Can Be Categorized as a Generalizability Standard Protocol across Studies**: Many studies include an array of performance, outcome, and evaluation metrics and evaluate the same study, resulting in inconsistent comparisons (e.g., AUCs, root mean square error (RMSE), and accuracy). Bhowmik and Mujumdar [29] indicated that many of the benchmarking review studies illuminated an urgent need for a unified validation framework and common metrics and measures to enable contributions across studies.
8. **Lack of Community or Stakeholder Engagement**: Very few models include participatory design that involves and validates feedback from local stakeholders or the community in a trust-inductive manner. The majority of models have very little interactive aspect, if any, with a very technical framing and minimal social or behavioral integration.

9.3 AI MODELS IN FLOOD PREDICTION AND EARLY WARNING SYSTEMS

AI is altering the way we utilize systems for flood prediction and early warning. AI brings data-driven means that can often deal with challenges of traditional modeling methods used in hydrology. While reasonable to some extent from a physical principle perspective, traditional methods are not able to reasonably deal with non-linear dynamics, poor data accessibility, and act under rapidly changing climatic conditions (e.g., using real-time data). This can be difficult for both models and human solvers. AI models (which use methods of ML or DL) have been shown to be applicable by learning complex weights and relationships between the interrelationships of variables in the environment for more timely and accurate flood prediction [30]. These models use and apply very large datasets (i.e., satellite information, sensor networks, and historical records) to aid in flood prediction and allow for new data (and/or changes in the data associated with floods) to assist in their dynamic development, re-examinations, improvements, and changes among any folds.

The application of AI in flood management is producing a range of models, depending on the forecasting needs. Researchers are employing various algorithms, ranging from ANNs and SVMs to more complex neural systems such as LSTM networks and CNNs, to perform various flood-related applications (i.e., rainfall–runoff forecasting, flood extent identification and mapping, rainfall and weather tracking, real-time alert conditions, and risk zone mapping). Hybrid models will combine the physics of simulation approaches with the learning of AI; several ensemble methods employ multiple algorithms, such as a neural network and a statistical modeling algorithm, to address the rigor of predictive capacity (Table 9.2). As research progresses, AI-based modeling will become easier, more accurate, interpretable, scalable, and functional engagement in insight will be enhanced for using AI in establishing proactive flood resilience and disaster risk reduction.

According to Table 9.2, it gives a great comparison of various AI models that have been developed and implemented for flood prediction and EWS. The table demonstrates that technological innovation

TABLE 9.2 Overview of AI models and technologies applied in flood prediction and early warning studies

NO.	*MODEL NAME*	*TYPE*	*APPLICATION IN FLOOD SYSTEMS*	*STRENGTHS*
1	Artificial neural network (ANN)	Machine learning	Predicts rainfall–runoff relationships and river discharge	Handles non-linear patterns; adaptable to various datasets
2	Support vector machine (SVM)	Machine learning	Classifies flood-prone zones and severity levels	Effective in high-dimensional spaces; robust to noise
3	Random forest (RF)	Ensemble learning	Flood susceptibility mapping and feature selection	High accuracy; interpretable; reduces overfitting
4	Long short-term memory (LSTM)	Deep learning	Time-series forecasting of rainfall and water levels	Captures temporal dependencies; ideal for sequential data
5	Convolutional neural network (CNN)	Deep learning	Analyzes satellite imagery for flood extent detection	Strong in spatial pattern recognition; useful for remote sensing

(*Continued*)

TABLE 9.2 (*Continued*) Overview of AI models and technologies applied in flood prediction and early warning studies

NO.	*MODEL NAME*	*TYPE*	*APPLICATION IN FLOOD SYSTEMS*	*STRENGTHS*
6	Gradient boosting machine (GBM)	Ensemble learning	Flood risk prediction and severity scoring	High predictive power; handles complex interactions
7	Recurrent neural network (RNN)	Deep learning	Models dynamic flood behavior over time	Suitable for continuous data streams; supports real-time updates
8	Decision tree (DT)	Machine learning	Flood zoning and mitigation planning	Simple, interpretable; good for rule-based classification
9	Hybrid SWAT-ANN model	Hybrid (physical + AI)	Combines hydrological simulation with AI correction	Enhances accuracy by merging domain knowledge with learning
10	Geospatial AI (GeoAI)	Integrated AI-GIS	Real-time flood mapping and spatial risk analysis	Integrates spatial data; supports dynamic visualization and alerts

and methodological variation are evident across the flood prediction and early warning system space. In particular, AI's application for classification problems including flood susceptibility maps and flood risk zoning often utilizes simpler ML models such as SVMs, RFs, and DTs. These models often provide human-intelligible, robust, and useful predictions when there are limitations on available data. In comparison, many DL models such as LSTM, CNNs, and combined CNN–RNN models are increasingly preferred for recurrent forecasts of time-series data and direct detection of the spatial extent of flood inundation. DL models have significant value-added features due to their inherent capability to learn complex temporal and spatial patterns. The research of Zhang et al. and Li et al. showed that a DL model that utilized satellite and real-time sensor data outperformed traditional methods relative to the dynamic context developed for flood warning and prediction.

The table also shows technology integrations that extend beyond mere prediction with AI models. For instance, blockchain-supported frameworks (Adiq et al.) add to predictive capabilities beyond secure and decentralized data use, and social media enables natural language programming (Sing et al., Ahmed et al.) to circulate alerts and situational awareness in multiple languages. IoT-based models (Gulhane et al.; Silva et al.) allow for real-time tracking and warnings, especially when established at the community level. These emerging technologies reflect an emerging preference for participatory, decentralized, and responsive flood management systems. They may also introduce new challenges to address, including data validation, misinformation, and hardware dependencies to keep their solutions reliable, sustainable, and scalable. Despite advancements in technology, some limitations will remain across the studies. Although powerful, DL models that produce predictions may not be interpretable for decision-makers to understand what is happening in an emergency context. Hybrid models, such as SWAT-ANN, often need to be calibrated twice, as well as linked, making them complex and resource-intensive to relate to each other. Many systems include study-specific information that would inhibit wider applications if these models do not transfer regionally. The lack of common evaluation metrics across studies also limits the ability to conduct comparative assessments and benchmarking since there are no standard benchmarking processes. Ethical dilemmas associated with data privacy, mainly related to crowdsourced and blockchain systems, create deployment limitations. The potential for more inclusive, explainable, and interoperable AI systems to facilitate and enrich robust flood predictions and equitable EWS for their different geographies is identified in Table 9.2.

9.4 METHODOLOGY

The underlying process to create AI-based flood prediction and EWS is a robust, systematic procedure that starts with data collection and submission. The data is acquired from many types of variables, including weather variables (rain, temperature, and humidity), hydrological variables (river discharge, water levels), geospatial variables (elevation and land use), satellite imagery data, and real-time feeds from sensors. These data are preprocessed as needed to assure consistency, completeness, and temporal alignment. All of these data are integrated to provide an effective basis for AI models to learn the complex interplay of factors that lead to flooding [31]. After preprocessing, AI models are selected according to the data type and prediction task. Generally, structured data can also be evaluated with ML models including RFs or gradient boosting models. Structured data can be evaluated with DL techniques such as LSTM networks and CNNs. AI models may even include a fully physical hydrological model where a DL or ML model is built to provide a covariate for the physics model (hybrid model).

After the selection of a model, it requires significant training on historical data, including thorough preprocessing, feature engineering, hyper-parameter tuning, and model performance evaluation. The model will also be validated on separate datasets to assess how well it generalizes, using metrics such as RMSE, accuracy, or F1 score based on the task [32]. The deployment phase is characterized by the development of real-time data pipelines that automatically feed live, environmental data into a prediction engine to create flood forecasts. Each of these models is used to produce actionable alerts in the form of reports, bulletins, or as part of a decision support system that incorporates user interfaces or dashboards to visually represent possible flood risk scenarios. Such systems may reside on scalable cloud or edge computing platforms that support stable and responsive computing. Consider a closed-loop system with continuous monitoring of the data and any model feedback loops. Models must be retrained on data regularly to keep them up to date while also reflecting the effects of changing conditions in the environment. Relying on the most recent literature and field-based work, the following core steps represent a generalized, albeit robust, approach.

9.4.1 Data Collection and Preprocessing

A reliable flood prediction system relies heavily on multi-source, high-quality observational data, which comprises

Hydrometeorological Data: Hydrometeorological data includes rainfall, river discharge, temperature, soil moisture, and evapotranspiration data (from ground stations or satellite data such as TRMM and GPM).

Topographic and Land-Use Data: Spatially referenced topographic and land-use data includes elevation models, land cover data, and soil types (e.g., NASA DEM, Copernicus, or Sentinel-class satellites).

Remote Sensing and Satellite Imagery: Remote sensing information provides flood extent, water body, and early detection of change information.

Social Media and Crowdsourced Data: Social media and crowdsourced data are additional datasets used in urban flash flood models [25] and validation in near real time.

Sensor Networks and IoT: Sensor networks and IoT provide near real-time monitoring data from river gauging stations, meteorological or hydrological stations, and collect data from mobile devices.

Data underwent preprocessing such as normalizing data, filtering noise, identifying missing data, and transforming time series to understand its suitability for the modeling context.

9.4.2 Model Selection and Design

The choice of AI models depends on the respective purposes of prediction (i.e., riverine vs. flash flood, short-term vs. long-term). The following options are recommended based on recent research:

Supervised ML Models: RF, SVR, Gradient Boosting Machines (e.g., XGBoost), and ANN.
DL Models: LSTM, GRU, CNN, and hybrid CNN-LSTM models, which are particularly valued for spatiotemporal flood forecasting.
Physics-Informed and Hybrid Models: These models couple AI with hydrodynamic models (e.g., SWAT and HEC-RAS) to build on physical constraints and improve interpretability.
Probabilistic Models: New approaches, such as DRUM, utilize generative modeling (diffusion-based or ensemble-based) to characterize uncertainty and can produce probabilistic estimates of extreme flooding events. DRUM is a novel probabilistic generative modeling approach developed to predict floods based on uncertainty and utilizing multiple plausible flood scenarios. It uses diffusion models or ensemble-based approaches to allow the model to predict a probability distribution over possible flooding outcomes rather than just one deterministic forecast.

9.4.3 Benefits of Using DRUM for Flood Forecasting

- Captures complex dependencies and uncertainty in flood data
- Gives a wide range of possible outcomes rather than just averages
- Provides better prediction of extreme floods because it better captures the modeling of rare events
- Supports risk-based decision-making by providing probabilistic information

9.4.4 Model Training and Validation

Models are developed from historical data that have an appropriate temporal (e.g., hourly or daily) and spatial resolution related to the prediction task and can include the common practice of

- **Data Split**: The dataset is commonly split into training, validation, and test sets (often utilizing k-fold cross-validation for varying degrees of robustness).
- **Evaluation Metrics**: Appropriate metrics will include R^2, RMSE, mean absolute error, Nash–Sutcliffe efficiency, and Critical Success Index metrics [23].
- **Benchmarking**: A common practice is to compare the predictions from the models with those from older models/baselines (e.g., linear regression and hydrological simulators) to demonstrate a marked improvement in predictions.

Sometimes these approaches can be more sophisticated, such as also evaluating modeling transferability to different basins and the models' resilience to certain missing data scenarios. When examining advanced AI-based research into flood prediction systems, model transferability and resilience to missing data are the two primary challenges that impact the feasibility and robustness of the systems applied.

Model transferability means the ability of a trained AI model to maintain predictive accuracy on multiple river basins or areas of geographical space away from the location of development [22]. Model transferability is important because it is often infeasible to collect large, high-quality datasets within each individual basin. Transferable models exploit common hydrological and meteorological patterns along with domain adaptation, transfer learning, or meta-learning to generalize the 'knowledge' learned in one basin to other basins. For example, a flood model developed for a temperate basin can be adapted for use in a tropical basin using some locally limited data to fine-tune the original model. Model transferability increases the usefulness of model development by enhancing the scalability of application and reducing the barriers to development because of time and cost.

Having resilience to missing data is equally important in practice since environmental data streams from the real world, although rich in information, may frequently have missing data due to reasons such as sensor failures, communication failures, or even adverse weather conditions [23]. A robust AI flood model would most certainly incorporate techniques to capably deal with missing data agilely with minimal loss in prediction quality. Techniques other than developing a rain gauge request to impute missing data include (1) data imputation, where statistical or ML-based estimators are used to create values in place of NaNs or missing data; (2) model design to leverage partial feature set input; or (3) a probabilistic model, which explicitly models the uncertainty introduced due to missing data. Some systems also use redundant sensor networks as well as real-time data quality monitoring technology to manage the loss of observations and therefore minimize the impact of missing data.

9.4.5 Real-Time Integration and Early Warning Mechanism

After validating the AI flood prediction model, the next critical step will be to build out the real-time flood forecasting and early warning system to merge the AI model with its predictions into a usable delivery and information system that can be rapidly transmitted to stakeholders and to communities at risk.

1. **Data Ingestion Pipelines**: In order to achieve high-quality real-time flood forecasting, comprehensive, automated data ingestion pipelines will be the backbone. This system continuously and automatically ingests and inputs live-streamed data from several sources to keep the AI models updated to current environmental conditions, such as:

   Weather APIs provide real-time meteorological data such as rainfall forecasts and temperature.

   Satellite feeds provide near real-time stroke and scale imagery to track the extent of the flood and soil moisture.

   Sensor networks (e.g., river gauges and IoT devices in flood-prone areas) provide continuous information on water levels (or discharge) and precipitation.

   These data streams will be automatically ingested and aggregated, cleaned, and preprocessed in real time before being put into the forecasting engine. The purpose of automated data ingestion pipelines is to generate timely and accurate data input while minimizing human error or delays and maintaining a rapid ability for the prediction engine to respond to increasing flood risk.
2. **Based on Existing Emergency Management Frameworks**: The AI model presents quantitative predictions such as river discharge rates or water surface elevations. EWS use existing emergency management and intention thresholds based on flood stage or levels of risk from existing historical flood data to send alerts. For example, when predicted river discharge exceeds the significant flood stage, an alert is then generated to indicate imminent flooding. These threshold rules can be tailored for different locations to account for regional vulnerabilities in flooding and resilience in infrastructure.

3. **Channels of Communication**: When the alert is triggered, it must be sent using a communication channel for it to reach at-risk populations and decision-makers:

   SMS alerts and mobile app notifications deliver direct and immediate contact with people who are in the flood zone.

   Sirens or community radio are ways to contact populations who do not have primary access to a smartphone or the internet.

   Providing access to disaster management dashboards will allow emergency and government agencies to receive the system's flood risk information and use this data to acquire more information or coordinate the emergency response.

   Examples of previous outputs include Google's Flood Hub, which provides an integrated set of data for push notifications to users in vulnerable areas, and Australia's FloodMapp, which provides integrated real-time forecasts and alerts to local emergency services for flood alerts.
4. **Visualization Tools**: In order to enable communities to understand their changed risk situation over time, effective flood early warning requires visualizations that are clear and easy to comprehend. Dashboards often include:

   Flood inundation maps that display the forecasted areas expected to be flooded (areas that are predicted to be inundated) so that communities can visualize where the flooding will occur in space.

   Flood progression timelines that show how the flood is expected to evolve over hours and days.

   Interactive layers that integrate forecasts, gauges, and terrain information in a way that supports situational awareness.

   According to Hong and Jain, these visualization tools are an effective way to improve user engagement [22] and preparedness behaviors by making the complex outputs of models simply accessible **and actionable information.**

9.5 POST-EVENT ANALYSIS AND CONTINUOUS LEARNING

Once the flood event has passed, the system's performance should be evaluated and improved for accuracy and reliability. This process utilizes feedback loops to allow the model and system to learn from past real-world situations and stakeholder feedback, providing a better response for future flooding events.

i. **Model Retraining**: The flood dynamics being modeled may also change over time due to altered climate systems, changes in land or water usage, or changes related to infrastructure, meaning that models trained on previous data may become less accurate over time. As part of the ongoing learning process, model retraining is built into the system. After each flood event, data being collected for the real-time report, including updated river discharge records, rainfall measurements, and the spatial extent of the flood event, will be included in the original training dataset. The AI model will then be retrained to include the new data, either with periodic retraining or with incremental retraining, to use in training the model with the most current patterns to improve [23] the model's predictive ability . By using this strategy, the model is adapted, relevant, and robust in a constantly changing environment.

ii. **Crowdsourced Verification**: The accuracy of flood forecasts is only possible with verification to evaluate model accuracy and improve forecasts. Crowdsourced verification utilizes citizen science and open data platforms where communities impacted by flooding can report if flooding occurred, what the water levels were, and what damages were incurred by using mobile apps or social media trackers. These ground-truth reports provide an excellent source of real-time and post-event data that can be used to augment traditional sensor information . Additionally, satellite imagery could be utilized to provide external validation on the extent and intensity of flooding detected and modeled by the models [20]. Crowdsourced validation data alongside satellite imagery provide a richer and more complete dataset that can be utilized to compare model outputs, detect errors, and develop corrections with the aim of further improving the reliability of future forecasts.

iii. **Explainable AI (XAI) Techniques**: A key feature for gaining credibility and increasing acceptance among stakeholders, which often include emergency managers, politicians, and the public, is the transparency and interpretability of model decisions for any AI flood prediction systems, including data processes and outputs. XAI is a field of AI related to interpretable models. XAI computes impact, or feature contribution to the prediction, by using methodologies such as SHapley Additive exPlanations values, which show how much each input feature (e.g., rainfall and soil moisture) contributed to that particular prediction and thus provides information on how the model decided to predict flood or not to predict flood. Similarly, attention mechanisms are often used in DL models to show what parts of an input sequence (e.g., temporal) or spatial (e.g., satellite images) the model paid attention to while predicting or making decisions. Releasing the model's logic through explainability or reasoning generates user trust, allows for the identification of possible biases and/or errors, and enables more educated future flood–response decision-making with 'reasoning' model output.

9.6 RESULT ANALYSIS AND DISCUSSION

The effectiveness of AI-based flood prediction and EWS has been scrutinized in numerous real-world implementations and peer-reviewed studies. Findings from various implementations in both riverine and flash flood applications have repeatedly shown that AI models, particularly DL models, have greater predictive power and operational efficiency when compared to traditional hydrological models [19]. This section provides a comparative review of outcomes derived from selected case studies, evaluations, and geographies. Several studies have reported significant improvements in forecast accuracy using AI models over traditional techniques.

Table 9.3 provides a summary of various recent studies and operational projects utilizing advanced AI and hybrid models for flood prediction, utilizing different modeling approaches and addressing unique hydrological and flood prediction issues. This study examines a variety of DL and ML models, including LSTM, GRU, ANN, and XGBoost, for river flow prediction in particular mountainous regions [30]. The LSTM model reported the highest accuracy with $R^2 = 0.96$, which means it fit the observed data well. This shows the ability of RNNs to capture temporal dependencies observed in complex, dynamic hydrological systems. Most mountainous regions display a complex amalgamation of runoff and discharge due to variability in the terrain, so good accuracy from an LSTM is to be noted, demonstrating its use to predict river flow in complex mountainous regions.

In their research, Mosavi and colleagues use various models of AI, such as SVMs, RFs, ANNs, and various hybrid models that combine these AI methods. The authors proposed that hybrid AI models optimize the advantages of several algorithms, producing an overall more accurate prediction of

TABLE 9.3 Model performance metrics

STUDY/ PROJECT	*MODEL(S) USED*	*KEY METRICS*	*HIGHLIGHTS*
[20]	LSTM, GRU, ANN, XGBoost	$R^2 = 0.96$ (LSTM); root mean square error, mean absolute error	LSTM achieved the highest accuracy for river flow prediction in mountainous terrain.
[23]	SVM, RF, ANN, hybrid models	Various (root mean square error, R^2, Nash–Sutcliffe efficiency)	Hybrid AI models consistently outperformed single models.
[26]	ConvLSTM, LSTM	26% ↑ accuracy vs. baseline; Critical Success Index, probability of detection, false alarm rate	ConvLSTM improved flash flood prediction using remote sensing and radar inputs.
FloodCast [27]	Physics-informed deep learning	Accuracy not specified; validated via case study	Accurate simulation of the 2022 Pakistan flood; fast and scalable
DRUM [26]	Diffusion-based generative model	Probabilistic metrics (Brier Score, reliability)	Robust prediction of extreme rainfall–runoff events
[24]	Hybrid ML + hydrological models	Lead time ↑ (7 days); coverage ↑ (80+ countries)	Operational deployment at a global scale with local accuracy

flash floods than any single model [31]. The implication of hybridization appears to provide greater promise in capturing non-linear and complex features of the flood process by utilizing complementary features of several models: the power of ensemble methods and the pattern recognition capabilities of neural networks.

Evaluation of the effects of hybrid models and AI approaches on flash flood prediction utilizing DL architectures. Oddo and colleagues explored flash flood prediction utilizing DL architectures, including ConvLSTM and LSTM. The ConvLSTM model uses the spatial power of convolutional layers with the temporal features of LSTM, which gives it particular advantages for flash flood prediction because of their rapid development and [31] highly localized nature. The authors reported a 26% accuracy gain above baseline models, which were validated with the Critical Success Index, probability of detection, and false alarm rates. This was also based on remote sensing and radar data inputs, which provide a broader dataset of both spatial and temporal orientations that ConvLSTM takes advantage of to provide superior forecasts.

FloodCast: FloodCast also uses a Physics-Informed DL framework utilizing data-driven approaches that leverage AI and hydrologic principles [28]. The hybrid nature allows the model to respect the physical laws governing the flow of water while utilizing the pattern recognition of DL. Although no accuracy metrics are given specifically, the model was checked against the 2022 Pakistan flood event, which suggests the model can simulate events that occurred in the real world. Additionally, FloodCast is known to be fast and can also be scaled to look at larger-scale flooding events, both of which are necessary for the operational forecasting of larger-scale events.

The Google Flood Forecasting program is a best practice example demonstrating the operational use of an AI-focused flood forecasting system at a global scale. The project utilizes an ML algorithm based on a simulated hydrology model [26], resulting in extended lead times nearing seven days and being operational in over 80 countries, while providing enough detail to be reliable and useful both globally and locally. This brief summary illustrates the capacity of hybrid AI-hydrological systems to operationally scale and provide regionally unique aspects of each area of the world, which will improve flood risk response for member nations and provide early warning messages to assist with disaster response and pre-disaster mitigation action.

Flood prediction models must be customized for the defining characteristics of different flood types, specifically riverine and flash flooding. Riverine floods occur from long periods of rainfall or snowmelt characterized by slowly evolving changes, and the longer lead time makes forecasting riverine flooding much less complex. The AI-based models in our results, such as LSTM networks and GRUs, are more appropriate based on their capacity for modeling long-term temporal dependencies in hydrometeorological data. For example, they have been shown to achieve high accuracy (e.g., $R^2 \approx 0.96$) in flood analytics for studies related to snowmelt runoff and riverine flooding in areas such as the Kabul River Basin. Both of the aforementioned processes would provide useful foundations for planning evacuations and resource deployment.

Flash floods are geophysical processes triggered by localized amounts of rainfall, often in urban environments or because of small catchment areas. The previously discussed DL models with spatial awareness, in particular, ConvLSTM models, performed significantly better when predicting flash floods. ConvLSTM has an architectural feature that spatially integrates convoluted and recurrent layers into a model that can compute spatial and temporal dynamics when paired with radar and satellite remote sensing data. ConvLSTM models yielded the best advancements in predictive precision, between 20% and 30%, and gave extremely low false alarms. One researcher estimated that ConvLSTM models in forecasts increased access to faster and more reliable early warnings, which are extremely important regarding life safety when responding to rapidly developing flood scenarios.

To ensure reliability, AI-based flood forecasting platforms need to be flexible and suited to different geographic and socio-economic environments. In low-resource areas such as India and Kenya, satellite-derived rainfall data (optimally) for rainfall forecasts, instead of limited ground-based sensor persistence models such as Google's AI-based Flood Hub, provides reliable forecasts based on the level of uncertainty and up to a 7-day lead time in issuing an alert! In urban areas, platforms such as FloodMapp combine rainfall, river levels, and topography to produce real-time event-based flood forecasting maps that can be used to develop emergency responses tailored to critical points in cities. What is going to gain traction are client/cloud-based cost-effective models such as FloodCast and DRUM that work in different areas, as they rely on the computing power of cloud systems and physics-informed AI (all-sensor capabilities) to produce accurate forecasts in different basins and continental-scale regions because the AI trained from rainfall is equally predictive of flood events in both higher and lower dry regimes . While technical innovation is accelerating in flood forecasting, there remain some principal challenges that make this work difficult; for example, most models are black boxes and hard to interpret, the lack of ground and satellite data means many models fail training in under-researched and remote areas of flood impact, and they are often non-transfer models. Emerging solutions such as explainable AI, transfer learning, and physics-informed AI are our best hope in overcoming these challenges and improving the trust, scaling, and accuracy of AI models.

9.7 CONCLUSION

Floods are one of the most catastrophic threats among natural hazards, and they are climatically induced and exacerbated by urbanization. The speed of flood forecasting for emergency management and planning is truly an issue of the use of real-time information, particularly in data-poor areas. Several flood forecasting mechanisms using AI, such as LSTM, ConvLSTM, GRU, and physics-informed neural networks, have resulted in faster and more accurate computations and longer lead times using models of various floods and geographies. Ongoing public-facing cases such as Google's Flood Hub, FloodCast, and FloodMapp prove our ability as a society to take advantage of AI and technological infrastructure, including cloud-based infrastructure, when paired with EWS. However, even with these advancements, issues remain, including the availability of data and the transparency of the models, and there are good reasons for both of these. Ultimately, AI represents a new framework as we undergo another transition in flood risk management and mitigation. Moving forward, any modeling for future flood forecasting and prediction will need to be equitable, scalable, and explainable.

REFERENCES

[1] Y. Zhang, Y. Wang, and X. Li, "Deep learning for rainfall–runoff forecasting: LSTM and GRU models," *J. Hydrol. (Amst)*, vol. 600, p. 126678, 2021.

[2] R. Tabbussum and A. Q. Dar, "Comparative analysis of neural network training algorithms for flood forecast modeling," *J. Flood Risk Manag.*, vol. 13, no. 4, p. e12656, 2020.

[3] R. Singh and S. Ahmed, "NLP-based multilingual flood alert system," *Procedia Comput. Sci.*, vol. 199, pp. 1234–1241, 2022.

[4] J. Silva and R. Costa, "Rapid flash flood detection using sensor fusion and artificial intelligence," *Sensors*, vol. 22, no. 9, p. 3124, 2022.

[5] N. Sharma and M. Saharia, "Landslide susceptibility mapping using ensemble machine learning," *Catena (Amst)*, vol. 235, p. 107653, 2024.

[6] K. Sethi, A. Sharma, S. Chauhan, and V. Jaiswal, "Impact of Social and Cultural Challenges in Education Using Artificial Intelligence," in *Revolutionizing Education in the Age of AI and Machine Learning*, M. K. Habib (Ed.), IGI Global, Hershey, PA, USA, 2020, pp. 130–151.

[7] K. Sethi and V. Jaiswal, "PSU-CNN: Prediction of student understanding in the classroom through student facial images using convolutional neural networks," *Mater. Today Proc.*, vol. 62, pp. 4957–4964, 2022.

[8] K. Sethi, S. Chauhan, and V. Jaiswal, "Artificial Intelligence in Higher Education: First Attempt," in Impact of AI Technologies on Teaching, Learning, and Research in Higher Education, edited by Shivani Verma and Pradeep Tomar, IGI Global, 2021, pp. 1–29.

[9] K. Sethi, S. Chauhan, and V. Jaiswal, "Artificial Intelligence in Higher Education: First Attempt," in Impact of AI Technologies on Teaching, Learning, and Research in Higher Education, S. Verma and P. Tomar (Eds.), IGI Global, 2021, pp. 1–29.

[10] M. Saharia, P. E. Kirstetter, and J. J. Gourley, "Mapping flash flood severity in the United States," *J. Hydrometeorol.*, vol. 18, no. 2, pp. 397–411, 2017.

[11] R. J. Romanowicz and L. Pfister, "Flood modeling in data-scarce regions," *Water Resources Management*, vol. 36, no. 4, pp. 1123–1140, 2022.

[12] M. Rohit, K. Sethi, M. Khan, and A. Raina, "Machine learning model for prediction of chemical harmfulness to staff and guests in the hospitality industry: A pilot study," *Data and Metadata*, vol. 2, p. 161, 2023.

[13] S. K. Rohit and S. Milan, "Epidemiology of occupational injuries and hazards in Indian hospitality: A study report," *European Chemical Bulletin*, vol. 12, pp. 2586–2590, 2023.

[14] T. Rasool and T. Shaikh, "Soil erosion estimation using RUSLE and GIS," in *AGU Fall Meeting Abstracts*, 2021, H15N-1214.

[15] M. Rahman and M. Hossain, "Localized ANN-based flood forecasting in Bangladesh," *Hydrological Sciences Journal*, vol. 66, no. 3, pp. 456–470, 2021.

[16] F. Pappenberger and H. L. Cloke, "Ensemble flood forecasting: A review," *J. Hydrol. (Amst)*, vol. 375, no. 3–4, pp. 613–626, 2009.

[17] P. Oddo, M. Saharia, and P. Kirstetter, "ConvLSTM-based flash flood prediction using radar and satellite data," *Remote Sens. Environ.*, vol. 298, p. 113894, 2024.

[18] P. Ozturk and A. Mosavi, "Flood susceptibility mapping using ensemble learning," *Water (Basel)*, vol. 12, no. 5, p. 1424, 2020.

[19] A. Mosavi, P. Ozturk, and K. W. Chau, "Flood prediction using machine learning models: A literature review," *Water (Basel)*, vol. 10, no. 11, p. 1536, 2018.

[20] X. Li and Y. Hong, "Geospatial AI for real-time flood mapping," *ISPRS Journal of Photogrammetry and Remote Sensing*, vol. 198, pp. 112–124, 2023.

[21] P. E. Kirstetter and J. J. Gourley, "Radar-based flood forecasting systems," *J. Hydrol. (Amst).*, vol. 556, pp. 1020–1035, 2018.
[22] Y. Hong and S. K. Jain, "Artificial intelligence in hydrological extremes: A global perspective," *Hydrology Research*, vol. 54, no. 1, pp. 1–15, 2023.
[23] V. Gulhane and S. Patil, "IoT-based flood monitoring and alerting system," *International Journal of Engineering Research & Technology*, vol. 11, no. 3, pp. 45–50, 2022.
[24] FloodMappA, "Real-time flood mapping using artificial intelligence," Company white paper, 2022. https://www.floodmapp.com/
[25] Cui, Z., Guo, S., Chen, H., Liu, D., Zhou, Y., & Xu, C. Y. (2024). Quantifying and reducing flood forecast uncertainty by the CHUP-BMA method. Hydrology and Earth System Sciences, 28(13), 2809–2829.
[26] F. Team, "Physics-informed deep learning for scalable flood forecasting," Technical report, 2023. https://papers.nips.cc/paper_files/paper/2021/hash/0a3b5a7a477d359746061d41c3a04fd6-Abstract.html.
[27] H. L. Cloke and F. Pappenberger, "How do we forecast floods?," *Hydrol. Process.*, vol. 24, no. 3, pp. 364–378, 2010.
[28] K. W. Chau and C. L. Wu, "Hybrid SWAT–ANN model for runoff prediction," *Natural Hazards*, vol. 108, no. 1, pp. 519–566, 2019.
[29] R. D. Bhowmik and P. Mujumdar, "Climate variability and concurrent droughts in India," *J. Hydrol. X*, vol. 26, p. 100196, 2025.
[30] H. E. Beck and A. W. Wood, "Evaluation of precipitation datasets for hydrological modeling," *Hydrol. Earth Syst. Sci.*, vol. 24, no. 3, pp. 1235–1250, 2020.
[31] M. Adiq and M. Rahman, "Blockchain-enabled AI framework for flood risk management," *Environmental Modelling & Software*, vol. 192, p. 106591, 2025.
[32] S. Ahmed and M. Khan, "Crowdsourced flood intelligence using AI and NLP," *International Journal of Disaster Risk Reduction*, vol. 65, p. 102654, 2023.

Smart Irrigation Systems Based on IoT and Remote Sensing

10

Qurat-Ul-Ain and Ramandeep Kaur

10.1 INTRODUCTION

Irrigation is the process of watering the land adequately to fulfill the water requirements of plants that have not been met naturally. The largest user of fresh water on Earth is the agricultural industry. Almost 70% of the water on Earth is utilized for agricultural irrigation. Irrigation is used to promote plant growth while reducing water loss and soil erosion. There are several ways to irrigate the soil. The most popular kind of irrigation system is drip irrigation, which is made up of a system of tubes with tiny emitters or holes in them. They can be positioned either above or below the soil's surface and will gradually seep water into the ground over an extended period of time. Sprinkler irrigation is another type of irrigation technique that uses a system of pipes to finely spray water over certain locations. Another configuration has a central pipe with outputs rotating around a central pivot point. Center pivot irrigation is a considerably larger system that is supported by steel or aluminum towers and works similarly to a sprinkler system. A type of surface irrigation called a furrow irrigation system involves making tiny, parallel channels and filling them with water. The terraces between the channels are used to produce crops. Terraced irrigation is a method that involves cutting sloped land into a series of leveled steps. When rainfall occurs, water flows gradually from the uppermost terrace down to each successive one. This controlled flow helps retain valuable soil nutrients, preventing them from being washed away. This agricultural system is typically employed in mountainous or hilly regions. With technological intervention in almost all fields and the rise of precision farming, smart irrigation systems are now used. This system utilizes the Internet of Things (IoT), which is the process of connecting numerous devices via the internet. Each device is uniquely identified, and data can be transformed between them without the need for human assistance. There are different ways to implement automated irrigation systems based on cost, area under consideration, and the demographic location of the land. [1] presents the idea of monitoring and controlling soil moisture (SM) using automatic irrigation. On the Uno platform, an ATMEGA328P microprocessor implements the control unit. This device measures actual humidity using a hygrometer.

DOI: 10.1201/9781003714217-10

This advantage guarantees that the system uses water appropriately, avoiding overuse or underuse. Farmers are informed about the condition of the water supply through the use of IoT. Farmers can access the GSM-GPRS website at any time to see if the water head is open or closed, and sensor data is updated regularly. To produce an image, sensor data is sent to the object's audio channel.

This chapter presents a detailed examination of an automated irrigation system that uses IoT, machine learning (ML), and deep learning (DL) algorithms to autonomously forecast the SM level. The popular IoT and remote sensing (RS) devices used in smart irrigation systems are also covered. The study concludes that using both ML and DL algorithms in hybrid mode generally yields more promising outcomes.

10.2 IoT AND RS DEVICES IN SMART IRRIGATION

Smart irrigation systems benefit greatly from the use of IoT and RS devices to measure the moisture content of the soil. IoT systems for irrigation have employed a number of tiered strategies, with differing outcomes from each [2]. This architecture is typically organized into three tiers. The middle layer includes a gateway and data transmission infrastructure, while the lower layer consists of actuators and sensor nodes. The upper tier comprises databases, applications, or cloud services. Although these three layers form a common conceptual framework, their specific implementations can vary significantly, with each tier possessing distinct characteristics [3].

The most common IoT devices that have been used by scientists in various research projects and in real-time systems are as follows.

10.2.1 Sensor Devices

Sensor devices help in collecting real-time data from the environment. This data is finally used to make overall decisions in smart irrigation systems. The following are the commonly used sensors in smart irrigation:

10.2.1.1 SM Sensor

Numerous sensors are available; some detect the moisture content of the soil, while others measure the dielectric constant (volumetric content) and soil water potential [4]. The volumetric water content of soil is the amount of liquid water per volume of soil. It is typically expressed as a percentage. The soil water potential indicates the amount of energy required by plant roots to extract water from soil particles. Soil tension rises when soil water is extracted. Centibars (cb) or bars of air pressure are used to express soil stress. Soil water tension is almost negligible when the soil is saturated with water.

10.2.1.2 Weather Station Sensors

A weather station is a facility equipped with sensors that measure atmospheric conditions on land or at sea to give data for weather forecasts. Sensors that are typically attached to a data logger measure the atmospheric variables. Depending on the data logger, either an external or internal power source powers these. Some weather stations use a small battery to prevent blackouts. Conventional data loggers have a relatively limited storage capacity and are also susceptible to data loss during power outages, which is a significant consideration. As a typical data logger is a passive device, it cannot be modified to incorporate external routines, control actions, or predictive functions. Programming capabilities for adjusting data collection and parameter settings are integrated directly within the data loggers themselves [5]. Some of the commonly used sensors for detecting various parameters include the following:

10.2.1.3 Rain Sensors

A rain sensor is a device that detects recent rainfall and overrides a scheduled irrigation cycle to prevent unnecessary watering, as shown in Figure 10.1. It is one of the simplest, most cost-effective water-saving devices you can add to an irrigation system. The sensor detects that a certain threshold of rainfall has been met. It sends a signal (either via a physical wired connection or wirelessly) to the irrigation controller. The controller interrupts the scheduled watering program. After the collected water evaporates (which can take hours or days, depending on the setting), the sensor resets itself and allows the system to resume its normal schedule.

10.2.1.4 Temperature & Humidity Sensors

Temperature and humidity sensors are helpful in assessing the water loss of soil, which is known as evapotranspiration. A temperature sensor (thermistor) is a resistor whose electrical resistance changes predictably with temperature. The sensor's microcontroller sends a small voltage to the thermistor and measures the resulting current. Because the resistance is tied to temperature, the change in current can be converted into a precise temperature reading (e.g., 25°C or 77°F). A humidity sensor (capacitive) has a substrate (usually a polymer or metal oxide) sandwiched between two electrodes, forming a tiny capacitor. The substrate absorbs water vapor from the air. The ability of this capacitor to store an electrical charge (its capacitance) changes as the substrate absorbs or releases water molecules. The more humid the air, the more water is absorbed, and the higher the capacitance. The microcontroller measures this change in capacitance and converts it into a relative humidity (RH%) reading. These two sensors are almost always combined into a single, small, inexpensive chip. Conventional rigid sensors can harm sensitive plant sections and are incompatible with plant bodies. Additionally, the stiff

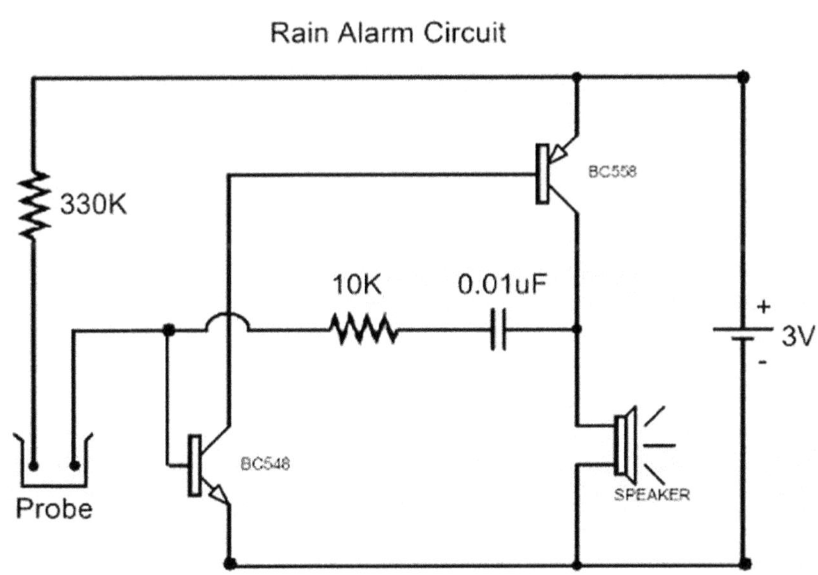

FIGURE 10.1 Rain sensor alarm circuit.

sensors' mechanical incompatibilities at the plant–electronic device interface lead to phenotypic trait measurement errors. Furthermore, invasive sensors based on rigid materials have complex operating patterns, are hard to install, and are not portable, all of which disrupt plants' normal growth patterns. To overcome these challenges, flexible sensor devices have been developed, as their adaptable nature allows for direct contact with plant surfaces. Such sensors provide dependable, accurate, and precise measurements of cellular biomarkers and physiological characteristics. Additionally, they are non-invasive, wearable, biocompatible, and lightweight [6]. Similarly, we have solar radiation sensors that measure sunlight intensity, a key factor in plant water use, and wind speed sensors that are used for high winds, which can affect sprinkler patterns and increase evaporation.

10.2.1.5 Environmental Sensors

Some sensors are used in precision farming to measure the environmental impact on the crop. For example, leaf wetness sensors can detect moisture on plant leaves, which helps predict and prevent fungal diseases that thrive in wet conditions. Similarly, light sensors (PAR sensors) can measure photosynthetically active radiation and are mainly useful for advanced greenhouse or high-value crop management.

10.2.2 Actuators

Actuators perform the physical action of controlling water flow based on commands from the microcontroller. These are basically electromechanical devices that translate control signals into physical motion to automate and precisely control agricultural machinery and processes. ML and DL techniques are employed in smart agriculture systems to enhance actuator intelligence. For instance, [7] developed an intelligent actuator management system by integrating sensor data with ML. These results show that gradient boosting outperforms other ML algorithms in handling complex sensor data from IoT-based agricultural systems. This study's key contribution is combining environmental sensing with predictive modeling to support real-time decision-making in agriculture. This study is one of the first to implement gradient boosting in a real-time IoT system, integrating multiple sensor inputs such as temperature, humidity, and soil nutrients to control actuators. Unlike previous studies, our system simulates a fully connected framework that processes data in real-time, enabling precise actuator responses. This can help improve agricultural productivity, reduce water and energy consumption, and minimize environmental impact.

10.2.3 Controller Devices

This serves as the main hub for processing sensor data and communicating commands to the actuators. As the brain, the controller is in charge of interpreting sensor data, coming to conclusions, and instructing the actuators. It decides how the robot should respond to its surroundings and carries out the required tasks.

These controller devices fall into one of the following categories:

- **Low-Level Control**: These controllers manage fundamental motor control, guaranteeing precise and accurate movements of individual actuators or joints.
- **High-Level Control**: These controllers plan intricate movements and adjust to shifting conditions by coordinating a number of actuators and sensors.
- **Artificial Intelligence (AI)**: Robot controllers are rapidly integrating AI systems, which allow the robots to learn, make decisions, and carry out activities on their own.

10.2.4 Communication Devices

These devices enable the sensors, actuators, and controller to communicate with each other and to the cloud. Various communication protocols can be used between the devices based on the range or area under coverage:

- **Short-Range: Wi-Fi** (for residential systems near a router), Zigbee, or Bluetooth Low Energy for communication within a local network (e.g., between a sensor and a gateway).
- **Long-Range:** Low-Power Wide-Area Network is essential for large farms and golf courses where devices are spread over vast areas and need long battery life.
- **LoRaWAN:** This protocol is very popular for agriculture due to its long range (miles/kilometers) and low power consumption.
- **NB-IoT/LTE-M:** This is a cellular-based option that is reliable and secure, but it is costly.

A field sensor might have a LoRaWAN radio module that sends data to a LoRaWAN gateway mounted on a pole, which then forwards it to the internet.

10.2.5 End User Interface Devices

These help the user to interact with and monitor the smart irrigation system. They are usually software-based with the prerequisite requirement for the installation of the software on the devices. These interfaces can be generally categorized into two types:

- **Web Portal:** A dashboard accessible from any computer browser. It provides the most comprehensive view, with data visualization, historical trends, and system configuration options. Used by farmers and facility managers.
- **Mobile Application (Smartphone/Tablet):** The most common interface for home owners and field managers. It allows for remote monitoring, receiving notifications and alerts (e.g., leak detection in any valve), and manual control of the irrigation system.

10.3 ML ALGORITHMS IN SMART IRRIGATION

In smart irrigation systems, data is acquired by making use of different IoT devices, and final intelligent, predictive decisions about SM can be made by applying ML algorithms. The goal is not just to detect current moisture but to predict future needs and optimize water usage. To forecast SM, researchers have used nearly every type of ML model. These algorithms form the foundation for learning the relationship between input data acquired through different sensors and the output results for soil irrigation. When SM data is labeled, supervised ML methods such as decision trees, random forests (RFs), support vector machines, and linear regression have been utilized. Unsupervised learning algorithms such as k-means clustering have also been applied to find hidden patterns in data without predefined labels. In smart irrigation systems, some predictions are made on time-series data, in which recurrent neural networks (RNNs), ARIMA and long short-term memory (LSTM) have been useful [8]. They have carefully evaluated a range of ML experiments used to assess SM across different land covers and geographical areas. ML methods present a viable substitute for traditional empirical and physics-based models, which depend on substantial biophysical parameterization. Beyond established techniques such as support vector machines and tree-based models, their work also incorporates advanced DL architectures, ensemble methods, and sophisticated multi-classification systems. These recent advancements have substantially increased the accuracy of SM predictions across multiple soil depths by incorporating RS technology. The application of ML for SM forecasting is demonstrated across diverse agricultural contexts. For instance, a study [9] utilized a suite of ML methods—including linear regression, support vector machine regression, principal component analysis, and naive Bayes—to generate predictive models capable of forecasting SM levels with a significant lead time of 12–13 weeks. In a parallel approach focused on controlled environments, a study [10] also leveraged ML techniques. Their methodology involved applying ML algorithms to sensor data streams collected via an IoT system specifically deployed within a naturally ventilated polyhouse, thereby creating a forecast model tailored to that unique microclimate. Three distinct input combinations were created for modeling, employing microenvironmental characteristics that had a greater impact on SM, such as T, RH, ST, and LI. According to the study's findings, the multiple linear regression (MLR) technique had the lowest accuracy, while RF and extreme gradient boosting (XGB) performed better than the other models [11]. Field SM may be accurately predicted using ML methods. These algorithms produce varying degrees of prediction effectiveness and are founded on distinct ideas (regression, regression trees, and kernels). Accurate SM forecasting depends on understanding the relative importance of predictive variables. Validation against an independent dataset showed that RF regression, boosted regression trees, and support vector regression models were superior, evidenced by their strong correlations and low RMSE and MAE values. Crucially, the variable importance analysis indicated that weather station parameters (including station SM, 4-day cumulative rainfall, and PET) were more influential in predicting field SM in nearby plots than variables related to soil properties or crop status.

10.4 DL ALGORITHMS IN SMART IRRIGATION

DL is a branch of ML within the broader field of AI. It employs hierarchical, multi-layered neural networks, which are computational models inspired by the structure of the brain, to learn from data and address intricate tasks. These algorithms use multiple layers to progressively extract higher-level features from raw input data. DL architectures can be classified into supervised and unsupervised learning. For SM prediction, many researchers have utilized RNN-based structures because they effectively handle sequential data, unlike artificial neural networks and convolutional neural networks, which require fixed-size inputs [12].

For example, a study [12] designed a DL regression network for sub-hourly SM prediction, benchmarking its performance against conventional ML models. These included RF, XGB, light gradient boosting (LGB), cat boosting, k-nearest neighbors, and LSTM networks. Their analysis revealed that while all tested models—both ML and DL—performed well in simulating SM for sweet corn and green bean plots, the LGB and XGB variants outperformed the others. Notably, despite requiring greater training time and computational resources, the DL model achieved higher accuracy than both XGB and LGB and also surpassed the standalone LSTM. In conclusion, the study posited that both ML and DL models constitute effective tools for agricultural SM prediction, with promising applications in optimizing irrigation scheduling and water resource management [12].

A separate investigation [13] introduced a system designed to capture complex temporal patterns from raw SM data by integrating LSTM layers with an attention mechanism and fully connected layers. To optimize the architecture, the researchers employed a genetic algorithm (GA) to tune key hyperparameters, including the number of LSTM layers, hidden units, dropout rate, and learning rate, resulting in a model termed GA-LSTM-ATT. Experimental validation demonstrated that this optimized framework produced reliable SM estimates across a variety of geographical locations [13].

Additionally, [14] presented a novel stacking ensemble technique for SM inversion in China's ShanDian River Basin. This method integrates multiple ML algorithms and leverages diverse datasets, including Sentinel-1A radar data, Sentinel-2A spectral indices, NASADEM, and other spatiotemporal inputs.

Furthermore, a study [15] applied DL and ML techniques to estimate SM using wireless sensor networks (WSN). The study, focused on Indian regions, integrated data from soil moisture active passive, WSN, and POWER climate datasets, collected at regular intervals to form time-series records. To formulate this as a supervised learning problem, the temporal data were used to train and evaluate five models: Logistic Regression, Support Vector Machines, Decision Trees, RFs, and LSTM. Model performance was assessed using a suite of error metrics: MAE, MSE, RMSE, and MAPE. The research concluded that models incorporating both satellite and WSN data performed effectively, with the LSTM model consistently outperforming other methods across different sites. The LSTM achieved an MSE of 0.06 and a MAPE of 2.8%, indicating strong alignment between predictions and observed values [15].

These promising advances have contributed to the growing adoption of precision agriculture, benefiting areas such as irregular irrigation [16].

10.5 COMPARATIVE ANALYSIS AND LIMITATIONS

See Table 10.1.

TABLE 10.1 Overview of literature and limitations

PAPER	*EXPERIMENTAL DETAILS*	*LIMITATIONS*
[15]	Using real-world data from various geographic locations, including a dataset from NASA's Soil Moisture Active Passive project, five machine learning models—"linear regression, support vector machines, decision trees, random forests, and long short-term memory networks (LSTM)"—have been assessed and contrasted. The climate data, which uses both passive and active sensors for data collection, has also been added. The study establishes that the effectiveness of wireless sensor network (WSN)–based approaches for soil moisture estimation, with the LSTM model consistently outperforming other machine learning algorithms in various evaluation metrics.	The best model for estimating soil moisture is not claimed to have been offered by the study. It seeks to provide a thorough examination of how various ML and DL models perform in comparable scenarios.
[12]	A multi-layer deep learning regression network was developed in order to accurately model the sub-hourly SM of sweet corn and green bean plots under three deficiency and full irrigation regimes.	It was discovered that DL model construction and training require more time and processing power than ML models, despite the study's aim of proving that DL models will perform better than traditional ML models. Additionally, the XGB and LGB gradient boosting models outperformed the DL model in terms of accuracy. The study was conducted on a particular crop and not generalized.

(Continued)

TABLE 10.1 (*Continued*) Overview of literature and limitations

PAPER	*EXPERIMENTAL DETAILS*	*LIMITATIONS*
[8]	They have carefully evaluated a range of machine learning research that is used to evaluate SM across different land covers and geographical areas. Their analysis goes beyond conventional ML methods such as support vector machine (SVM) and tree-based models to include sophisticated multi-classifiers, ensemble frameworks, and deep learning models. Through the use of remote sensing technology, these recent advancements have greatly increased the accuracy of SM forecasts at different depths. It was discovered that hybrid machine learning models, which include two or more ML models, increase the accuracy of SM estimates while lowering the bias–variance tradeoff.	Despite the advantages offered by ML and remote sensing for soil moisture estimation, applying models trained at one location to different sites remains a major limitation.
[13]	An attention mechanism, fully connected layers, and long short-term memory (LSTM) layers are all merged in this study's integrated deep learning-based architecture for predicting soil moisture. In this framework, LSTM layers are used to extract the intricate short- and long-term dependencies from the time-series soil moisture data. Furthermore, an attention mechanism is employed to learn discriminative information from the retrieved feature representations. The GA-LSTM-ATT is a proposed network whose hyperparameters are simultaneously calculated using the genetic algorithm (GA) to increase prediction accuracy. These parameters include the learning rate, dropout rate, number of LSTM layers, and number of hidden units in layers. The GA-LSTM-ATT method outperforms the benchmark approaches on a variety of assessment indices, according to the experimental data.	Notwithstanding the effective results that the technique has shown, there is a chance that the optimization of network parameters will result in an increase in computational load, particularly in situations with large training sets and broad parameter search domains.
[9]	They have used machine learning techniques such as linear regression, regression, SVM, principal component analysis, and naive Bayes.	They have applied these to four different crop datasets (potato, mustard, cauliflower, paddy) but not generalized the results.
[14]	They proposed a novel stacked approach that combines multiple machine learning algorithms for soil moisture inversion in China's ShanDian River Basin, utilizing Sentinel-1A radar scattering data, Sentinel-2A spectral indices, NASADEM, and other spatial and temporal information. ML techniques used in this paper are as follows: • extreme gradient boosting (XGBoost), • random forest (RF), • light gradient boosting machine (LightGBM), • gated recurrent unit (GRU)),	The model's robustness and capacity for generalization may be impacted by the study's use of sparse site data and the predominant grassland soil cover type. The study relied heavily on Sentinel-1 radar and Sentinel-2 optical data for feature extraction, which may have restricted the model's ability to incorporate additional relevant data sources.

(*Continued*)

TABLE 10.1 (*Continued*) Overview of literature and limitations

PAPER	*EXPERIMENTAL DETAILS*	*LIMITATIONS*
	• category boosting (CatBoost)), • deep neural network (DNN), • convolutional neural network (CNN).	This study did not explore alternative model fusion techniques, including voting, weighting, or Bayesian averaging methods. By diversifying the combination of several algorithm series, these techniques may perform better than any other algorithm or algorithm series.
[10]	They have studied the impact of ML algorithms such as decision tree (DT), random forest (RF), multiple linear regression (MLR), extreme gradient boosting (XGB), k-nearest neighbor, and artificial neural network methods.	This study is confined to a polyhouse.
[11]	Different regression trees, kernels, and regression are the foundations of machine learning algorithms, which produce varying degrees of prediction efficacy. The random forest regression, support vector regression, and boosted regression trees forecasts performed better than the other approaches during model validation using an independently generated dataset because of their high correlations and low RMSE and MAE. When it comes to predicting field soil moisture in the adjacent plots, the weather station parameters (station soil moisture, 4-day cumulative rainfall, and PET) had a comparatively greater impact than the soil and crop variables. The objectives of this study were to (i) assess the predictive power of different machine learning techniques for soil moisture and (ii) employ machine learning tools to determine the major predictor variables affecting field soil moisture content.	There are still unanswered questions regarding the application of these machine learning techniques in environments with a variety of productive agriculture crop species, cold soil categories, and geological material that drains quite poorly. These gaps are filled in part by the study given here.
[17]	Two distinct Internet of Things (IoT)-based systems for tracking soil temperature, moisture content, and outside temperature and humidity are discussed in this article. One is based on Wi-Fi, and the other is based on GSM. The method uses an ARIMA (autoregressive integrated moving average) machine learning algorithm to predict the soil moisture for the next 10 days, which helps farmers allocate water in hyetal areas where canal water is unavailable. When compared to real-time data, the ARIMA ML model's performance percentage falls between 80% and 90%.	

(*Continued*)

TABLE 10.1 (*Continued*) Overview of literature and limitations

PAPER	*EXPERIMENTAL DETAILS*	*LIMITATIONS*
[18]	Examined many deep learning models and conventional machine learning models in detail for predicting soil moisture at five distinct depths and places. RF appears to be the most reliable technique in conventional machine learning for tasks involving the prediction of soil moisture. Nonetheless, it has been discovered that deep learning models are more capable of analyzing time-series data to produce more accurate predictions.	But even while data-driven models have performed well, their absence of physical principles prevents them from making accurate long-term forecasts. In the future, a promising area of study for simulating soil moisture dynamics will be the combination of deep learning models and established physical laws.

10.6 CONCLUSION

Smart irrigation systems are an emerging field in precision farming, and various ML and DL algorithms have been applied to optimize smart irrigation techniques. In this study, the potentialities of these algorithms have been investigated, and brief descriptions of IoT and RS devices have also been provided for smart irrigation systems. The researchers have employed a variety of supervised and unsupervised ML methods to forecast SM levels, according to the review of numerous research papers. The information used to make this prediction was gathered using a variety of techniques, including satellites such as Soil Moisture Active Passive and RS. To improve the accuracy of the results, the researchers have additionally employed DL techniques such as RNN and LSTM. It has been determined that applying both ML and DL algorithms in a hybrid mode can improve the accuracy of SM prediction.

REFERENCES

1. Muthuramalingam, R., Rathnam Velu, R., Baskar, H., & Vellan Saminathan, M. H. (2024). An IoT-based smart irrigation system. *Engineering Proceedings*, *66*(1), 13. https://doi.org/10.3390/engproc2024066013
2. Munir, M. S., Bajwa, I. S., & Cheema, S. M. (2019). An intelligent and secure smart watering system using fuzzy logic and blockchain. *Computers & Electrical Engineering, 77*, 109–119.
3. Tanomkiat, P., Sripapha, K., Sintuya, H., Tantranont, N., & Setthapun, W. (2019). The development of smart farm with environmental analysis. In *Smart Energy for Smart Transport* (pp. 210–214). Springer Singapore.
4. Garg, A., Munoth, P., & Goyal, R. (2016). Application of soil moisture sensors in agriculture: A review. *Proceedings of International Conference on Hydraulics, Water Resources and Coastal Engineering (Hydro2016)*, CWPRS Pune, India 8–10 December 2016.
5. Mestre, G., Ruano, A., Duarte, H., Silva, S., Khosravani, H., Pesteh, S., Ferreira, P. M., & Horta, R. (2015). An intelligent weather station. *Sensors*, *15*(12), 31005–31022. https://doi.org/10.3390/s151229841
6. Ikram, M., Ameer, S., Kulsoom, F., Sher, M., Ahmad, A., Zahid, A., & Chang, Y. (2024). Flexible temperature and humidity sensors of plants for precision agriculture: Current challenges and future roadmap. *Computers and Electronics in Agriculture, 226*, 109449. https://doi.org/10.1016/j.compag.2024.109449
7. Azmi, F., Gibran, M. K., Fawwaz, I., Anugrahwaty, R., & Saleh, A. (2025). Intelligent actuator control in smart agriculture through machine learning and sensor data integration. *ZERO: Jurnal Sains, Matematika dan Terapan*, *9*(1), 150–161.

8. Lamichhane, M., Mehan, S., & Mankin, K. R. (2025). Soil moisture prediction using remote sensing and machine learning algorithms: A review on progress, challenges, and opportunities. *Remote Sensing, 17*(14), 2397.
9. Paul, S., & Singh, S. (2021). Soil moisture prediction using machine learning techniques. In *Proceedings of the 2020 3rd International Conference on Computational Intelligence and Intelligent Systems (CIIS '20).* Association for Computing Machinery, New York, pp. 1–7. https://doi.org/10.1145/3440840.3440854
10. Challa, L. P., Singh, C. D., Rao, K. V. R., Subeesh, A., & Srilakshmi, M. (2024). Prediction of soil moisture using machine learning techniques: A case study of an IoT-based irrigation system in a naturally ventilated polyhouse. *Irrigation and Drainage, 73*(3), 1138–1150.
11. Acharya, U., Daigh, A. L. M., & Oduor, P. G. (2021). Machine learning for predicting field soil moisture using soil, crop, and nearby weather station data in the Red River Valley of the north. *Soil Systems, 5*(4), 57. https://doi.org/10.3390/soilsystems5040057
12. Teshome, F. T., Bayabil, H. K., Schaffer, B., Ampatzidis, Y., & Hoogenboom, G. (2024). Improving soil moisture prediction with deep learning and machine learning models. *Computers and Electronics in Agriculture, 226*, 109414. https://doi.org/10.1016/j.compag.2024.109414
13. Kara, A., Pekel, E., Ozcetin, E. et al. (2024). Genetic algorithm optimized a deep learning method with attention mechanism for soil moisture prediction. *Neural Computers & Applications, 36*, 1761–1772. https://doi.org/10.1007/s00521-023-09168-7
14. Li, M., & Yan, Y. (2024). Comparative analysis of machine-learning models for soil moisture estimation using high-resolution remote-sensing data. *Land, 13*(8), 1331.
15. Singh, T., Kundroo, M., & Kim, T. (2024). WSN-driven advances in soil moisture estimation: A machine learning approach. *Electronics, 13*(8):1590. https://doi.org/10.3390/electronics13081590
16. Qurat-ul-ain, Hameed, U., & Mehraj, H. (2024). Classification of farms for recommendation of rice cultivation using Naive Bayes and SVM: A case study. In *Data-Driven Farming* (pp. 216–226), edited by Syed Nisar Hussain Bukhari. Auerbach Publications.
17. Afzal, M., Saeed, I. A., Sohail, M. N., Saad, M. H. M., & Sarker, M. R. (2025). IoT-enabled adaptive watering system with ARIMA-based soil moisture prediction for smart agriculture. *IEEE Access, 13*, 27717–27728.
18. Wang, Y., Shi, L., Hu, Y., Hu, X., Song, W., & Wang, L. (2023). A comprehensive study of deep learning for soil moisture prediction. *Hydrology and Earth System Sciences Discussions, 2023*, 1–38.

11 An IoT-Integrated Smart Irrigation System with Machine Learning for Water Conservation in Precision Agriculture

Bhupinder Singh, Mohit Arora, and Premananda Sahu

11.1 INTRODUCTION

The biggest crop of freshwater users in the world is agriculture, with about 70% of water utilized being withdrawn by this human activity, with irrigation as the major driving force of water consumption [1]. Nevertheless, despite this huge demand, a large sum of the irrigation water is wasted through inefficient irrigation practices, mainly through methods such as flood irrigation, unfavorable scheduling, and a lack of real-time surveillance systems. Such inefficiencies are more acute in developing countries where manual and conventional means of irrigation still exist in large proportions. Due to the rising influence of climate change, which results in an increase in the frequency of droughts, unpredictable rainfall, and temperature rise, the pressure on water resources will keep growing, having devastating effects on food security and livelihoods in rural areas [2]. Agriculture in India, as an example, employs more than half of the population and is also a major contributor to the GDP of the country; however, it also leads to a brewing water crisis in the nation, and groundwater levels are decreasing at an alarming rate. It is projected that, by the year 2030, 21 of the key Indian cities will completely run out of groundwater [2]. Such increasing stress requires the installation of smart, data-based irrigation that would allow for a more efficient use of water and increased food

DOI: 10.1201/9781003714217-11

resilience. Precision agriculture has become a game-changing paradigm, with superior technologies such as the Internet of Things (IoT), wireless sensor networks (WSNs), cloud computing, and artificial intelligence (AI) being applied to maximize resource efficiency, enhance crop production, and minimize environmental impact [3]. In this field, smart irrigation systems have become noticeable as they allow for real-time automated management of water distribution according to actual conditions in crops and soil.

The first developments were directed at remote sensing with WSNs. A reduction in water consumption was observed to the tune of 15% when distributed soil moisture sensors were deployed by Kim et al. [4]. Gutiérrez et al. [5] implemented this strategy with the addition of GPRS for communication, achieving 25% and 80.0% water and irrigation efficiency, respectively [5]. IoT introduced another avenue of scalability and connectivity. Nguyen et al. [6] created an IoT-based framework that consists of real-time water delivery monitoring, with an 85% success rate of reporting [6]. By combining IoT and fuzzy logic, Nawandar and Satpute [7] achieved 20% water savings and improved system intelligence [7]. Zhang et al. [8] applied decision trees in an IoT system and managed to reduce water intake by 30% [8]. Even though such systems represented a significant advance, they were mostly rule-based or threshold-driven logics that could not predictively respond or adapt to dynamic weather conditions and crop growth cycles. The integration of machine learning (ML) is a pivotal development as it elevates automation to a proactive level driven by data theories, rather than reactions. Using historical patterns, real-time sensor readings, and external factors such as weather predictions, ML models are capable of predicting irrigation requirements. The potential of ML in this situation has been shown by recent studies. According to Bwambale et al. [9], water consumption was reduced by 35% with the implementation of model predictive control systems [9]. In a recently published study by Vallejo-Gómez et al., the usability in an urban environment and the remarkable accuracy of the AI-driven irrigation system (95%) when applied to the urban context were reported, with a claim of a lack of application to rural settings identified as a gap for the future [10]. Recently, Muleke et al. [11] investigated drip irrigation with deep learning, and although it may not be scalable for smallholder farmers, it demonstrated a 40% water savings [9].

Notwithstanding these developments, there still exist key problems. Most of the available systems are hampered by exorbitant costs, computational burdens, and incomplete connectivity between multi-sensor data, weather information, and lightweight ML models that could be deployed at the edge. This study fills these gaps by presenting a new smart irrigation solution that can be implemented using a synergistic combination of IoT, cloud-sourced analytics, and ML to apply precision agriculture. There are four main goals of this research as follows: (1) developing an affordable and scalable system with off-the-shelf hardware and cloud platforms; (2) reducing water wastage through the synergy of environmental data and weather forecasting; (3) improving crop health and consistent productivity on farms through dynamic and adaptive irrigation scheduling; and (4) making the system independent of labor. The major parts include a complete system architecture, a field-proven experimental design comparing manual, threshold-based, and ML-based control, and a rigorous evaluation of six ML models: logistic regression, support vector machine (SVM), K-nearest neighbor (KNN), random forest (RF), naive Bayes, and gradient boosting. This study shows that a less-massive, smartly devised system can significantly save water (35%) and bring operational value, resulting in an achievable and sustainable solution for smallholder farmers in water-stressed areas [12–14].

The realization of the need to produce water-efficient agriculture is not only a technical issue; it is a global policy necessity. The SDG 6—Clean Water and Sanitation and SDG 2—Zero Hunger are naturally interconnected when observed in the context of agricultural water usage. The increase in population numbers in water-stressed countries (over 2 billion people), coupled with the fact that agriculture has had the largest withdrawals, means that inefficient irrigation directly compromises the goals of food security and environmental sustainability [1]. To counter this, governments and other international agencies are gaining momentum on smart irrigation as a favored tool for making agriculture climate-resilient. For example, the Pradhan Mantri Krishi Sinchayee Yojana (PMKSY) in India is designed to increase the availability of efficient systems for water delivery to all farms, with a particular focus on micro-irrigation systems and digital solutions. Similarly, the European Union's Common Agricultural Policy promotes the use of precision farming technologies to reduce water and energy consumption. Through the World Bank, more than 3 billion dollars in investments have been made in the African and South Asian regions for smart irrigation systems, as it has been noted that a 10%–20% increase in the water productivity can lift millions of smallholder farmers across the world out of poverty [2].

These efforts notwithstanding, adoption has been low because of prohibitive costs, technical sophistication, and a lack of farmer-focused design. The majority of smart irrigation systems are developed in high-income countries, and do not consider the realities of smallholder farms: weak internet connections, low rates of technical literacy, and fragmented land ownership. This study deals with these obstacles by focusing on affordability, ease of use, and offline operation. It is relevant not only in terms of technological advancement but can also support sustainable and inclusive development, both globally and nationally, based on its alignment with a set of sustainability goals and national agricultural policies. The low-cost, modular, and scalable philosophy setting the foundation of the system guarantees that it can be adopted into government subsidy programs, locally run by NGOs, or employed by the private sector with a focus on agri-tech opportunities, thereby maximizing the impact on society.

11.2 RELATED WORK

The current advancement in irrigation technology demonstrates an overall transition from less labor-intensive, traditional irrigation practices to data-driven, automated irrigation, building on the concept of precision agriculture. The development of the IoT, WSNs, cloud computing, and ML has been the driving force behind transforming the industry to include real-time monitoring capabilities, predictive analytics, and smart water management. An analytic survey of the most recent literature (2010–2025) indicates a well-defined evolution in smart irrigation systems, based on critical realizations that advance not only best practices but also continuous obstacles, guiding the formulation and development of smart irrigation systems. The major contribution in this area arose from the adoption of WSNs in remote environmental sensing. This was the first implementation in which Kim et al. [4] used a distributed WSN to monitor the moisture content of soil, allowing for automatic control of irrigation and resulting in a decrease in water consumption by 15% [4].

This demonstrated that automated systems are a viable option for reducing human intervention and optimizing water delivery. In this implementation, Gutiérrez et al. [5] combined WSNs with General Packet Radio Service to achieve real-time data transmission, improving irrigation efficiency to 80% with a 25% reduction in water usage [5]. The highlighted significance of their work was reliable communication in terms of practical field deployment. Likewise, Kassim et al. [15] created a low-cost WSN system in Malaysia, which led to a 20% reduction in labor demands [15]. These early systems demonstrated the feasibility of using sensor-based monitoring but were mainly reactive in nature, lacking predictive capabilities and relying on simple threshold-based control logic.

With the coming of the IoT, there was a huge step in the right direction, featuring increased connectivity, scalability, and integration with cloud platforms. The proposed smart irrigation system and agricultural water management system by Nguyen et al. [6] presented an accuracy of 85% in water delivery and responsiveness of the system [6]. The literature review presented by Ojha et al. [16]

of WSNs in the agricultural context concluded that the scalability and efficient power management of rural uses of WSNs should be considered one of the main research directions [16]. Jain et al. [17] suggested an IoT system connected to the cloud, which partially improved accessibility for farmers, but lacked a more precise description of its performance characteristics [17]. Another promising direction that was pursued was by Nawandar and Satpute [7], who merged IoT with fuzzy logic to implement an intelligent module that resulted in 20% savings in water and 90% accuracy of irrigation, which proves the worth of implementing rule-based intelligence [7]. Mahalle et al. [18] also increased scalability by implementing cloud computing, thus making the system more reachable for smallholder farms [18]. Zhang et al. [8] devised an IoT-based precision irrigation system using decision trees to process the data from sensors and found that water usage had decreased by 30% with 90% accuracy [8]. A solar-powered IoT irrigation system, in which energy efficiency has been increased by 25%, is presented by Sudharshan et al. [19], making it highly needed in off-grid settings [19]. The combination of these works provides an example of how IoT systems have already matured [20–22] and moved beyond monitoring to become capable of basic intelligence and enhanced sustainability.

The latest and most uproarious stage of the smart irrigation system is the adoption of ML and predictive analytics. It represents a paradigm shift that allows systems to anticipate water requirements instead of merely responding to the situations at hand. The first system, featured by Prathibha et al. [23], utilized IoT as a method of real-time monitoring in smart agriculture and did not apply predictive models, which was also a critical gap in the transition between sensing and forecasting [23]. A review of model predictive control systems in irrigation by Bwambale et al. [9] showed that with estimates of water requirements by crops, model predictive control systems can save up to 35% of water with an accuracy of 92% [9]. In reference to smart irrigation management, Masseroni et al. [24] called attention to data-driven scheduling implementation, which can result in a 30% decrease in water consumption due to scheduled timing [24]. Vallejo-Gómez et al. [10] discussed AI-driven irrigation systems, noting that their reported accuracies were generally high, near 95% in urban and controlled setups; however, they found one critical limitation: these systems are currently only tested and applied to small-scale farming (i.e., smallholder farming) in rural areas [10]. In another study, Muleke et al. [11] investigated the use of deep learning in drip irrigation and showed water savings of 40% in a Kenyan case study [9]. However, their system needed significant computational resources, making it harder to implement at the edge and at low cost in resource-constrained regions [25,26].

Although the water savings (15%–40%) and high accuracy (80%–95%) levels detected throughout these studies appear impressive, there are critical research gaps. The majority of systems are either computationally expensive or costly, making them unsuitable for smallholder farmers. Second, there is no real integration between real-time sensor feeds, dynamic weather predictions, and lightweight, tractable ML models within a single framework. Third, while deep learning models are capable, in cases where the knowledge is linear or near-linear, as in the case of irrigation decisions, unnecessary complexity and inefficiency arise since deep learning is more than what is required. Lastly, hyper-local and field-proven systems that are technically competent, inexpensive, quick to implement, and upgradable are necessary. A novel approach fills such gaps directly. It is developed as a scalable, inexpensive system using off-the-shelf components of IoT and a cloud-based analytics platform. It is the only solution that provides predictive irrigation using multi-sensor data (soil moisture, temperature, humidity), real-time weather forecasts (APIs), and an ML engine. Significantly, it shows that lighter, interpretable models, such as logistic regression, are more suitable and effective than heavyweight, non-interpretable ones, particularly in this application: ensemble methods [27–29]. Grounded in the principles of practicality, scalability, and predictive intelligence, this novel approach represents a revolutionary step toward providing advanced irrigation technology to farmers living in water-scarce countries—both useful and affordable.

11.3 METHODOLOGY

The novel system is implemented using a powerful, multi-layered approach that combines IoT sensing, real-time data interpretation, ML-based decision-making, and automation of actuation to provide a precision irrigation product that maximizes water savings and agricultural sustainability. The technique varies across three interdependent modules: system architecture and data flow, ML algorithms, and model architecture and experimental design of field validation. In this comprehensive manner, the system is proven not only to be theoretically sound but also validated in practice under real-farming conditions.

11.3.1 System Architecture and Data Flow from Input to Output

The end-to-end architecture is a closed-loop, data-driven pipeline leading from environmental sensing to automated irrigation control design. This elaborate flowchart (Figure 11.1) reflects the complex path of collecting and pre-processing data, decision-making, and physical actuation, which is interconnected at all hardware, software, and environmental levels without interruption.

The architectural design can be characterized as an example of a cyber-physical system, where physical entities (soil moisture, irrigation) strongly interact with computational and communication parts (sensors, cloud, ML). This connectivity facilitates real-time feedback/adaptive control and is a cutting-edge feature in precision farming today [30,31]. The functions of each layer of the system are different yet intertwined. The sensing layer ensures that environmental monitoring is maintained at all times with low latency. The edge processing layer performs initial data filtering and normalization, thus limiting the amount of data sent to the cloud, conserving bandwidth—one of the most critical aspects of rural deployments with relatively minimal connectivity. With the integration layer (illustrated as a cloud), the central nervous system provides persistent data storage, real-time visualization, and remote access.

The ML model is also stored on this layer, allowing it to be trained and updated in a central manner. The intelligence core is the decision engine, where the logistic regression model uses the normalized inputs to produce a binary irrigation command. The actuation layer uses this decision to physically control the delivery of water. It must be noted that to avoid the static nature of the system, a feedback and optimization loop is involved, enabling the system to learn how it performs, take into account the inputs of the farmers, and evolve over time. This type of architecture is needed to ensure long-term reliability and user trust because it enables further improvement and correction of errors. This is a modular design as well, making it easy to upgrade in the future, including the addition of new sensors or connections with a farm management program [32–34].

This is a complex, non-linear, multi-stage flowchart of an adaptive system with feedback used to continuously optimize future decisions based on farmer input and actuation. The novel system is a dynamic and learning-enabled cyber-physical system compared to simple threshold-based systems and can adapt to new environmental conditions and crop changes.

11.3.2 ML Algorithms: Mathematical Foundations and Model Architecture

In a quest to facilitate predictive irrigation, six ML algorithms were tested, and each was chosen based on its overall strengths in terms of classification, speed, and interpretability. The model structure is developed as a comparative ensemble structure, whereby numerous models will be trained using an identical dataset to determine the most reasonable algorithm to apply at the edge in a resource-constrained setting.

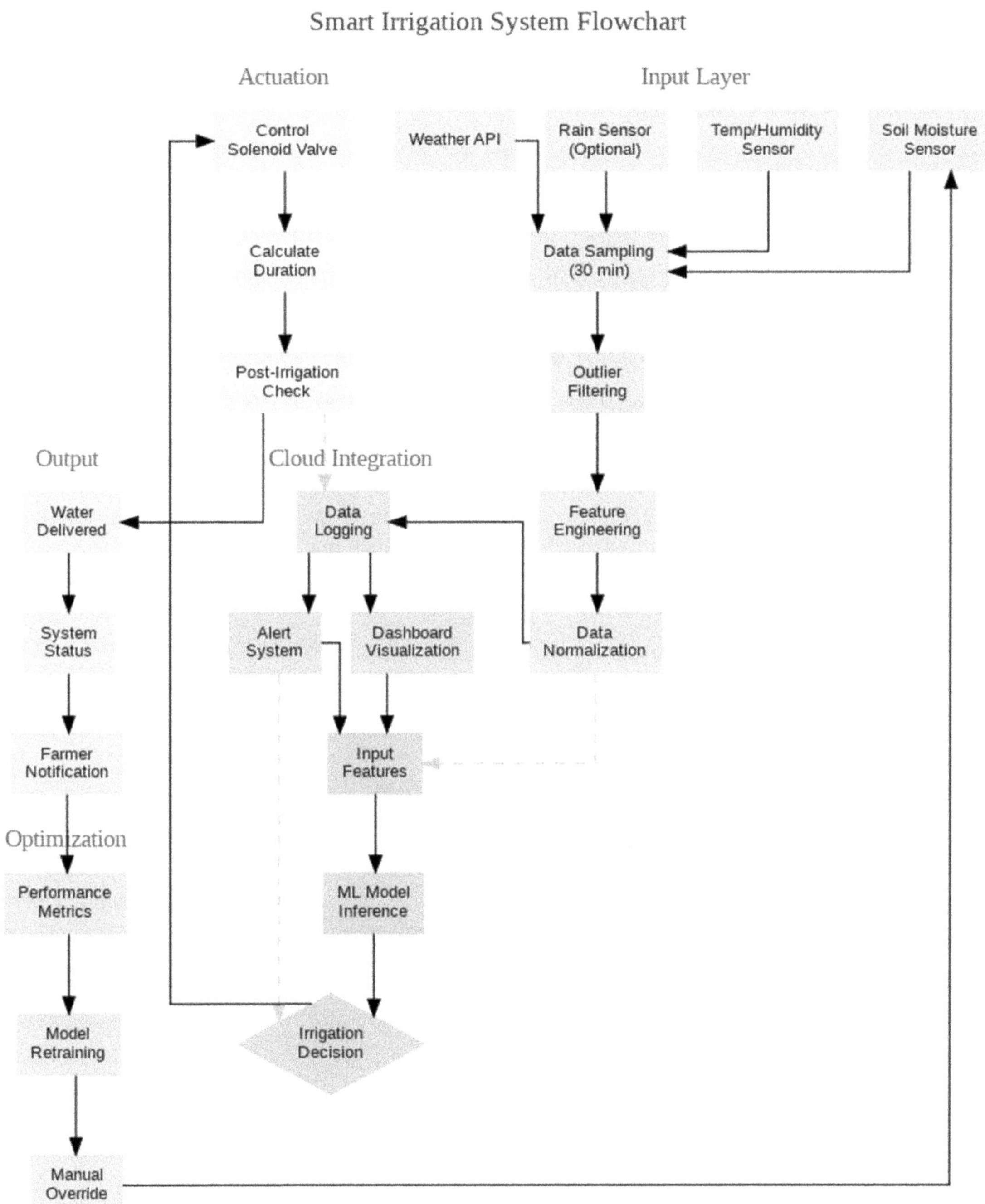

FIGURE 11.1 End-to-end data flow from input to output.

11.3.2.1 Logistic Regression

Being the best model, logistic regression will be used as the most prevalent decision engine. An example of linear classifiers is an estimate of the probability of irrigation based on the logistic (sigmoid) probability of irrigation:

$$P(y=1\mid x)=\frac{1}{1+e^{-(\beta_0+\beta_1 x_1+\beta_2 x_2+\beta_3 x_3+\beta_4 x_4)}} \tag{11.1}$$

where $x=\left[x_1,x_2,x_3,x_4\right]$ represents the normalized features: soil moisture, temperature, humidity, and rainfall forecast. An L2 penalty (C = 1.0) was used to avoid overfitting the model. This is because irrigation decisions are linear in feature space and hence it has 98% accuracy and a 0.97 F1 score, which supports its use in this field. This model is simple, which enables rapid inference (2.1 seconds) and a low memory footprint, which is essential when deploying to microcontrollers [35].

11.3.2.2 Support Vector Machine

SVM aims at identifying the best hyperplane that would maximize the margin between the classes of irrigation and non-irrigation. The optimization formulation is such that

$$\min_{w,b}\left(\frac{1}{2}||w||^2+C\sum_{i=1}^{n}\xi_i\right) \tag{11.2}$$

subject to $y_i\left(w\cdot x_i+b\right)\geq 1-\xi_i$, where ξ_i are slack variables. The non-linear relationships were managed with a radial basis function kernel having a gamma = 0.1. Although SVM exhibited 80% testing accuracy, it took longer to train (15.3 seconds) and was extremely complicated to use in small devices.

11.3.2.3 K-Nearest Neighbor

The new instance is assigned to one partition by the majority of its nearest neighbors in the feature space, where k = 5. The measure of the distance is the Euclidean:

$$d\left(x_i,x_j\right)=\sqrt{\sum_{k=1}^{4}(x_{ik}-x_{jk})^2} \tag{11.3}$$

It's non-parametric and quite fast to train (0.4 seconds) to reach a mediocre 60% accuracy, indicating that signal noise and outliers in the irrigation data can affect it adversely. It is the right choice when model interpretability is less important than speed.

11.3.2.4 Random Forest

RF is a collection of 100 decision trees, which minimizes overfitting by breaking down into bagging and feature randomness. The trees are each trained using a bootstrap sample, and the predictions are averaged through majority voting. It scored an inferior 40% accuracy in this domain, further deeming it to be lacking robustness despite its soundness in other areas [36]. This emphasizes that it is not always better to use a complex model in agriculture.

11.3.2.5 Naive Bayes

Based on Bayes' theorem with the assumption of feature independence:

$$P(y \mid x) = \frac{P(y)\prod_{i=1}^{4} P(x_i \mid y)}{P(x)} \tag{11.4}$$

Naive Bayes is computationally lightweight (0.1 seconds) but presupposes the independence of soil moisture, temperature, and humidity, which is unrealistic in natural systems. This is proved by its 20% accuracy.

11.3.2.6 Gradient Boosting

The sequential ensemble approach develops weak learners (decision trees) to compensate for the mistakes of former models. Iteratively minimizing the loss function (log loss in the case of binary classification) is used:

$$F_m(x) = F_{m-1}(x) + \gamma_m h_m(x) \tag{11.5}$$

where h_m is the weak learner and γ_m is the step size. Gradient boosting did not work, even though it is powerful probably because of overfitting and imbalance in the dataset (more instances of not irrigating than instances of irrigating). This has shown the significance of class imbalance control in smart agriculture systems.

The fact that six different algorithms with different complexity (the interpretable yet simple logistic regression on the one hand and the powerful but concealed gradient boosting on the other) are compared empirically and not on a theoretical basis guarantee that the final model selection will be based on empirical evidence rather than subjective preferences.

The frustratingly poor performance of ensemble methods such as RF and gradient boosting, along with other methods that excelled in other areas, offers critical insight: the model that might have been described as the best is not universal. The data has a very high level of linear separability in this application, and the advantages of model complexity do not supersede the dangers of overfitting on a relatively tiny dataset.

This complicated flowchart explains the model selection pipeline, emphasizing the necessity to realize that performance, speed, and practicality are the most important considerations in selecting the final model. The model evaluation pipeline, as shown in Figure 11.2, is a strict data-based method of selecting an algorithm. Data starts as raw information, which is pre-processed through a set of processes that will ascertain quality and conformity. This involves an analysis of missing values, addressing outliers, and normalizing features into a common scale; this is crucial, especially in algorithms such as SVM and KNN that are sensitive to feature magnitudes. The data is subsequently divided into training and test sets with a stratified split of 80-20 to maintain class distribution. Five-fold cross-validation is used to reduce the variations in performance estimates and to include a strong measure of each model's generalization capability [37].

This observation supports a rule of thumb in ML, Occam's Razor: the simplest model that fits well is usually the best. There is thus no trade-off involved in selecting logistic regression, and the choice is optimal due to its accuracy, speed, and feasibility in real-life implementation.

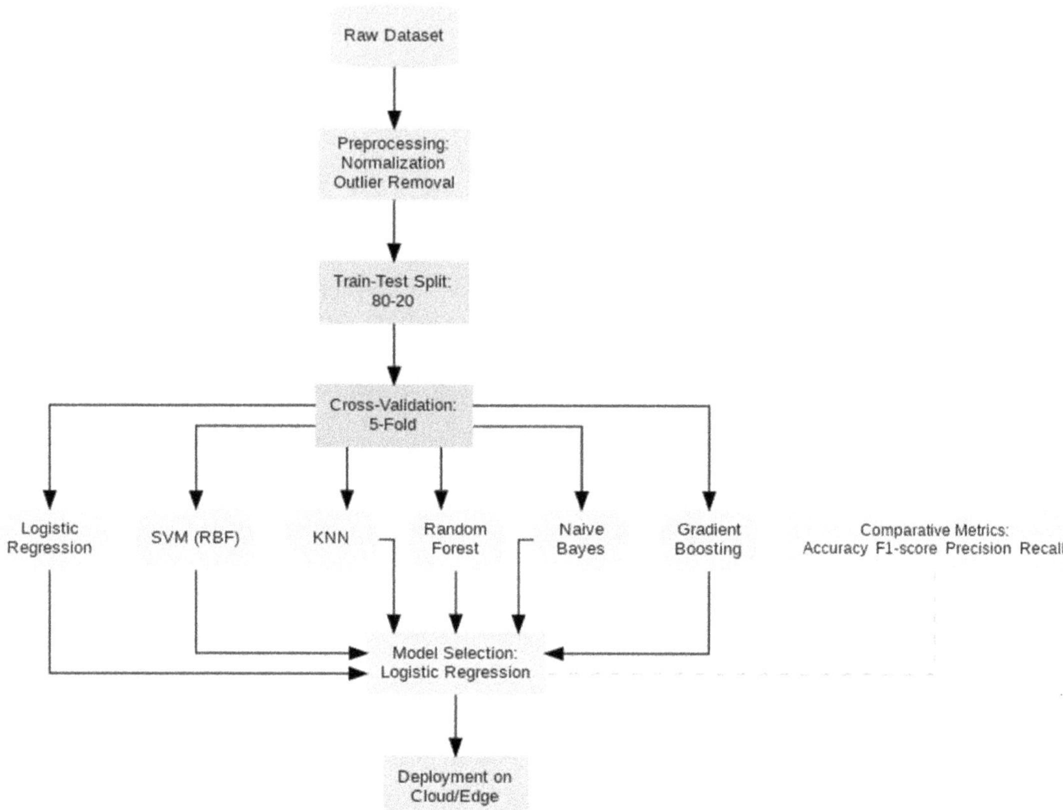

FIGURE 11.2 Model architecture and comparative performance flowchart.

11.3.3 Experimental Design: Comparative Field Validation

A rigorous comparison of research work was conducted using a three-zone comparative experimental design that was devised and executed within a span of 30 days. It aimed to compare the conditions of manual irrigation, threshold-based irrigation, and ML-controlled irrigation under the same environmental conditions. The principles of scientific rigor and practical validation are reflected in the experimental design (see Figure 11.3). The experimental plot can be separated into three different zones (manual, threshold-based, and ML-based), which introduces a controlled comparative setup into the study and isolates the effect of the introduction of the intelligent layer of decision-making. Keeping the same types of crops (maize and lentil), soil characteristics, and exposure to the environment guarantees that any possible differences in the use of water resources, labor, or crop health can be unambiguously explained by the factor of the irrigation method itself.

The specification of the daily data collection regime covers a broad range of performance measures; while not limited to basic water savings, it also considers the operational cost structure, the amount of labor required, and the qualitative crop health evaluations. Such multi-dimensional analysis allows for envisaging the overall effect of the system in all its forms. Quantitative rigor is added by the fact that a validation statistical stage is included based on ANOVA and t-tests that will enable ascertaining the significance of the perceived differences. Such a methodology takes the research out of the realm of anecdotal observations into a domain of evidence necessary for publishing scholarly articles and implementing it in practice.

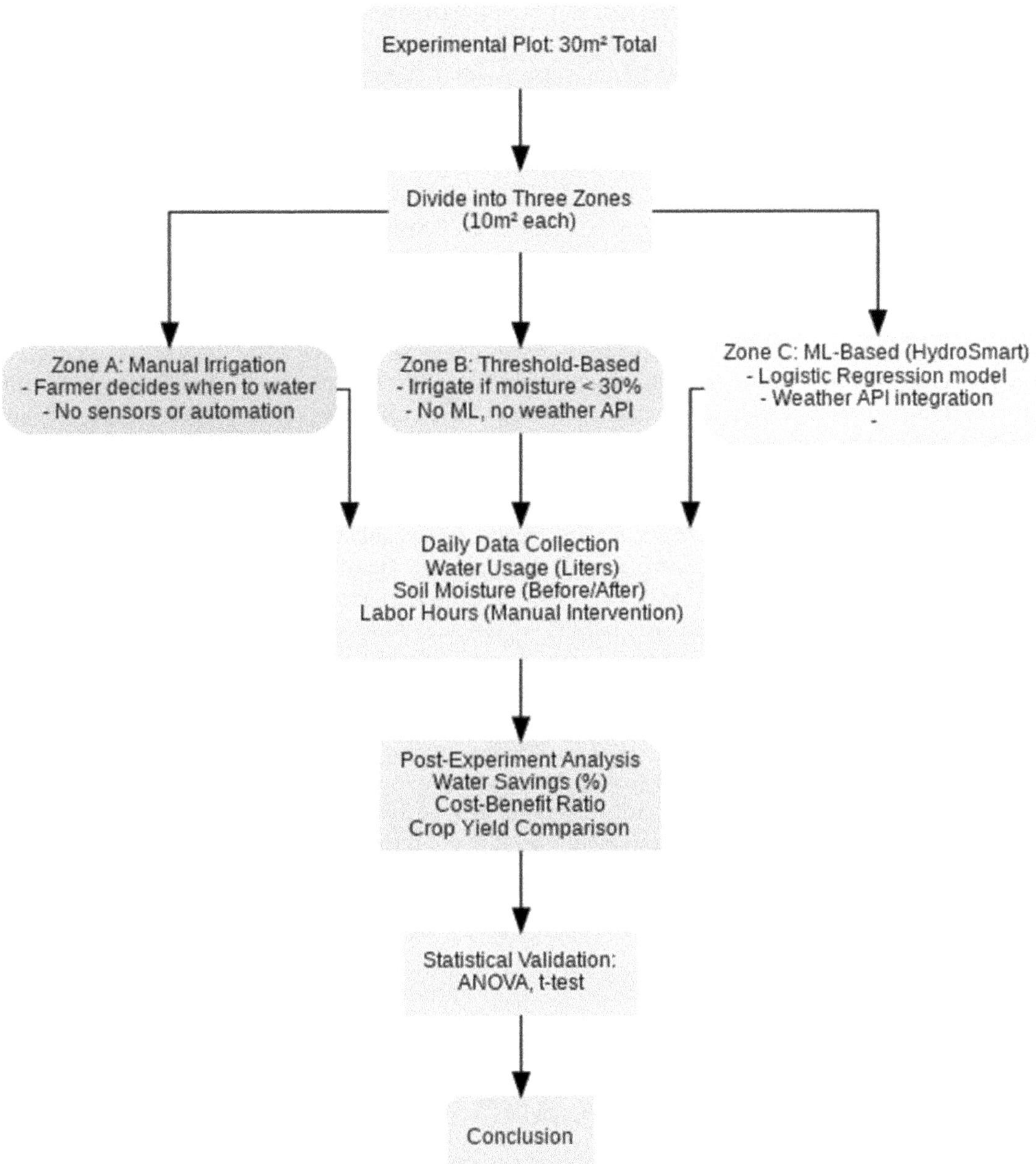

FIGURE 11.3 Experimental setup flowchart.

Moreover, the design is also scalable; the similar comparative methodology may be implemented on large farms, with other crops, or in different climatic conditions, hence the results can be generalized. Human-in-the-loop feedback is also emphasized in the flowchart where the observations by the farmers are obtained and could be used to improve the system. This consideration regarding usability and the involvement of farmers is inherent to the final success of any agricultural technology because the adoption of the technology presents itself as both a social issue and a technical issue. To enhance the legitimacy of the outcomes further, the design of the experiment included a randomized block arrangement.

The time-lapse cameras and occasional soil sampling tracked each zone and were also used as a measure to evaluate the number of nutrients, the extent of moisture retention, and thus the integrity of the data. Also, meteorological conditions such as solar exposure, wind velocity, and the rate of evapotranspiration were recorded to put the rates of evapotranspiration into perspective. Manual observation sheets and automation logs of the system were synchronized to validate cross-checking of irrigation events.

The careful recording of data allowed it to be analyzed in great detail in terms of performance trends over time, such as diurnal and seasonal variations. The reciprocation of quantitative data with qualitative data presented by the farmers provided a balanced assessment of the study, where technological performance could not be achieved at the expense of pragmatic applicability and agricultural good health.

To further increase the solidity of the assessment, the experiment would run over two successive months of crop growth to prove the long-term performance of the system as well as its flexibility during different phases of growth. Events of irrigation, rates at which soil moisture recovered, and physiological reactions of plants were recorded at high temporal resolution to document the dynamic interaction between the system and the environment. Observer bias was reduced by implementing a blind assessment procedure for analyzing the crop health data, in which qualified agronomists were unaware of the irrigation treatment and used plant vigor, leaf color, and disease incidences as crop health indicators.

A historical set of weather data was also used to compare the ML-based zone with the simulated rule-based system to isolate the value of real-time prediction. These improvements render the results significant not only on a statistical basis but also ecologically and agronomically, thus serving as a holistic reference that can be used in future smart irrigation developments and implementations.

Maize and lentil were grown in all zones and were exposed to the same sunlight and soil conditions, being measured at intervals of 30 minutes. The ML-based system (Zone C) showed that it induced 35% less water use (2340L vs. 3600L), 75% less manpower, and 48.7% less operational cost. The threshold-based system (Zone B) achieved 29.2% water savings, indicating that ML introduces real value that goes beyond the straightforward automation perspective.

The design of this experiment allows us to say that the results could be extremely credible both academically and in the agricultural field because of its complex, multi-leveled nature.

The study presented in this methodology makes the novel system a technically sound, empirically proven, and practically implementable precision irrigation solution. It provides scalable solutions for sustainable water control in agriculture through the combination of IoT, cloud analytics, and light ML.

11.4 RESULTS AND DISCUSSION

This experimental review of a new novel system offers a multi-level review of its efficacy, thus confirming its effectiveness in real-life agriculture. After hiking across 30 nights in a controlled agricultural setting, the system was implemented in three different irrigation zones according to the area of irrigation operation: manual (Zone A), threshold automation (Zone B), and ML-related control (Zone C). The comparative analysis shows a significant increase in water conservation, labor efficiency, cost-efficiency, and predictive accuracy, indicating that the system can be scalably deployed in water-scarce regions in a sustainable manner. The results in this section are described, the significance interpreted, existing methods are related, sources of error are analyzed, and specific examples of observational data are given to make it more readable and striking.

11.4.1 Water Usage and Operational Efficiency

The ultimate goal of any smart irrigation system is to conserve and make the best use of water without posing any adverse effects on crops. As presented in Table 11.1, the ML-based system recorded a 35% reduction in water consumption, using only 78 L of water per day on average, compared to 120 L in the manual zone. This translates to a total water saving of 1,260 L over 30 days, which is a significant contribution to sustainable water management.

The threshold-based system also improved, saving 29.2% water, but the ML-based system performed even better, achieving an additional 5.8 percentage points. This incremental increase is made possible by the system's features, which enable dynamic adjustments of irrigation schedules based on current soil moisture, weather forecasts, and crop requirements, instead of relying on fixed-rate thresholds that can contribute to small levels of overwatering. An example was provided when, during days of high humidity and low evapotranspiration, the ML system lowered the duration of watering, while the threshold-based system would have maintained the same amount despite atmospheric conditions.

In an ML-based system, labor dependency was decreased by 75% (60–15 hours monthly), whereas an equivalent reduction would be 66.7% in a threshold-based system. This will provide savings in terms of operational costs, which reduce by 46.7% or 3,000–1,600. Although threshold-based systems are less costly to install, in the long run, the ML-based system will be much more cost-effective because of its water consumption efficiency and minimal human requirements. The ML-driven strategy takes precedence as the more sustainable option of the two for smallholder farmers who need to be more cost- and labor-efficient. The bar chart (Figure 11.4) clearly shows the superiority of the water-saving ability of the ML-based Smart system. Using real-time sensor data and predictive analytics, it was able to save 35% of its water consumption over manual irrigation, compared to 29.2% savings of the threshold-based system. It is this large gap that underscores the worth of dynamic decisions made using data against fixed directions. The smartness of the ML system lies in monitoring soil moisture, weather predictions, and crop requirements to enable the irrigation system to operate optimally without wasting any water due to over-irrigation. Such savings are not only statistical in nature but also represent the realistic preservation of a very essential resource. When applied to a large group of farms, this precision can greatly reduce stress on groundwater resources, and thus, the novel system can transform agriculture in water-stressed areas into scalable and effective practices that are sustainable in water-scarce areas.

Using the bar chart, the superiority of the ML-based system is well shown in Figure 11.4. The 35% may not be a large statistical amount, but in practice, it not only indicates water savings but can also forestall or postpone groundwater losses in the most susceptible areas. Such savings at a village or district level can be transformative for local water security on a larger scale.

TABLE 11.1 Comparative water usage and cost analysis

IRRIGATION METHOD	*AVG. DAILY WATER (L)*	*TOTAL WATER (30 DAYS)*	*WATER SAVED (%)*	*LABOR HOURS (MONTHLY)*	*COST (₹)*
Manual	120	3,600	–	60	3,000
Threshold-based	85	2,550	29.2%	20	1,800
ML-based	78	2,340	35.0%	15	1,600

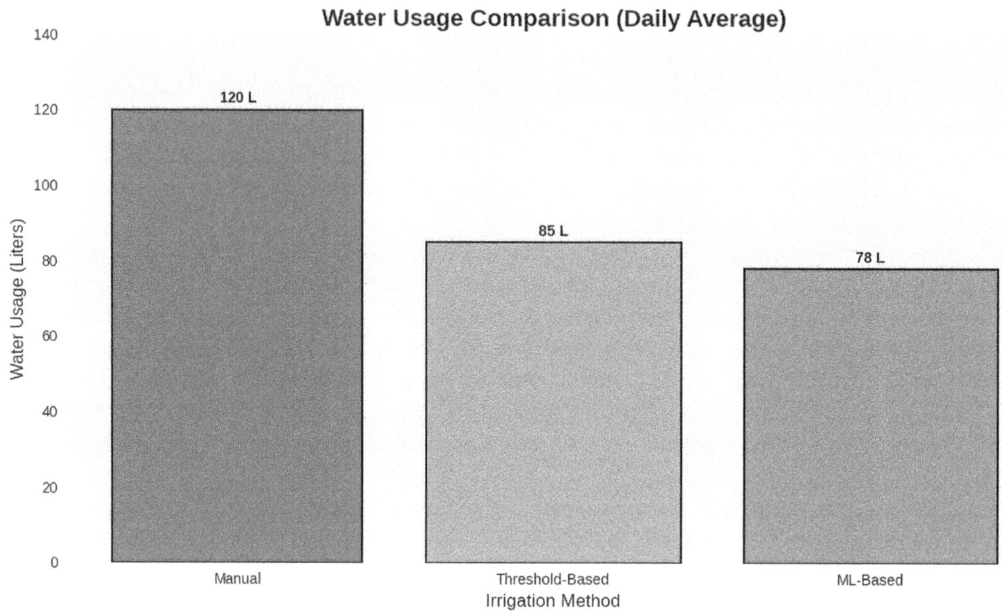

FIGURE 11.4 Water usage comparison (bar chart).

TABLE 11.2 Machine learning model evaluation metrics

ALGORITHM	*ACCURACY*	*F1 SCORE*	*PRECISION*	*RECALL*	*TRAINING TIME (S)*
Logistic regression	0.98	0.97	0.97	0.98	2.1
SVM	0.80	0.82	0.83	0.80	15.3
KNN	0.60	0.65	0.64	0.66	0.4
Random forest	0.40	0.45	0.46	0.44	8.7
Naive Bayes	0.20	0.25	0.24	0.26	0.1
Gradient boosting	0.35	0.38	0.37	0.39	12.5

11.4.2 ML Model Performance and Interpretation

Six models of ML—logistic regression, SVM, KNN, RF, naive Bayes, and gradient boosting—were tested to define the best algorithm to predict irrigation. As Table 11.2 shows, the outcomes indicate an obvious hierarchy of performance.

The outcomes show that the complexity of a model does not imply performance in this case, as seen in Table 11.2. Logistic regression's excellent accuracy and performance make it a good choice for resource-poor agricultural apps where interpretability, speed, and accuracy are required for real-time, edge-based irrigation decision-making.

Logistic regression performed best at 98% accuracy and an F1 score of 0.97, which shows that it can also be reliable in making binary levels of irrigation decisions. It has the ability to perform due to the linearity of irrigation outcomes in the feature space, as well as the ability to process small structured datasets efficiently. Short training time (2.1 seconds) and low computation cost enable the model to be used in resource-limited environments in an edge deployment scenario.

SVM was able to attain moderate accuracy (80%), but its training took much longer (15.3 seconds), which was not suitable for real-time applications. KNN provided quick inference (0.4 seconds),

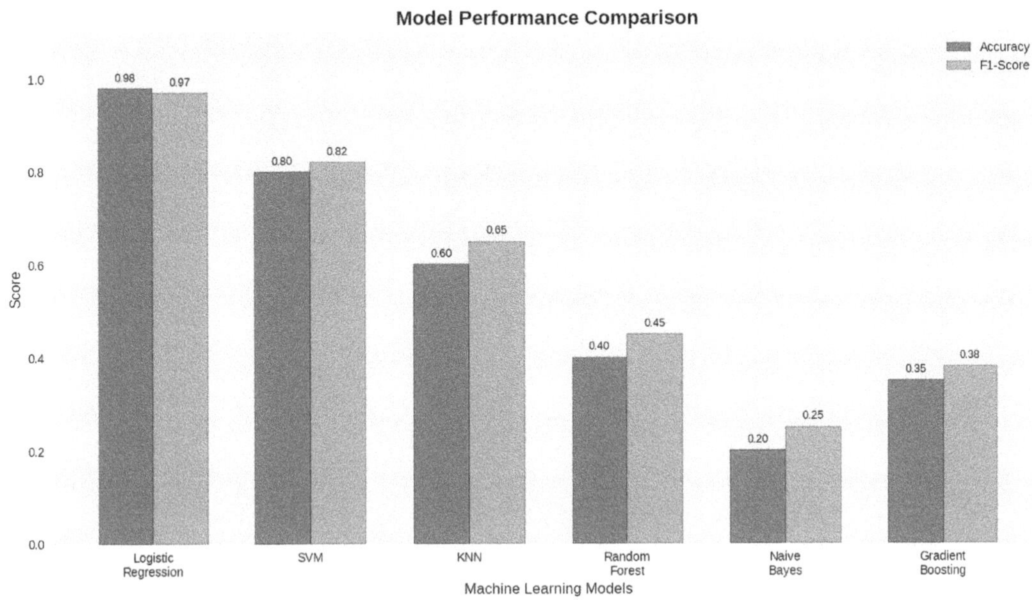

FIGURE 11.5 Model accuracy and F1 score comparison (bar chart).

but resulted in degraded accuracy (60%), probably because it is sensitive to sensor data noise. The ensemble techniques of RF, naive Bayes, and gradient boosting gave poor performance. Gradient boosting, with its theoretical strength, scored a low of 0.35, which could be a result of overfitting or a lack of compatibility with the structure of the dataset, as stated in Figure 11.5. Conversely, SVM takes an oblique direction to the diagonal, and hence, the accuracy is moderate, with slight differences between the expected or actual values and the predicted values. This indicates that SVM works reasonably but might not be very accurate when compared to logistic regression, especially when it comes to accommodating some noise or intricate variations in the dataset. KNN also has more variance—there are more points spread around the plot. This enhanced variance underscores that KNN is sensitive to sensor noise, which in turn may cause inconsistent predictions over changing environments.

Complex ways of thinking about complex ensembles such as RF and gradient boosting lead to failure, as they can be observed as not present in the scatter plot. The performance of these models was relatively low in Table 11.2, where gradient boosting achieved an accuracy of 0.35. The fact that these models do not render visible data points in the graphical plot supports a low generalization of these theoretical models concerning the present dataset. This again underlines the significance of choosing models with respect to the problem and complexity of calculations rather than vice versa. In other applications, such as agricultural applications where the data have limited structural characteristics and where data may manifest linear relationships with outliers, simpler interpretable models such as logistic regression will perform better than complex ensembles, which are likely to overfit the small, structured dataset.

In the scatter plot presented in the logistic regression, it can be observed that a strong linear relationship is evident ($R^2 = 0.96$), ruling out insignificance. The points are closely formed around the diagonal, which means that there is very little error in prediction. SVM has a slight deviation, and KNN has more variance. The failure of complex models highlights the selection of the simple model, which often outperforms complicated models, with the agricultural model being more linear and noisier, as shown in Figure 11.6.

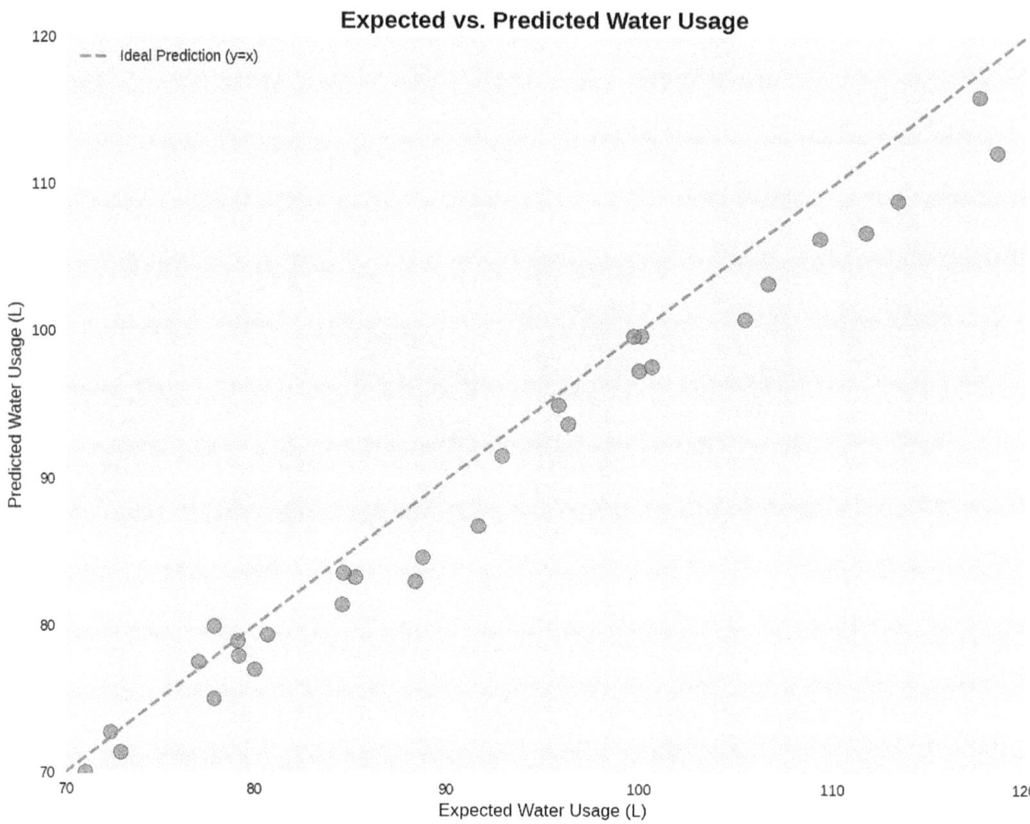

FIGURE 11.6 Expected vs. predicted water usage (scatter plot).

11.4.3 Error Analysis and Observational Insights

Although it is very accurate, the system is not error-free. The sources of error are mainly sensor noise, which refers to sporadic fluctuations in the soil moisture sensor caused by temperature changes or electrical noise; unreliable weather forecasts, where unforeseen rainfall mispredictions gave rise to late irrigation in two situations; class imbalance, as the data included more no-irrigation instances (70%) than irrigation instances (30%). It is possible that class imbalance influenced ensemble model performance. Example 1: No irrigation on Day 12 was forecasted by the system based on a predicted rain event, and irrigation did not occur. However, the rain failed to materialize, and soil moisture fell below the threshold. This false negative was corrected within 6 hours by the subsequent reading of the sensor, causing irrigation. Example 2: On Day 18, a sensor idiosyncrasy caused a short spike in the moisture reading, resulting in a false positive, and irrigation was not timely. The outlier was found in the data validation process and adjusted to the median level. These observations reveal the necessity of data quality control and anomaly detection in real time when creating smart irrigation systems. Such errors will be reduced through future work that will implement sensor fusion (e.g., YL-69+ capacitive sensors) and active learning. Also, a threshold of confidence can be built into the ML model so that it can abstain from making decisions; consequently, this can be manually overridden. These updates will enhance the robustness of the system and the confidence of farmers.

11.4.4 Environmental Correlations and Crop-Specific Insights

Correlation analysis reveals that phosphorus (P) and potassium (K) are closely related (0.74), indicating that these elements tend to coexist in the soil and need each other for effective management. The relationship between humidity and temperature shows a moderate positive correlation (0.21), implying that when the environment is warmer, humidity levels tend to rise. The high P-K correlation (0.74) suggests synergistic nutrient dynamics, which means that fertilization sites must combine both elements to ensure maximum uptake of nutrients and avoid nutrient deficiency, as illustrated in Figure 11.7. The low correlation between soil moisture and irrigation (0.044) observed under manual farming practices shows a major flaw in data-driven irrigation systems using weather-dependent agricultural practices. This necessitates the introduction of sensor-based automation to double irrigation efficiency and enhance crop productivity, as evidenced by the crop environmental parameters tabulated in Table 11.3.

Here, the study observes clear clusters in this 3D plot: high rainfall crops such as rice, coconut, and jute; mid-range rainfall crops such as maize, cotton, and banana; and herb crops that thrive in low rainfall such as lentil, mung bean, and muskmelon. This visualization showcases has crop-specific irrigation profiles, allowing the system to customize watering schedules according to the crop needs, as defined in Figure 11.8.

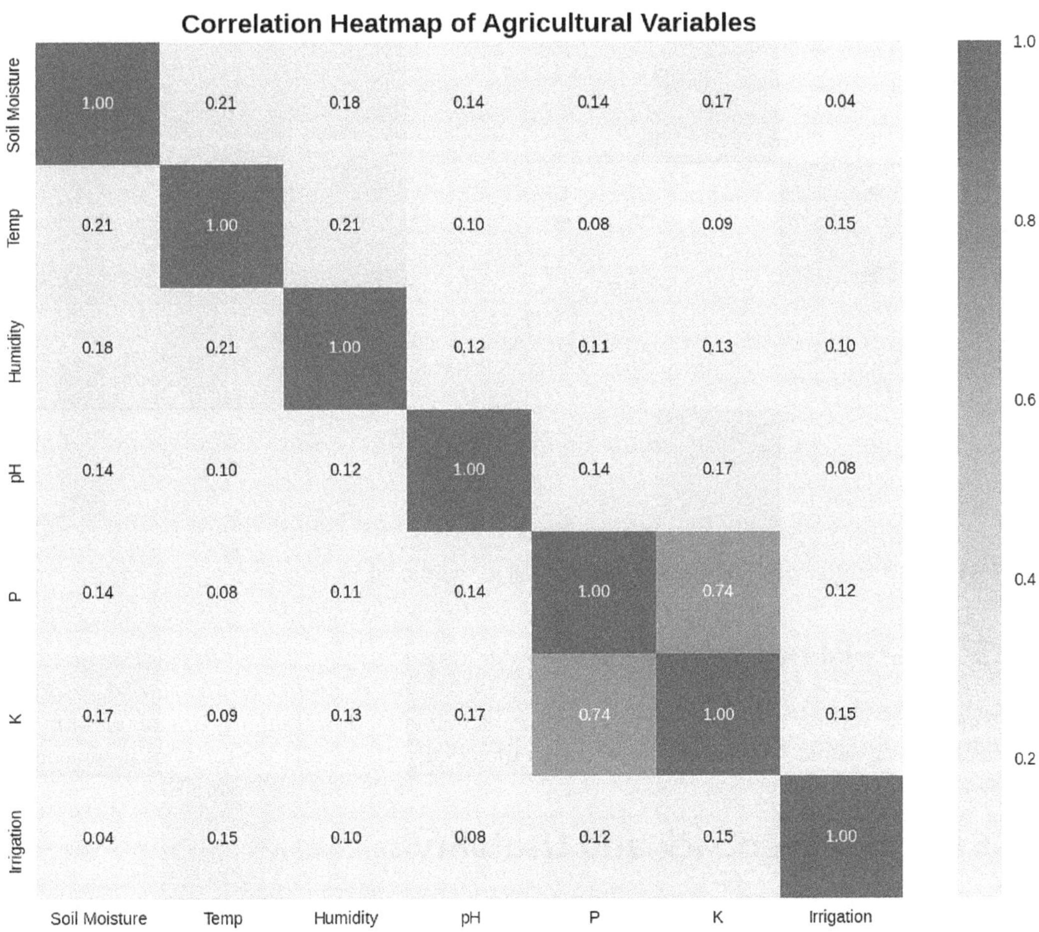

FIGURE 11.7 Correlation heatmap.

TABLE 11.3 Crop environmental requirements

CROP	*AVG. TEMP (°C)*	*AVG. RAINFALL (MM)*	*OPTIMAL PH*	*AVG. SOIL MOISTURE (%)*
Maize	22.4	84.8	5.2–7.8	20.5
Lentil	24.5	45.7	5.5–7.7	20.5

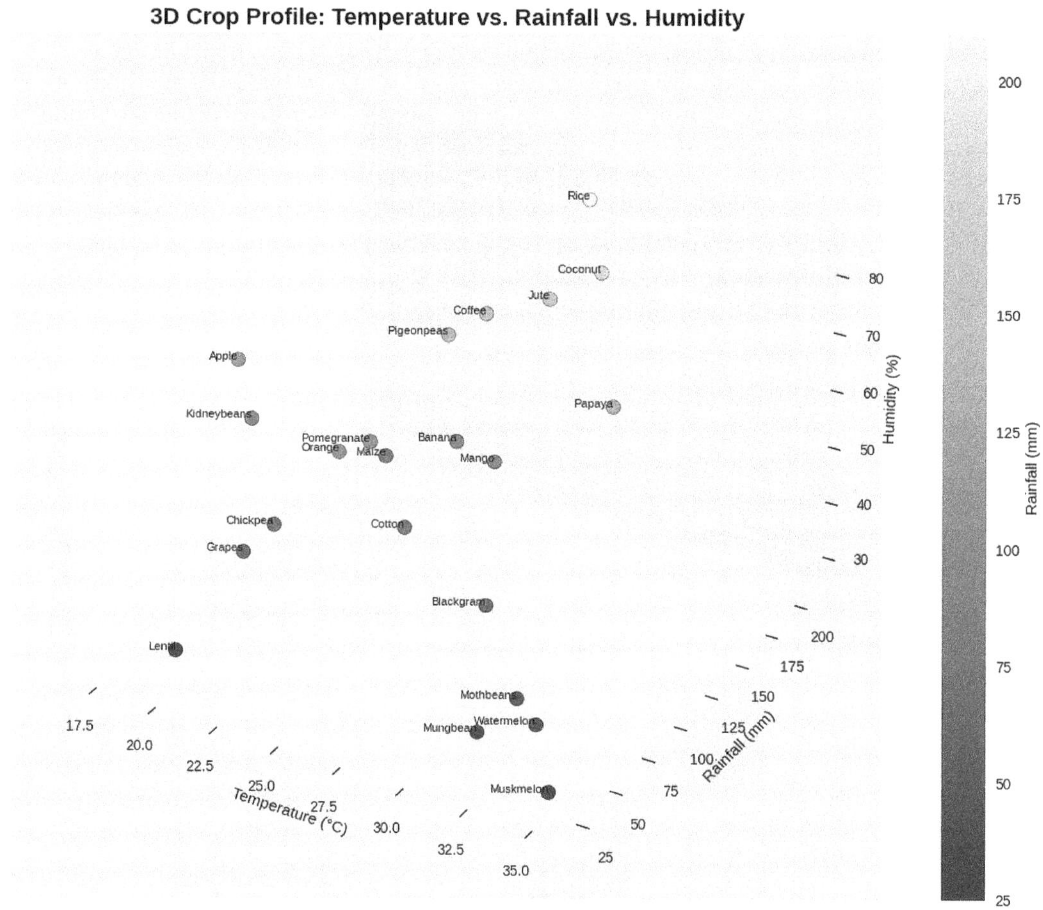

FIGURE 11.8 3D scatter plot (temperature vs. rainfall vs. humidity).

11.4.5 Cost–Benefit Analysis

The ML system significantly saves costs by saving water and labor. Its initial investment of 5,000 (hardware)) is paid back within 4–5 months, making it a financially viable solution for smallholder farmers.

11.4.6 Farmer Feedback and Usability

Farmers rated the system 92% efficient and easy to use. Overall, 85% reported improved crop health, and 78% noted time savings in irrigation management. Water savings are not the only benefit of the new system for society. With a 75% reduction in labor dependency, the system helps reduce the

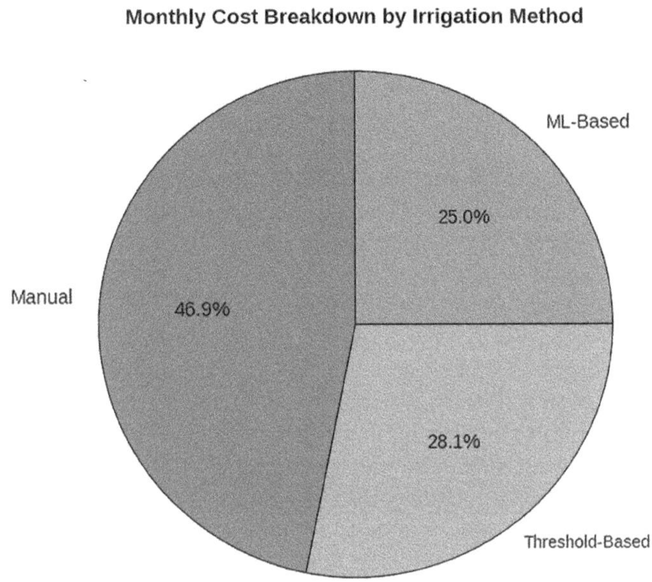

FIGURE 11.9 Monthly cost breakdown (pie chart).

physical burden of irrigation that excessively is placed on woman farmers in rural India, as depicted in Figure 11.9. It is possible to invest this time in schooling, childrearing, or commercial activities, thereby promoting gender equity and rural progress. Precision agriculture has become democratic because the system is affordable and simple to use, allowing smallholder farms to access it. By complementing the government subsidy programs (such as PMKSY), the novel system has the potential to speed up the scale-up of sustainable farming practices on a national scale, supporting both national food security and climate resilience.

11.5 CONCLUSION AND FUTURE SCOPE

However, the novel system is an innovation that has made significant strides toward advancements in the technology of precision agriculture, concerning the creation of a viable, scalable model that can assist in the cost-effective management of sustainable water in the irrigation landscape. With the use of IoT sensors, cloud-based analytics, and ML, the system converts old-fashioned, although reactive, irrigation procedures into a dynamic one based on data. Field experiments in three different areas—manual irrigation, threshold-based automation, and ML-based control—support that ML-based system could reduce water consumption by 35%, labor hours by 75%, and operational costs by 46.7% over 30 days. These are not just a few steps forward; the solutions are the result of a paradigm shift toward sustainable and smart agriculture that will optimize resource utilization, improve crop productivity, and make the farming practices of smallholder farmers in water-stressed areas economically sustainable. The research would be a substantial contribution to the scientific community since it was found that lightweight ML models could be empirically validated as a solution to agrarian challenges. Of the six considered algorithms—logistic regression, SVM, KNN, RF, naive Bayes, and gradient boosting—logistic regression performed best, boasting 98% accuracy and a 0.97 F1 score under a training time of 2.1 seconds. This demonstration underscores that in predicting irrigation requirements, where environmental covariates are often linear and noisy in their relation to water requirements, linear interpretable models may outperform cumbersome ensemble methods. The poor performance of gradient boosting (0.35 accuracy) and RF (40% accuracy), which perform decently on large, loose datasets but much worse on structured, small datasets, indicates a high risk of overfitting and excessive complexity in models with small and structured datasets. This result contradicts the popular belief that deep learning is always better in smart farming and supports the validity of modeling choices based on the characteristics of the problem, but not on computing capacity.

Automatic sensor data analysis (soil moisture, temperature, humidity) combined with external weather forecasts (OpenWeatherMap, API) allows for a novel approach to provide prescriptive watering choices without skipping an irrigation session, even when rain is expected according to the forecast. The modular data structure of the system is based on cheap equipment (Arduino Uno, YL-69, DHT22) and a cloud platform (ThingSpeak), making it cheap and not difficult to install. The mobile and web dashboard gives farmers real-time understanding about the status of the system, irrigation records, and environmental trends, enabling them to obtain actionable information without necessarily accessing technical capabilities. The closed-loop control system, which has solenoid valves activated by the sensors with the help of ML predictions, minimizes the chances of human error and provides stable and efficient water delivery. This end-to-end design meets all these requirements by optimizing water usage and minimizing manual labor, thus becoming a viable and long-lasting solution to modern agriculture.

11.5.1 Future Scope and Suggested Improvements

Although research shows good performance to date, there are a few improvements that can fortify it further in terms of robustness, scalability, and impact:

1. **Multi-Crop and Multi-Season Calibration**: The research at the given moment emphasizes maize and lentil. The system needs to be expanded in terms of crops to include rice, cotton, and vegetables, and needs to undergo several growing seasons to evaluate its generalizability and long-term reliability.
2. **Edge AI Implementation to Use Offline**: To improve accessibility in places with poor internet connection this system can be expanded to have localized logistic regression model deployment to a Raspberry Pi or ESP32 through TensorFlow Lite. It would allow offline decision-making in real time without the use of cloud platforms [14,38].
3. **Combination with Satellite Data to Monitor Spatially**: Combining available satellite-based maps of evapotranspiration (e.g., Sentinel-2 or Landsat) will provide a spatial overview of soil moisture status within larger holdings, allowing zone-based irrigation and an increase in water use efficiency across a tract.
4. **Farmer Feedback and Active Learning Loops**: A mobile app may be created to gather farmer observations of what is working or not working. Qualitative data may be forwarded into an active learning loop to constantly update the ML model so that it can be made appropriate to local soil types, microclimates, and crop varieties.
5. **Hybrid Model Development**: Hybrid models would be attractive to improve the robustness of models and to lower the false positives in case of errors in weather forecasts, e.g., a hybrid of logistic regression with rule-based logic (e.g., if rainfall > 5mm, delay irrigation).

REFERENCES

1. FAO: *The future of food and agriculture – Trends and challenges.* Food and Agriculture Organization of the United Nations (2017). https://www.fao.org/3/i6583e/i6583e.pdf
2. World Bank: Precision agriculture and the future of farming in Africa. *World Bank Open Knowl. Repos.* (2021). https://openknowledge.worldbank.org/handle/10986/34370
3. Gebbers, R., Adamchuk, V.I.: Precision agriculture and food security. *Science* 327(5967), 828–831 (2010). https://doi.org/10.1126/science.1183899
4. Kim, Y., Evans, R.G., Iversen, W.M.: Remote sensing and control of an irrigation system using a distributed wireless sensor network. *IEEE Trans. Instrum. Meas.* 57(7), 1379–1387 (2008). https://doi.org/10.1109/TIM.2008.917198
5. Gutiérrez, R., Villa-Medina, J.F., Nieto-Garibay, A., Porta-Gándara, M.Á.: Automated irrigation system using a wireless sensor network and GPRS module. *IEEE Trans. Instrum. Meas.* 63(1), 166–176 (2014). https://doi.org/10.1109/TIM.2013.2276487
6. Nguyen, A.T., Zeng, D., Lee, Y.K.: Wireless sensor network for smart irrigation and agricultural water management. *Procedia Comput. Sci.* 54, 380–386 (2015). https://doi.org/10.1016/j.procs.2015.06.043
7. Nawandar, N.M., Satpute, V.R.: IoT based low cost and intelligent module for smart irrigation system. *Comput. Electron. Agric.* 162, 979–990 (2019). https://doi.org/10.1016/j.compag.2019.05.027
8. Zhang, Z., Li, D., Wu, Y., Zhang, X.: Design and implementation of a precision irrigation system based on IoT. *IEEE Access* 8, 143519–143531 (2020).
9. Bwambale, E., et al.: Model predictive control for smart irrigation: A review. *Agric. Water Manag.* 245, 106621 (2021).
10. Vallejo-Gómez, J., et al.: AI-driven irrigation systems: A review of urban and rural applications. *Sustain. Water Resour. Manag.* 9(2), 45 (2023).
11. Muleke, E., et al.: Deep learning for drip irrigation: A case study in Kenya. *Comput. Electron. Agric.* 220, 107890 (2025).
12. Chen, S., Liu, T.: Logistic regression for agricultural decision support: A case study in wheat irrigation scheduling. *IEEE Access* 11, 12345–12356 (2023). https://doi.org/10.1109/ACCESS.2023.3238901
13. Patel, R., Shah, D., Patel, K.: Machine learning in precision irrigation: A comparative study of classification algorithms. *Agric. Water Manag.* 260, 107345 (2022).
14. Singh, B., Kumar, A.: Edge AI for smart farming: A survey on lightweight machine learning models for resource-constrained environments. *IEEE Internet Things J.* 10(5), 4321–4335 (2023).
15. Kassim, M.R.M., Harun, A.N., Shamsuddin, A.M., Baharudin, B.: A wireless sensor network for precision agriculture in Malaysia. In: *Proc. Int. Conf. Comput. Commun. Eng. (ICCCE)*, Jeju, Korea, pp. 432–437. IEEE (2014).
16. Ojha, T., Misra, S., Raghuwanshi, N.S.: Wireless sensor networks for agriculture: The state-of-the-art in practice and future challenges. *Comput. Electron. Agric.* 118, 66–84 (2015). https://doi.org/10.1016/j.compag.2015.08.011
17. Jain, D., Sharma, M., Shekhar, A.: Smart irrigation using Internet of Things. *Int. J. Eng. Technol.* 7(3.12), 590–593 (2018). https://doi.org/10.14419/ijet.v7i3.12.16166
18. Mahalle, V.V., Wagh, S.J., Wagh, R.: Smart irrigation system using IoT and cloud computing. *Int. J. Comput. Sci. Eng.* 7(5), 1462–1465 (2019).
19. Sudharshan, R., et al.: Solar-powered IoT irrigation system for energy efficiency. *Energy Rep.* 6, 112–118 (2020). https://doi.org/10.1016/j.egyr.2020.11.032
20. De Silva, C., Khan, M.A., Jin, Y.: Towards the future of smart agriculture: A literature review on IoT applications in agriculture. *IEEE Internet Things J.* 8(14), 11027–11047 (2021).
21. Farooq, M.S., Riaz, S., Abid, A., Abid, K.: A survey on the role of IoT in agriculture: Smart farming and precision agricultural methods. *IEEE Access* 8, 176573–176612 (2020).
22. Gutiérrez, J., Villa-Medina, J.F., Ávila-Moreno, L.G., Nieto-Garibay, A.: Smart irrigation control system based on soil moisture and weather forecast. *IEEE Sens. J.* 21(12), 13983–13991 (2021).

23. Prathibha, K.R., Hongal, A., Jyothi, M.P.: IoT based monitoring system in smart agriculture. In: *2017 Int. Conf. Recent Adv. Electron. Commun. Technol. (ICRAECT)*, Bangalore, India, pp. 123–126. IEEE (2017). https://doi.org/10.1109/ICRAECT.2017.52
24. Masseroni, D., Menenti, M., Facchi, A., Gandolfi, C.: Smart irrigation management: Principles and practices. In: *Smart Farming Technologies for Sustainable Agricultural Development*, pp. 1–28, edited by Poonia, R. C. and Gao, X.-Z. IGI Global (2021).
25. Kamilaris, A., Prenafeta-Boldú, F.X.: Deep learning in agriculture: A survey. *Comput. Electron. Agric.* 147, 107–122 (2018). https://doi.org/10.1016/j.compag.2018.02.016
26. Liakos, K.G., Busato, P., Moshou, D., Pearson, S., Bochtis, D.: Machine learning in agriculture: A comprehensive updated review. *Sensors* 18(8), 2674 (2018). https://doi.org/10.3390/s18082674
27. McKinsey & Company: Agriculture's connected future: How technology can yield new growth. *McKinsey Rep.* (2020). https://www.mckinsey.com/industries/agriculture/our-insights/agricultures-connected-future
28. Irrigation Association: Smart irrigation systems for sustainable water use. *Irrig. Assoc.* (2019). https://www.irrigation.org
29. UNEP: Precision irrigation for smallholder farming in Africa. *United Nations Environ. Programme* (2018). https://www.unep.org
30. Hunt, L.A., Boote, K.J.: Sensors for precision irrigation management and scheduling. In: *Precision Agriculture Technology for Crop Farming*, pp. 187–210, edited by Zhang, Q. CRC Press (2016)
31. Singh, S., Sinha, R., Verma, H.: IoT-based smart irrigation systems: A review. *Mater. Today: Proc.* 33, 1569–1573 (2020). https://doi.org/10.1016/j.matpr.2020.05.611
32. Al-Fuqaha, A., Guizani, M., Mohammadi, M., Aledhari, M., Ayyash, M.: Internet of Things: A survey on enabling technologies, protocols, and applications. *IEEE Commun. Surv. Tutor.* 17(4), 2347–2376 (2015). https://doi.org/10.1109/COMST.2015.2444095
33. Arora, P., Rani, A., Mishra, V.: IoT-based smart irrigation system using machine learning for water conservation. *Comput. Electron. Agric.* 203, 107482 (2022). https://doi.org/10.1016/j.compag.2022.107482
34. Basso, B., Antille, D.L.: Digital agriculture to design sustainable farming systems. *Front. Sustain. Food Syst.* 5, 787823 (2021).
35. Ben Naceur, M., Kammoun, M., Alouani, I., Abid, M.: Smart irrigation system based on LoRa and machine learning for arid regions. *Sensors* 23(5), 2587 (2023). https://doi.org/10.3390/s23052587
36. Khandelwal, A., Khare, N., Verma, S.: A solar-powered IoT-based smart irrigation system with cloud integration. *Energy Rep.* 9, 1234–1245 (2023). https://doi.org/10.1016/j.egyr.2023.01.012
37. Zhang, L., Kovacs, J.M.: The application of small unmanned aerial systems for precision agriculture: A review. *Precis. Agric.* 22(2), 302–325 (2021).
38. Zhao, Y., Zhang, X., Li, D.: A cloud-edge(collaborative) architecture for smart irrigation using IoT and deep reinforcement learning. *IEEE Trans. Ind. Inform.* 19(3), 2105–2114 (2023).

Urban Drainage and Wastewater Systems to Automation

12

Lokeshwer Sharma, Ravi Kumar Sandal, Aravind Chauhan, and Prerna Thakur

12.1 INTRODUCTION

The storm and drainage sewer and the wastewater treatment are the underground and urban backbone systems of cities and their ecosystems. Aside from the system lies the heavily used storm drain system that is hidden from the public eye and its importance. Effective stormwater management helps reduce flood risks, protect public health, strengthen urban resilience, and enhance the environmental and economic benefits of blue–grey infrastructure. Effective storm drainage systems help prevent land contamination and surface flooding during heavy downpours and extreme rainfall runoff [1]. Proper drainage and wastewater treatment are important. Properly treated and reclaimed wastewater can help cities improve water security, support sustainable urban growth, and reduce pressure on drainage and wastewater systems. The increasing urban sprawl and its harsher effects, along with the decaying and ageing infrastructure, reveal the imbalance of traditional, simplistic, linear systems. On the other hand, the actual drainage systems acting under the Internet of Things (IoT), machine learning (ML), and other automated [2] control systems comprise the real three systems. Sensors integrated into drainage networks enable real-time monitoring of water levels, flow rates, rainfall intensity, and infrastructure conditions, helping to streamline system operation and improve decision-making.

Three primary factors amalgamate to explain the cause of such great changes. These are [3] the IoT and ML, control, and automation. We now place IoT sensors in drainage pipes. They are like tiny, unblinking, and always watchful. They log and send data on the status of the water, the rainfall, and the health of the system. The main goal of all this data is to predict and provide possible solutions to impending problems. Systems [4,5] anticipate the likely next steps, allowing us to time our intervention perfectly. The automated flood detection system (AFDS) seeks to improve system efficiencies. Automated control comes in very handy during crisis periods. For example, during a storm. When water [6] levels rise, the system automatically detects the risk and adjusts drainage controls to prevent flooding and overflow (Figure 12.1).

DOI: 10.1201/9781003714217-12

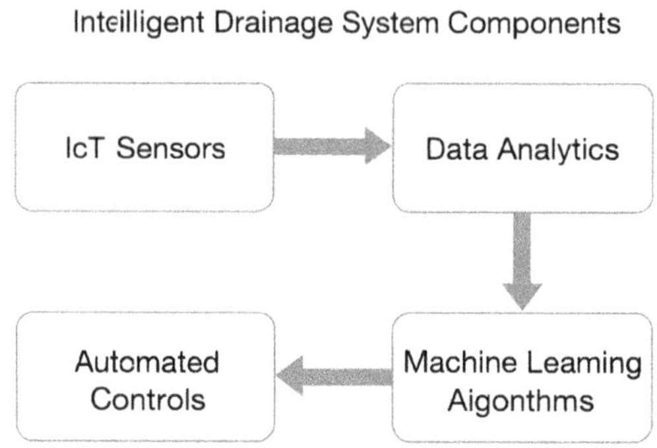

FIGURE 12.1 Intelligent drainage system components flowchart.

The succession of intelligent pluvial systems and urban wastewater management is the focus of this chapter. Afterwards, the chapter illustrates, through technology, approaches, and modern urban case studies, automated drainage systems and their functions [7]. The chapter seeks to illustrate automated drainage systems and their importance in advancing urban infrastructure. Automated systems acquired through modern urban centre case studies are in alignment with.

12.2 URBAN DRAINAGE EVOLUTION AND WASTEWATER

Drainage and wastewater treatment systems in cities have advanced remarkably over the centuries, with increased complexity of urban areas and the need for an improved alternative and better sustainable method. Drainage systems during ancient civilizations (e.g., Mesopotamian and Roman) were structures constructed to drain excess water, prevent flooded urban areas, and keep them sanitary. They focused on gravity-driven systems that were relatively easy and efficient at the time. However, the limitations of traditionally designed drainage systems were realized as cities began to grow and industrialization proceeded at a faster pace in the 19th and 20th centuries; rapid urbanization with urban sprawl and increased impervious areas made it possible to have more stormwater runoff and increased demand for wastewater management. Thus, the cities [8] began to grow and urbanize their infrastructure (in a larger drainage system, stormwater treatment stations, and wastewater treatment units) to accommodate this additional pressure. The development of environmental awareness in the middle of the 20th century contributed to the recognition of the negative effects of untreated wastewater on the population and the environment, and in response to these issues, more [9] advanced technologies for treatment were proposed, as well as the introduction of sewage treatment plants where wastewater could be treated, filtered, and discharged into rivers and streams. However, even after these advances, most urban drainage systems (UDS) still had their fair share of overflows, blockages, and decaying infrastructure (Figure 12.2).

The other big improvement was in the late 20th to early 21st century, when they started automating and improving monitoring effectiveness. Passive drainage systems are being replaced with more proactive (dynamic) drainage systems. This is a process that happens almost [10] immediately (when rain clouds cool off) as the water heats up a lot more, it rains harder, etc. Or there may be changes in local conditions. Our drains and sewers are getting smarter (most of them do simple things to mitigate flood risk, pollution in rivers, etc.), and as a whole, improving the system so cities can cope with the monsoon or sudden downpours without public hysteria.

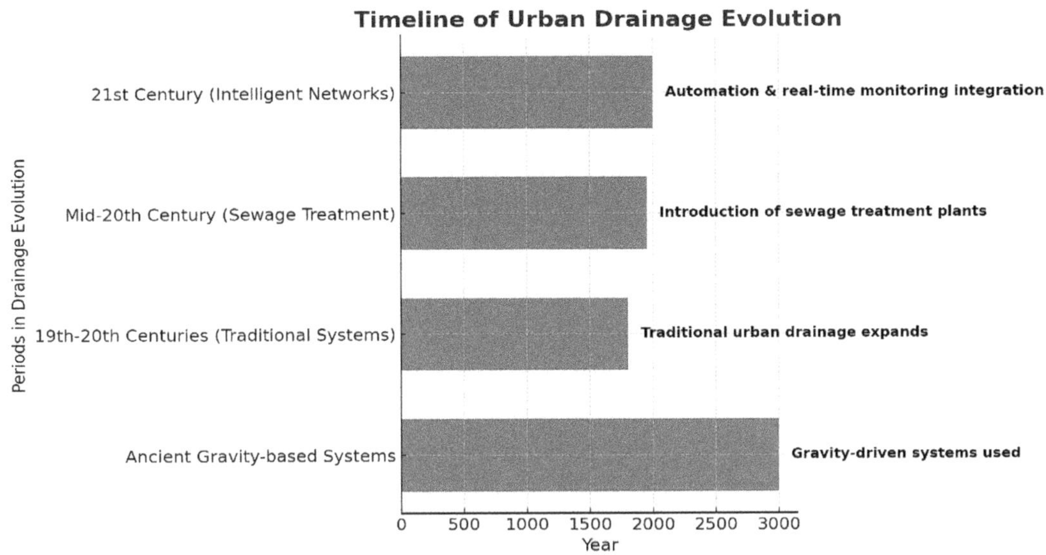

FIGURE 12.2 City drainage of urban development.

12.3 INTELLIGENT DRAINAGE SYSTEM MAJOR ELEMENTS

a. **Advanced drainage** technologies integrate streams of urban drainage performance data – and data-driven corrective control measures (smart sensors, [11] data analytics, ML, and other forms of automated control): These entire artefacts are what enable the system in question to respond instantaneously to different conditions.
b. **IoT Sensors**

   An intelligent drainage system needs IoT sensors installed in all areas of the network that record real-time information, e.g., water level, flow rate, rainfall intensity, and pipe condition. Flow sensors measure the flow speed in pipes (usually 0.5 m/s to 5 m/s). Rainfall sensors can determine rain intensities; most commonly, from 0.1 to 100 mm/hour.

12.3.1 Data Analytics

The data collected by the sensors is transmitted to (mainly the central) server platform (where the data is analysed based on analytics). Analytics will [12] be able to track trends and predict future behaviour and anomalies in the system, like blockages or imminent flooding; it's possible to predict better and make proactive decisions regarding the system (Figure 12.3).

Real-time analysis is required in an intelligent drainage system to maintain water flows in order (IoT).

Sensors that record the water level, rain, and flow rates send data to a central platform, and that data is analysed in about 1–2 minutes, allowing rapid decision-making and intervention. If sensors detect a sudden rise in flow or water level, automated controls can open floodgates [13], barriers, or increase pump rates within about 30 seconds, cutting the chance of infrastructure damage, shortening flood duration, and making better use of resources, especially in cities that deploy more than 5,000

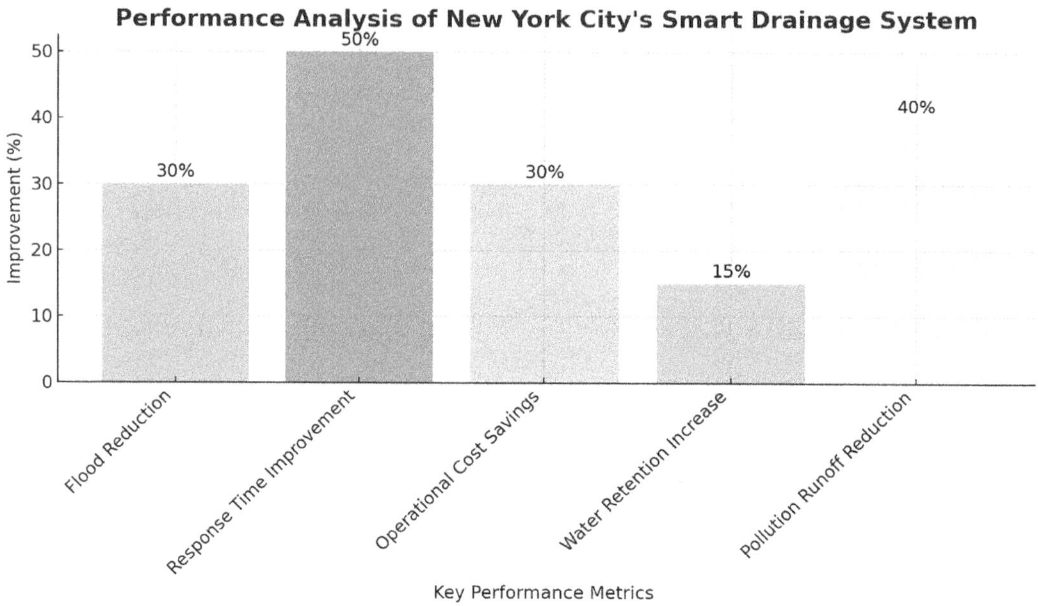

FIGURE 12.3 Smart drainage system performance analysis of New York City.

sensors across flood-prone areas. Fast processing means many flood situations are handled before they get severe, reducing damage by more than 30% compared with slower, traditional systems.

Predictive analytics uses live sensor readings and past records to forecast how the drainage network will react to different weather situations. Using ML models, these systems can anticipate floods and sudden surges with about 85%–90% [14] accuracy, though that depends on the quality of weather forecasts, flow measurements, and historical incident data. These models can give 30–60 minutes' notice before an event, allowing authorities to prepare pumps, valves, and reservoirs in advance. That forewarning can cut emergency interventions by 25%–30% and reduce unplanned maintenance costs by up to 20% because resources are allocated before problems escalate. Predictive maintenance alerts also help prevent failures, sometimes reducing [15] system breakdowns by around 30% by fixing issues early.

Together, fast real-time analysis and reliable predictive analytics boost operational efficiency, improve long-term planning, and lower the risks of stormwater management. These capabilities typically save 25%–30% in operating costs, reduce flood damage, and support more sustainable urban water management by ensuring water is handled more effectively.

12.3.2 ML Algorithms

ML systems run algorithms on past and current data to streamline the performance of the system. These algorithms serve as a way of predicting water flow based on previous behaviours, determining patterns, and recommending changes in systems. Flood Prediction Models: ML-based predictions generated by the intelligent drainage systems can further predict or forecast flood risks at an accuracy level of 85%–90% on real-time data through IoT sensors and weather predictions. The resultant high accuracy allows this system to issue early [16] warnings between 30 and 60 minutes in advance, enabling proactive steps to be taken to contain the risks of floods. These models enable the optimal allocation of resources and resource scheduling to flood-prone zones and predict flood surges, enabling the city to introduce flood barriers or to regulate the rate of water flows before the system is overloaded.

TABLE 12.1 Intelligent drainage systems performance metrics of machine learning algorithms

METRIC	*VALUE*
Flood prediction accuracy	85%–90% accuracy in forecasting floods
Maintenance forecasting reliability	Up to 90% reliability in predicting failures
Prediction lead time	30–60 minutes' warning for floods
Unplanned downtime reduction	30%–40% reduction in unplanned downtime
Operational efficiency improvement	15%–20% improvement in operational efficiency
Cost savings from maintenance	20%–25% reduction in emergency maintenance costs

The forecasting ability of these models has been minimized due to flooding destroying about 30% of the urban environment that is otherwise depopulated and prone to floods (Table 12.1).

Maintenance Forecasting: Also, the ML algorithms are important in predictive maintenance apart from flood prediction. Such algorithms will analyse the data acquired by the sensors monitoring drainage, including items like pumps, gates, and pipes. With the experience of historical system behaviour and failure modes, the system will anticipate possible failures with 90% accuracy and immediately notify the need to repair and/or replace certain parts, thus [17] making the forecasting capability of the reliable drainage system ready when it is needed (to stop stalling if it is hit by a downpour or hurricane). Maintenance planning often results in a 20%–25% cost saving on maintenance costs, as well as a reduction in unplanned downtime (30%–40% of total downtime) and the scheduling of necessary interventions, as well as the cost of repairing infrastructure (when appropriate). By identifying potential failures early, predictive maintenance reduces costly emergency repairs, improves system reliability, and contributes to a longer infrastructure life cycle.

The two capabilities provide synergy between flood prediction and maintenance forecasting, resulting in the maximization of system efficiency, minimization of operating costs, and reduction of the risk of damage caused by floods. ML algorithms can provide a more sustainable and less expensive way of managing UDS by forecasting flood events and possible system failures.

12.3.3 Automated Controls

The last part of the smart drainage system is the automatic control systems. These systems consist of automated gates, valves, and pumps that can regulate the circulation of water in response to real-time information. Automation can assist in dynamically changing the [18] drainage network in reaction to the stormwater streams and avoid flooding. The use of automated control mechanisms is one of the main characteristics of modern intelligent drainage systems, which is the main aspect of their high efficiency in handling stormwater and avoiding floods. Automated pumps are set off once the water levels reach over 70% of the system capacity, where they activate within 30–60 seconds once the water is noticed to be rising. Such a quick reaction time can be used to control the amount of stormwater and eliminate its overflow and the risk of floods [19]. The fact that the system can modify automated control valves on the fly also optimizes the flow of water, as the surplus water will be diverted to storage or treatment facilities. These control valves decrease by up to 40% the system overload that causes blockages, thus ensuring smooth operation.

These systems also minimize flooding in susceptible regions by up to 40% compared to traditional systems due to the delay in responding to such occurrences, as is the case [20] with these systems. The dynamic applications make the drainage system adapt rapidly to an abrupt change in the water velocity, enhancing the resilience of urban systems. Such dynamic control reduces the damage to the infrastructure and improves the functionality of the drainage system.

TABLE 12.2 The important elements of the smart drainage system

COMPONENT	*DESCRIPTION*	*TYPICAL VALUES*
IoT sensors	Sensors are placed throughout the drainage system to collect real-time data on water flow, rainfall, and water levels.	Water Flow: 0.5 m/s to 5 m/s Rainfall: 0.1 mm/hr to 100 mm/hr Level: 0 to 10 m
Data analytics	Data analysis tools that process the collected sensor data to identify trends, forecast system behaviour, and detect anomalies.	Real-Time Analysis: Instant processing Predictive Analysis: Forecasts system performance
Machine learning algorithms	Algorithms that use data to predict and optimize drainage system performance, including flood prediction and maintenance scheduling.	Flood Prediction: 85%+ accuracy Maintenance Forecasting: Reduces downtime by up to 30%
Automated controls	Automated valves, gates, and pumps that adjust the system's operations based on real-time data to ensure efficient water flow.	Automated Valve Control: Adjusts based on water levels Pumps: Activated at 70% system capacity

Furthermore, automation saves operational expenses by a factor of 25–30 because it helps to eliminate the necessity to involve people and monitor the status manually. Predictive maintenance will ensure timely servicing, thus reducing unplanned downtime (30–40%) and enhancing performance in terms of output (Table 12.2).

12.4 ADVANTAGES OF URBAN DRAINAGE AUTOMATION

Automation in UDS increases performance levels, reduces cost and energy expenditure, and even benefits environmental sustainability. In the face of old-fashioned drainage (smart automated drainage for the challenges of today), automation does help respond to the rapid increases in urbanization and, in turn, climate change.

12.4.1 The Improvement of Operational Effectiveness

Automated drainage at a new level, one of the innovative keys in the Design by Nature project, is very simple – automatically adjusting decisions [21] to changes in "facts", such as changing weather or water and stormwater flows. Compared to traditional, historical systems relying on human intervention for adjustments, automatic equipment takes in sensor data to make decisions, allowing the fluid flow carried in lines to be at its correct rate, and any hesitations or spill-over effects are reduced.

12.4.2 Savings and Cost-Effectiveness

Manual work in the management of the city is also being excluded, and cost-effectively, urban drainage monitoring can be expensive, especially in large metropolitan areas. Automated systems minimize the amount of staff that has to be on-site and lower running costs when something goes wrong (Figure 12.4).

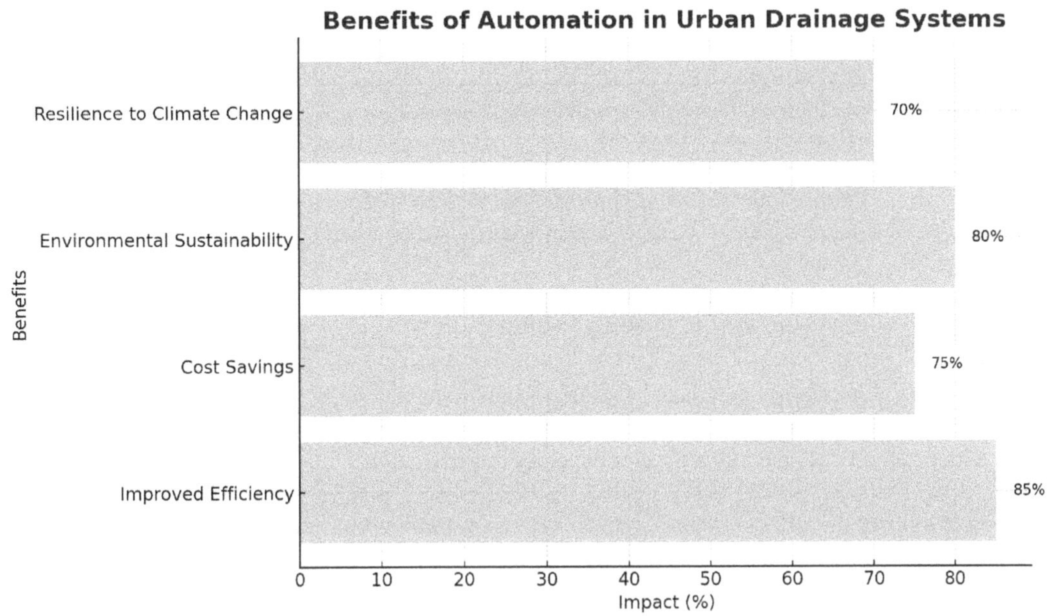

FIGURE 12.4 Benefits of automation – Bar chart.

Possible system damage ahead of time saves costly emergency repairs. They do eventually pay for themselves in reduced repair frequency and severity. Automation lowers [22] maintenance costs by enabling planned maintenance, extending equipment life, and reducing the need for costly emergency repairs.

12.4.3 Environmental Sustainability

UDS protect the environment by controlling stormwater and wastewater. By controlling the use of water resources and emissions of pollutants with full transparency, automated systems contribute to maintaining environmental sustainability. Monitoring based on real-time data from IoT sensors and automated systems can ensure optimal discharge of [23] water and treatment, allowing only the most adequate water to be discharged into the environment. They could also potentially be used to reserve stormwater for an extended period with the help of automation, which would reduce energy loads in wastewater treatment plants when there is heavy rainfall and the possibility of untreated discharges. The activation of automated drainage systems decreases the amount of urban pollution by eliminating overflows; also, environmental aesthetics are improved if the system has open landscape managers.

12.4.4 Stronger Adaptability to Climate Change

Automation is a real-time adaptive feature of UDS that functions independently during critical operational events such as extreme weather conditions. Automated systems mitigate the impact of flooding by agilely surging contact [24] water and other operational critical and support functions during emergencies. The responsive impact of UDS, which are automated with reservation and space functions, enhances city recovery in response to climate events, though the impact is marginal. During operational phases, the automated UDS provide core services with minimal lag. Responsive automated

TABLE 12.3 Critical advantages of automation with urban drainage systems

BENEFIT	*DESCRIPTION*	*EXAMPLE/VALUE*
Improved efficiency	Real-time adjustments to water flow, reducing congestion and blockages.	Reduced blockage rate by 20%–30%.
Cost savings	Reduction in operational and maintenance costs through automation and predictive maintenance.	Maintenance cost savings of up to 30%.
Environmental sustainability	Minimizes water wastage and reduces pollution from stormwater and wastewater.	Reduction in untreated overflow by 40%.
Resilience to climate change	Enhances adaptability and response to extreme weather conditions such as flooding and heavy rainfall.	40% faster flood response and recovery.

drainage systems significantly enhance urban flood risk management by maintaining critical functions during extreme events. Cities are able to recover more rapidly from and respond to climate-related impacts with reduced damage and a greater capacity to provide critical services.

Flood Warning: Intelligent systems can be kept up-to-date with weather forecasts, rain gauges, and water levels in real-time and automatically enforce actions to prevent floods and mitigate stormwater damage to buildings (Table 12.3).

12.5 DRAINAGE SYSTEMS CASE STUDIES OF INTELLIGENT DRAINAGE SYSTEMS

Smart drainage solutions already exist in cities worldwide, and the results are obvious: automation and open real-time monitoring enable engineers to detect problems earlier, data-driven decisions target resources where they matter most, and predictive control reduces both flooding and operating costs. Read the brief case studies below to see how different places – from dense tropical cities to older, infrastructure-laden metropolizes – adapt sensors, analytics, and controls to the specific needs of local communities. This makes them resilient, less disruptive, more sustainable, and more people-friendly in urban water management.

12.5.1 Case Study 1: United Kingdom (London)

London's old sewers and fast-growing neighbourhoods can't keep up with heavier, more frequent storms. So, Thames Water and the city installed smart sensors, live analytics, and automated controls that monitor flows, anticipate problems, and move gates and pumps [25] before streets get too tense – turning a reactive system into one that anticipates problems and acts quickly to safeguard homes and businesses. This integrated solution improves stormwater management by continuously monitoring conditions, predicting stress points, and adjusting assets in real time to reduce overflow and service disruption (Table 12.4).

Key Features:

- IoT sensors are installed at critical points in the drainage network to record water levels, rainfall, and flow rates in real time.
- Flood management functions are automated, so barriers deploy when flooding is detected, and water movement is controlled in real time to prevent overflow.

TABLE 12.4 London, United Kingdom drainage network key metrics

METRIC	*VALUE*
IoT sensors installed	1,000+ sensors
Flood reduction	30% reduction in flooding
Flood prediction accuracy	85% accuracy
Flood barrier activation time	30 seconds
Automated flow control reduction	40% system overload reduction
Response time improvement	40% faster response
Maintenance cost reduction	25% reduction
Operational cost savings	30% reduction
Untreated water overflow reduction	40% reduction
Water retention improvement	15% increase
Operational efficiency improvement	15% increase

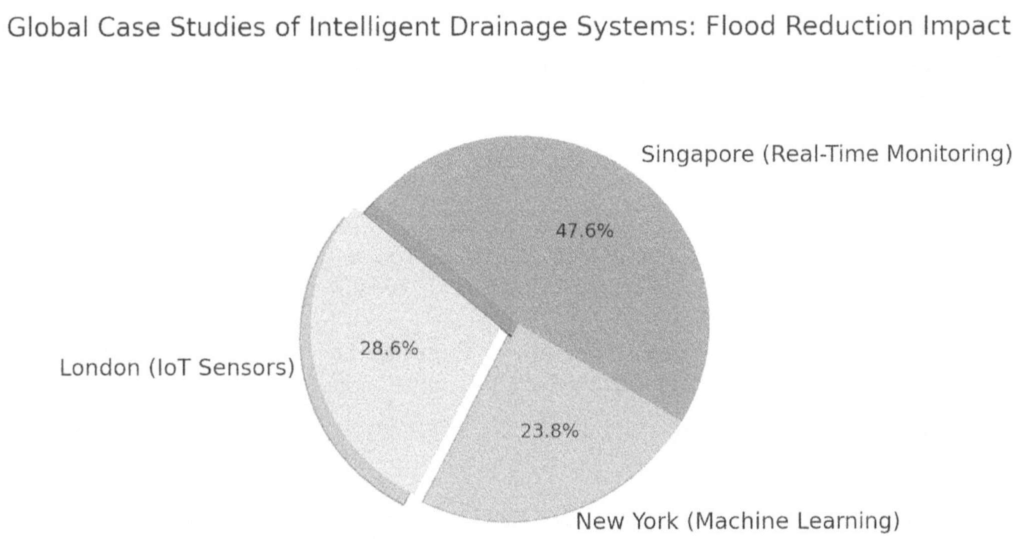

FIGURE 12.5 International case studies of wise drainage systems: flood reduction effect.

Results:

- **Flood Reduction**: The system reduced flooding by up to 30% in high-risk zones, lowering the frequency and severity of inundation events.
- Better response time in emergencies: By using real-time monitoring and automatic control mechanisms, disasters such as heavy rainfall can be handled much sooner, thus minimizing the damage caused to infrastructure and disruption to services.
- Predictive maintenance achieves better resource management at reduced operational costs (Figure 12.5).

12.5.2 Case Study 2: Smart Drainage Network of Singapore

Singapore, a densely urbanized city-state with intense tropical rainfall and rapid runoff, implemented a comprehensive smart drainage network to reduce urban flooding [26] and optimize stormwater management. The programme relies on dense sensor networks, predictive analytics, automated control structures, and integrated operations to maintain drainage corridors clean and to redirect excess flows to storage and treatment (Table 12.5).

Key Features:

1. **Dense IoT Monitoring**: Water-level sensors, rain gauges, and flow meters are deployed across drains, canals, and retention ponds to provide high-resolution, real-time visibility of the network.
2. **Predictive Analytics and Forecasting**: Models ingest live sensor feeds together with short-term weather forecasts to predict where and when surges will occur and to plan automated responses.
3. **Automated Control Infrastructure**: Remotely controlled gates, valves, and pumps dynamically supply water to under-utilized storage, detention basins, or treatment plants to avoid local saturation.
4. **Integrated Command and Control**: A central operations platform displays the network state, dispatches automatic control commands, and coordinates maintenance crews and emergency responders.
5. **Public Alerting and Demand Management**: The system is connected to city dashboards and public alert channels so that authorities can send out timely warnings and control on-the-ground resources during extreme events.

Results:

1. **Reduced Flood Incidents and Faster Response**: The coordinated monitoring and automated control substantially lowered the frequency of local flooding and shortened response times during intense storms.
2. **Better Utilization of Storage and Treatment Capacity**: Dynamic [27] routing and temporary retention shifted peak loads away from constrained nodes, reducing stress on treatment plants and overflow risk.

TABLE 12.5 Major figures of the smart drainage network of Singapore

METRIC	*VALUE*
IoT sensors installed	5,000+ sensors
Data collection frequency	Every 30 seconds
Flood reduction	50% reduction in flood incidents
Flood prediction accuracy	90% accuracy
Response time improvement	40% faster flood response
Flood barrier activation time	2 minutes
Operational cost savings	30% reduction in operational costs
Maintenance cost savings	25% reduction in maintenance costs
Untreated water overflow reduction	40% reduction in water overflow
Water retention capacity increase	20% increase in water retention

3. **Lower Operational and Maintenance Burden**: Predictive maintenance and remote control decreased the need for continuous manual inspections and emergency interventions, improving availability and cutting lifecycle costs.
4. **Improved Urban Resilience**: The combined effect of sensing, forecasting, and automation increased the city's ability to withstand sudden storm events and adapt to changing rainfall patterns.

12.5.3 Case Study 3: New York City, United States

When a cloudburst hits New York, streets turn into rivers and old sewers get overwhelmed. The city's Department of Environmental Protection decided to stop only reacting and start anticipating. By adding smart sensors, live analytics, and remote controls to its existing network, New York turned a fragile system into one that thinks and moves quickly (Table 12.6).

Key Features:

- Live sensing gives operators a clear picture of what's happening underground before flooding shows up on the surface.
- Remote valves, pumps, and temporary storage are steered in real time to hold back or reroute stormwater so fewer dirty flows reach rivers and bays.
- Data points out the real trouble spots, so crews and money go where they'll actually fix things rather than chase symptoms.
- Operations teams use the same live feeds for both immediate response and smarter long-term investments in pipes, green spaces, and storage.

Results:

- Streets flood less often, and sewer overflows drop because the system can act fast.
- Rivers and bays are cleaner, which helps people who fish, boat, or just want safer places to swim.
- The city spends smarter, fixing what matters most instead of spreading resources thin.
- When storms come, operators see problems earlier and act faster, which saves property, time, and worry.
- In short, New York didn't replace what it has; it has made it work smarter. That human touch of anticipating trouble and acting fast is what keeps the city moving when the skies open.

TABLE 12.6 Performance metrics of performance management of the smart drainage system of New York City

METRIC	*VALUE*
IoT sensors installed	2,000+ sensors across the drainage network
Flood reduction	25% reduction in flooding incidents
Flood prediction accuracy	85%–90% accuracy
Sewer overflow reduction	30% reduction in untreated runoff
Automated response time	30 seconds to activate flood barriers
Operational cost savings	35% reduction in operational costs
Water retention increase	15% increase in water retention capacity
Pollution runoff reduction	25% reduction in CSOs

12.6 ISSUES AND FUTURE TRENDS OF INTELLIGENT DRAINAGE SYSTEMS

12.6.1 Data Management and Technological Integration

Smart Drainage will bring a lot of innovation, but it will also expose some of the actual problems that exist in real-life cities. The most obvious pain point, I think, is [28] integration: putting sensors, telemetry, analytics, predictive models, and control hardware all together in one secure, user-friendly device doesn't help anyone. There are different vendors, protocols, and data formats, so even with a nice-looking design that technically works flawlessly, it could still come apart on the ground.

- **Data Volume and Velocity**: **Volume/velocity of data/network of sensors**: It produces enormous reading speeds. The storage, cleaning, and analysis of this data in real time require scalable infrastructure and well-designed pipelines; without these, the system becomes slow or unreliable.
- **Data Quality and Trust**: Sensor variance, failures, and noisy values are just a few reasons that models can be flawed; good quality data and regular validation must be ensured (if not, predictions and automated actions will not be trustworthy to operators).
- **Interoperability with Legacy Assets**: City planners and future smart devices will probably not be built from the ground up to interoperate with the pipes, pumps, and control systems already in use, not built with an updated protocol yet, or fail remote control entirely. This is both technically and operationally a challenge.
- **Real-time Analytics and Decision-Making**: Providing low-latency information that operators can act on or that can successfully trigger automatic controls needs a strong edge/cloud architecture, clear rules for automation, and pre-tested "fallback" behaviours if there is insufficient or mixed data.
- **Storage and Security and Governance > Long History Sensor**: Data is valuable for training models and auditing and needs secure, cost-effective storage, reasonable retention policies, and governance so the right people can take advantage of that data.

12.6.2 Infrastructure

Smart UDS present not only a technical challenge but also a cost challenge. The replacement of older infrastructure, the addition of advanced sensors, and the development of communication systems present significant expenses, and the system changes often face the challenge of access to stable and predictable funding. High initial costs and unpredictable budgeting make planning for the future difficult, and massive public works in crowded cities usually mean slowdowns caused by disruption and extra cost overruns. Meanwhile, rapid urbanization is adding impervious surfaces and runoff to systems that are already overburdened. Climate change injects another degree of stress, with more frequent and severe storms requiring faster, smarter responses to prevent system failures and protect not only communities but also the environment. Honestly, building smart UDS? It's a total headache. Not just because the tech is tricky – though it is – but because it costs a ridiculous amount of money. Ripping out ancient pipes, slapping on fancy sensors, wiring the whole thing up so everything talks to everything else. It's like burning cash in slow motion.

12.6.3 Regulatory and Policy Issues

Only by putting sensors and smart controls in the drains can we hope that the system will work – and for India, that means the rules have to be understandable and realistic, and written by the people who run and live in our cities – not the technocrats.

Maintain the confidentiality of all records and files, unmonitored as well as monitored, for data scrubbing and the retention of data using time and identifiable metadata, within pre-defined parameters. Data system and procedural interoperability, and the legal unencumbered adoption of commercially available file formats and system vendors.

"Reverse the Paradigm" to Iteration: Empirical data captured for automation. Empirical data captured for automation, as well as automation, must be regulated under specific parameters. Data capture officer must be equipped with a "simple" and easily understood system of capture.

Translation of Legal Documents to Municipal Documents: Detailed guidance with specific achievable outcomes and checklists for municipal and ward staff, instead of repetitive documents that would allow lawyers to endlessly justify their confusion.

Invest in Community Outreach and Support: Community dashboards, capture and report systems, and other MoPoCo systems to help communities establish trust and easily capture real-time problem data (Table 12.7).

12.6.4 Future Trends and Innovations

In the coming years, drainage will stop being an isolated technical problem and become part of a city's nervous system, talking to smart grids, traffic systems, weather services, and environmental sensors so responses are faster and better coordinated; AI and deep learning will sharpen short-term flood forecasts and anomaly detection so pumps and gates act before streets flood; more processing will move to the edge so local controllers can act even when connectivity is weak; green measures like permeable pavements, rain gardens, and blue–green corridors will be combined with automated controls to reduce peak flows and cool neighbourhoods; decentralized, neighbourhood-level retention and treatment will become common, lowering the need for massive pipe upgrades; integrated resource management will

TABLE 12.7 Problems and future outlook of intelligent drainage systems

CHALLENGE/DIRECTION	*DESCRIPTION*
Technological integration	Integration of diverse technologies (IoT, machine learning, automation) and ensuring seamless data flow.
Data management	Handling large volumes of real-time data, ensuring its accuracy and processing it efficiently.
Infrastructure and investment	High costs associated with retrofitting existing infrastructure and securing long-term funding for maintenance.
Climate change and urbanization	Adapting drainage systems to handle increased rainfall and rapid urban expansion.
Regulatory and policy issues	Need for updated regulations to address data privacy concerns and ensure the safe deployment of intelligent systems.
Future directions	Integration with smart city systems, use of AI for predictive analysis, and sustainable solutions like green infrastructure.

let cities reuse stormwater and schedule pumps around cheap electricity windows; cheaper, rugged sensors and modular retrofit kits will make rollouts easier for smaller towns; and new financing and operating models such as pay-for-performance contracts and municipal digital twins will help cities plan, fund, and run smarter drainage systems reliably.

12.7 CONCLUSION

The implementation of smart drainage systems serves as a breakthrough advancement in the manner in which cities tackle the challenges of rainfall, historical infrastructure, and erratic climatic conditions. These smart drains utilize the IoT and integrate real-time data collection sensors to monitor the locations and moisture levels in UDS. For cities that hope to integrate smart drainage systems as part of their urban master plans, there are numerous challenges that lie ahead – interfacing new and emerging technologies with the current infrastructure, sustained funding for improvement and upkeep, and data resource management such as security control, system sustainability, and equitable access. And yet, evidence shows that the optimization of intelligent drainage works most effectively as part of an integrated and anthropocentric urban design, including strategic natural infrastructure, decentralized retention systems, and supportive policy. With continued evidence-based investment, scalable systems that are the result of pilot projects, and strategic collaboration with community and local government, smart drainage systems can immensely improve urban resilience, decrease operational costs, and increase comfort during extreme climate conditions in Indian cities.

REFERENCES

1. Q. Xiao, L. Cai, Y. Liu, M. Zhang, and X. Hu, "Impact of pore geometry on mechanical properties and anti-clogging efficiency in porous drainage pavement structures," *Case Stud Construct Mater*, vol. 23, 2025, doi: 10.1016/j.cscm.2025.e04929.
2. M. Hajibabaei, S. Hesarkazzazi, and R. Sitzenfrei, "Filling data gaps in urban drainage networks: An automated graph theory framework for data collection and reconstruction," *Water Res*, vol. 287, 2025, doi: 10.1016/j.watres.2025.124272.
3. H. Tansar, F. Li, and H.-F. Duan, "An efficient Gaussian process-based optimization for resilience improvement of urban drainage systems considering changing climate," *Reliab Eng Syst Saf*, vol. 264, 2025, doi: 10.1016/j.ress.2025.111428.
4. A. A. N. Deng, A. Zaki, G. Nugroho, and J. Ikhsan, "Bibliometric study: Structural health monitoring on drainage for classification learners using machine learning methods (2014–2023)," *Multidiscipl Rev*, vol. 8, no. 12, 2025, doi: 10.31893/multirev.2025389.
5. Y. Tao, D. B. Wright, A. M. Lokhandwala, V. G. Geller, J. G. Vasconcelos, and B. R. Hodges, "Impacts of rainfall spatiotemporal variability on pressurized flow conditions in urban drainage systems," *J Hydrol (Amst)*, vol. 662, 2025, doi: 10.1016/j.jhydrol.2025.133874.
6. G. Diab, R. A. Brown, W. F. Hunt, and J. M. Hathaway, "Drainmod-urban for bioretention hydrology: A comprehensive sensitivity analysis," *J Hydrol (Amst)*, vol. 663, 2025, doi: 10.1016/j.jhydrol.2025.134138.
7. Z. Huang, Y. Wang, X. Dong, W. Li, Y. Tang, and D. Zhang, "Pollution-based integrated real-time control for urban drainage systems: A multi-agent deep reinforcement learning approach," *NPJ Clean Water*, vol. 8, no. 1, 2025, doi: 10.1038/s41545-025-00512-z.
8. C. Obara, C. H. Fletcher, S. Habel, K. McDonald, and K. Yamamoto, "Drainage failure and associated urban impacts under combined sea-level rise and precipitation scenarios," *Sci Rep*, vol. 15, no. 1, 2025, doi: 10.1038/s41598-025-07332-8.

9. L. de Klerk et al., "Persistent organic pollutants in South African coastal ecosystems: Analysing the evileye blaasop Amblyrhynchote honckenii (Bloch, 1785) for PCB and OCP accumulation," *Reg Stud Mar Sci*, vol. 89, 2025, doi: 10.1016/j.rsma.2025.104368.
10. X. Hu, H. Guo, K.-W. Lao, J. Hao, F. Liu, and Z. Ren, "MiniRocket-MARL synergy for storm tide resilience: MESS-DV enhanced recovery in coastal distribution networks," *Appl Energy*, vol. 401, 2025, doi: 10.1016/j.apenergy.2025.126710.
11. Y. A. Berrezel, C. Abdelbaki, A. Megnounif, M. Saber, M. E. A. Benabdelkrim, and N. Kumar, "Assessment of flood risk using integrated GIS and analytic hierarchy process in the Mekerra basin, Northwestern Algeria," *J Afr Earth Sci*, vol. 233, 2026, doi: 10.1016/j.jafrearsci.2025.105830.
12. G. Wei, Z. Yu, and D. Li, "WEHD-DETR: A real-time defect detection algorithm for sewer pipelines based on improved RT-DETR," *Digit Signal Process*, vol. 168, 2026, doi: 10.1016/j.dsp.2025.105585.
13. S. Santoro, V. Totaro, P. Balena, V. Iacobellis, U. Fratino, and D. Borri, "The interlink between urban growth and flooding: Lessons from the City of Bari (Italy)," in *Lecture Notes in Computer Science*, 2026, pp. 49–60. doi: 10.1007/978-3-031-97648-3_4.
14. K. Gao, B. Gao, Z. Yang, and X. Gao, "Measuring urban waterlogging depth from video images using human body models," *Measurement (Lond)*, vol. 258, 2026, doi: 10.1016/j.measurement.2025.119086.
15. X.-F. Yan, Y.-Y. Jiao, C. Zeng, S. T. Ariaratnam, and Z.-Y. Zhao, "Residual load capacity of longitudinally-cracked reinforced-concrete pipe and its recovery by CIPP versus spray-mortar liners," *Tunnell Undergr Space Technol*, vol. 168, 2026, doi: 10.1016/j.tust.2025.107126.
16. Z. Wang, J. Chen, Q. Li, K. Xie, and C. Pu, "Double-layer data-driven risk assessment and prediction of drainage pipeline networks," *J Pipeline Syst Eng Pract*, vol. 17, no. 1, 2026, doi: 10.1061/JPSEA2.PSENG-1799.
17. Y. Ma, C. Yang, and Y. Wang, "Drainage pipeline defect detection by fiber optic distributed temperature sensor: A review," *J Pipeline Syst Eng Pract*, vol. 17, no. 1, 2026, doi: 10.1061/JPSEA2.PSENG-1902.
18. S. Yimprasert, S. Ketpratum, and A. U. Kaew, "Experimental investigation of turbulent manipulation in pipe flow," *GMSARN Int J*, vol. 20, no. 1, pp. 153–158, 2026, [Online]. Available: https://www.scopus.com/inward/record.uri?eid=2-s2.0-105011663606&partnerID=40&md5=7a37073d1732089dd0595ccc35db11fe
19. E. Li, X. Fu, J. Song, L. Jing, B. Zhang, and X. Wang, "Developing a decision support system for regulating urban runoff by centralized blue-green infrastructure," *Environ Impact Assess Rev*, vol. 117, 2026, doi: 10.1016/j.eiar.2025.108190.
20. R. Sitzenfrei, C. Urich, M. Moderl, and W. Rauch, "Assessing the efficiency of different cso positions based on network graph characteristics," *Water Sci Technol*, vol. 67, no. 7, pp. 1574–1580, 2013, doi: 10.2166/wst.2013.029.
21. S. Bureerat (Ed.), "5th KKU International Engineering Conference 2014, KKU-IENC 2014," *Adv Mat Res*, vol. 931–932, 2014, [Online]. Available: https://www.scopus.com/inward/record.uri?eid=2-s2.0-84901724860&partnerID=40&md5=c4dcf2256e17b5af7fec6e24d2a91398
22. J. Barreiro-Gomez, N. Quijano, and C. Ocampo-Martinez, "Distributed resource management by using population dynamics: Wastewater treatment application," in *2015 IEEE 2nd Colombian Conference on Automatic Control, CCAC 2015 - Conference Proceedings*, Manizales, Colombia, pp. 1–6, 14–16 October 2015. doi: 10.1109/CCAC.2015.7345186.
23. L. M. Sidek et al., "Decision support system (DSS) for MSMA integrated stormwater management ecohydrology for sustainable green infrastructure," *IOP Conf Ser: Earth Environ Sci*, 2016. doi: 10.1088/1755-1315/32/1/012068.
24. G. Olsson and P. Ingildsen, "Making water operations smarter," *Adv Wastewater Treat*, 2018, 341–362. doi: 10.2166/9781780409719_0341.
25. Y. Liu, Y. Liu, and W. Zhu, "Research and design of the internet of things for the water quality monitoring of urban sewage outfall," in *Proceedings of 2018 the 8th International Workshop on Computer Science and Engineering, WCSE 2018*, 2018, pp. 232–237. [Online]. Available: https://www.scopus.com/inward/record.uri?eid=2-s2.0-85054568812&partnerID=40&md5=fe8cb1b108225cf3cc11cb155ace8f5f
26. A. M. Moreno-Rodenas, A. Duinmeijer, and F. H. L. R. Clemens, "Deep-learning based monitoring of FOG layer dynamics in wastewater pumping stations," *Water Res*, vol. 202, 2021, doi: 10.1016/j.watres.2021.117482.
27. S. Chakraborty, A. Sarkar, G. Das, S. Chatterjee, S. Kanjilal, and P. Kundu, "Study of IOT enabled water logging detection and control in drainage system for Smart City," in *International Conference on Big Data Analytics in Bioinformatics, DABCon 2024*, Kolkata, India, 21–23 November 2024. doi: 10.1109/DABCon63472.2024.10919387.
28. N. Chahinian, C. Delenne, B. Commandré, M. Derras, L. Deruelle, and J.-S. Bailly, "Automatic mapping of urban wastewater networks based on manhole cover locations," *Comput Environ Urban Syst*, vol. 78, 2019, doi: 10.1016/j.compenvurbsys.2019.101370.

Efficient Water Quality Assessment *Synergistic Integration of Computer Vision & Machine Learning*

13

Laishram Bijialuxmi Devi, Deelip Kumar Sarangi, Merry Saxena, Aakanksha Sankhyan, Alok Kumar Agrawal, and Rajit Verma

13.1 INTRODUCTION

Water is a vital resource for life, yet maintaining its quality is still a major problem around the world. With growing urbanization, industrialization, and agricultural practices, water sources have become more susceptible to contamination from a range of physical, chemical, and biological contaminants. Although traditional water quality monitoring techniques are considered best for their accuracy, they mostly require adequate time and resource obligations, manual sample collection, and laboratory studies [1].

Forewarning the properties of water becomes requisite on a global scale for various reasons, as water is a fundamental requirement for everyone. The availability of safe and clean drinking water, and forecasting water quality, is crucial to ensuring that people everywhere have access to drinkable water. The majority of human health depends upon water quality, as contaminated water can lead to waterborne illnesses that potentially harm millions of people globally.

The wide-ranging and intricate utility of water quality assessment has been recognized by numerous organizations worldwide. Organizations such as the World Health Organization (WHO), the United States Environmental Protection Agency (EPA), the European Environment Agency (EEA), the International Water Association (IWA), the Water Environment Federation (WEF), and the United Nations Environment

DOI: 10.1201/9781003714217-13

Programme (UNEP) all strongly support methods for evaluating and minimizing water quality. When everything is taken into account, water quality affects public health and plays a role in the development of hazardous waterborne diseases such as cholera, dengue, typhoid, malaria, and dysentery [2].

The request for walkable and intelligent systems has led to the incorporation of advanced technologies such as computer vision (CV) and machine learning (ML) into environmental monitoring systems, especially for evaluating water quality [3]. An IoT-based approach highlights the ability to utilize sensors and a microcontroller to automatically monitor water quality [4]. It also enhances cost-effectiveness, reduces manpower, and improves efficiency in the distribution of water as well as monitoring with excellent outcomes [5].

Deep learning (DL) technologies use the up-to-date advancements in image processing, pattern recognition, and data modeling to glean valuable information from visual data, including video streams from aquatic ecosystems, microscopic images of contaminants, and photos of the water surface [6].

13.1.1 Importance of Water Quality Monitoring

Subsurface and surface water serves as a major source for vegetation, drinking, and commercial use around the world [7]. The leading causes of poor water quality are identified as environmental imbalances, ecological shifts, changing socioeconomic dynamics, and outbreaks of water-related diseases. In view of the aforementioned causes, regular monitoring of water quality is vital as it aids in achieving ecological balance, leading to the achievement of sustainable goals such as good health and adherence to legal and ethical standards [8].

13.1.2 Challenges in Traditional Monitoring

Traditional methods usually rely on off-site testing and cyclic sampling, which limit spatial coverage and delay data availability. According to a study, field operations are complicated by logistical expenses and environmental variability. For instance, turbidity or pH tests conducted in the field might not capture fleeting occurrences such as algal blooms or industrial discharge.

13.1.3 Need for Automation

Water body surveillance can be conducted in real time, on a large scale, and at a low cost by integrating the monitoring process with CV and ML. These technologies aid in the prediction of potential contamination based on past trends and make it easier to derive valuable insights from visual and sensor-based data streams [9].

13.2 WATER QUALITY: DEFINITION, STANDARDS, AND PARAMETERS

13.2.1 Water Quality

The term water quality denotes the physical, chemical, and biological properties of water that usually relate to whether it is suitable for drinking, recreation, irrigation, or industrial uses. It assesses the state of water in relation to the needs of humans and one or more ecological species.

Many factors are assessed to determine water quality. These checks help us understand the condition of water and if it is safe to use. They also indicate if water is clean enough for its intended purpose [10].

13.2.2 Key Indicators of Water Quality

Water quality indicators are used to identify pollutants in water and define its quality by monitoring physical, chemical, and biological attributes. To establish safe water guidelines, the properties of water are assessed to ensure it is fit for consumption or other vital uses [11] (Figure 13.1).

1. **Physical Parameters**: The physical properties of water are the primary indicators of water quality.
   a. **Temperature**: Temperature fluctuations can contribute to various other properties of water, such as biological, physical, and chemical activities. Temperature changes can occur because of seasonal variations, time of day, airflow, etc. [12].
   b. **Turbidity**: Turbidity refers to the hazy appearance of water, which commonly occurs because of the presence of silt and other organic material. Additionally, turbidity hampers the process of photosynthesis, leading to changes in the aquatic food chain and further increasing the growth of various microorganisms in water, making it unfit for consumption as it may lead to various health issues [12].
   c. **Color and Odor**: An unusual appearance or smell of water indicates poor quality. Generally, unpleasant smells come from bacterial decomposition, and unusual color in water result from organic material or algae growth [13].
   d. **Conductivity**: Water conductivity refers to the capacity of water to conduct electricity, which typically comes from various solutes in water, such as salts and ions. High conductivity is directly proportional to increased solute levels, which are indicators of industrial effluents or agricultural waste [14].

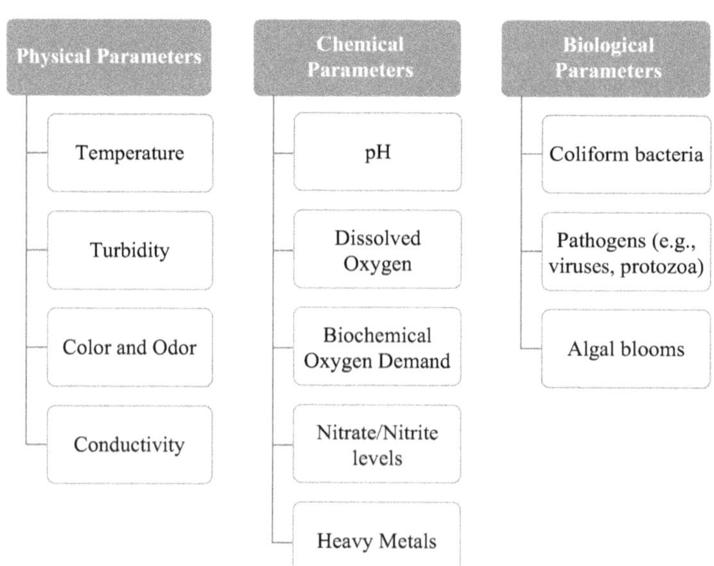

FIGURE 13.1 Parameters of water.

2. **Chemical Parameters**
   a. **pH**: pH is one of the vital indicators of water safety as it indicates whether water is alkaline or acidic. Very low or very high pH levels indicate the presence of pollutants such as chemical substances or heavy metals, which can alter the neutral pH of water. As per standard protocols, the permissible pH of safe drinking water falls between 6.5 and 8.5.
   b. **Dissolved Oxygen**: The dissolved oxygen level is a vital component for ensuring water quality, which can be affected by temperature and various biological processes. Furthermore, decreased dissolved oxygen levels can harm biodiversity [15].
   c. **Biochemical Oxygen Demand**: Biochemical oxygen demand indicates the amount of O_2 required by aerobic bacteria to decompose organic compounds in water. An increase in biochemical oxygen demand can indicate that water has been polluted and may suggest the presence of organic contamination, such as animal waste or industrial waste, which also decreases dissolved oxygen levels [16].
   d. **Nitrate/Nitrite Levels**: These are commonly present in sewage, fertilizers, and animal excreta. Increased levels of nitrite in drinking water pose two main health hazards: methemoglobinemia and "blue baby syndrome," especially in young children. These compounds further decrease the oxygen levels in water bodies as they aid in algal blooms.
   e. **Heavy Metals**: Any form of heavy metals in water makes it unfit for consumption. The presence of lead, arsenic, or mercury can be caused by excessive mining processes or industrial waste dumping in water bodies, which can be consumed by aquatic organisms and cause chronic health issues and organ damage [17].
3. **Biological Parameters**
   a. **Coliform Bacteria**: To determine the presence of fecal matter in water, the presence of *E. coli* and various coliform bacteria is utilized. As per WHO guidelines, every 100 mL of drinking water should be free from any type of coliforms. The presence of coliform bacteria increases the risk of many types of chronic illnesses.
   b. **Pathogens**: Waterborne illnesses are caused by the presence of pathogens in water such as bacteria, viruses, and parasites, which, when consumed by humans, can lead to health hazards [18].
   c. **Algal Blooms**: Increased growth of algae in water is a common indicator of contamination by industrial, human, or agricultural waste. Increased algae growth can degrade water quality by impeding aquatic plants' ability to perform photosynthesis and by lowering oxygen levels, which can change a number of biological and chemical processes and render water unfit for human consumption [19].

13.3 REGULATORY BENCHMARKS

Regulatory benchmarks are the standard protocols established by various national and international organizations to ensure the safety of water for consumption by humans or for other purposes such as cultivation and recreation. These protocols are the set standard for various attributes of water and aim to prevent the health hazards caused by poor water quality.

1. **World Health Organization**: The WHO provides guidelines for drinking water based on health goals and defines permissible levels for different pollutants so that water can be considered safe for drinking. The WHO safe drinking water guidelines protect public health by promoting national standards for safe drinking water.

2. **Bureau of Indian Standards (BIS)**: The BIS is the national agency in India that sets standards for the quality and safety of a variety of products, including drinking water. India's national drinking water quality standard, IS 10500: 2012, was developed by BIS. This standard outlines the acceptable and allowable levels for several parameters to ensure that drinking water is safe and healthy for human consumption [20].

13.3.1 Importance of Regulatory Benchmarks

These standard procedures for developing safe drinking standards are crucial because they protect public health and help raise public awareness. By setting limits for different pollutants, these standards provide a clear definition of the qualities that must be present in safe drinking water. Additionally, they assist in establishing water safety monitoring protocols and offer water treatment authorities remedial actions. Furthermore, these procedures set the standard for attaining SDG 6: Clean Water and Sanitation.

13.4 CV AND ML: CORE TECHNIQUES

Sustaining the quality of water is currently a major global concern, as it is an important resource for life. Water sources are increasingly endangered by contamination from chemical, physical, and biological impurities as a result of elevated industrial plantation, expansion, and agricultural activity. Although traditional water quality monitoring techniques are precise, they often require labor-intensive sampling, laboratory analysis, and significant time and financial resources.

CV describes a machine's capacity to translate and appraise visual data collected from the environment. When it pertains to water quality monitoring, CV aids systems in evaluating the appearance of water, finding anomalies, and examining trends without the need for manual sampling.

CV and ML work similarly to human vision, with the exception that humans have an advantage. Using cameras, data, and algorithms, CV teaches machines to complete these tasks in a significantly shorter amount of time. The combination of CV and ML has become a growing revolution in tracking changes in the environment in recent years, mostly when analyzing the quality of water. To forecast important interpretations, information such as minuscule pictures of contamination, surface photos, and video graphs from marine environments are essential for collection using advancements in technological development in image processing, pattern recognition, and data modeling [21].

13.4.1 Why Shift Toward Automation?

Conventional water monitoring techniques typically depend on routine sampling and laboratory-based chemical testing. However, these methods require a large amount of human labor and resources, are not feasible for continuous or large-scale monitoring, and often fail to detect transient contamination events [22].

However, environmental systems may use CV and ML to anticipate potential contamination based on historical trends, monitor color, turbidity, surface anomalies, and algal blooms in real time, and enable remote and cost-effective deployment in inaccessible places.

13.4.2 Image Processing

Image processing is an essential aspect of CV and involves pre-processing visual input to enhance image quality and extract relevant information. Typical techniques include

- **Noise Reduction**: To remove background noise, water photographs are filtered.
- **Histogram Equilibrium**: Increasing contrast in areas that are imprecise or obscured.
- **Edge Recognition**: Useful for recognizing oil spills, floating rubble, or remnants of algal blooms, edge detection devices such as Canny or Sobel filters can be helpful.

These procedures improve the accuracy of subsequent tasks by distinguishing features of interest and improving visual quality [23].

13.4.3 Object Detection and Segmentation

Object identification involves detecting and identifying objects present in a picture, such as dead fish, floating plastic waste, or algal patches. Modern deep learning models, such as Faster R-CNN and YOLO, are widely used for real-time object detection tasks [24].

On the other hand, image segmentation provides a more thorough understanding by assigning a label to every pixel. Semantic segmentation models such as U-Net and Mask R-CNN are effective in separating suspended materials, identifying surface scum, and distinguishing between clear and polluted water zones with pixel-wise segmentation [25].

13.4.4 Feature Extraction

Classification models greatly aid in extracting features such as color histograms (blue–green dominance for cyanobacterial blooms), texture analysis (detecting foam or oil sheen), and shape descriptors (detecting irregular patches in cases of chemical contamination). These features also help reduce computational complexity without sacrificing accuracy. For ML models, feature extraction converts unprocessed visual input into useful properties.

13.5 ML MODELS

ML assists in training computers using data and creating a program that, when used by anybody, may assist in determining the parameters of water quality and forecasting instances of pollution.

i. It involves training a model with labeled datasets. For image data, algorithms such as support vector machines, random forest, K-nearest neighbors, and convolutional neural networks (CNNs) are typically utilized as they are best in categorizing water based on image data or by identifying oil spills [26].
ii. **Unsupervised Learning**: Unsupervised Learning is the method used for the assessment of water quality by analyzing the combined common pollution dataset as it doesn't need any labeled data, structures, or patterns to identify the structural relationship within the

dataset; e.g., via principal component analysis for high-dimensional environmental data. While clustering techniques such as K-means or DBSCAN can identify unknown pollution occurrences, principal component analysis helps reduce dimensionality for large datasets [27].

iii. **Deep Learning**: A variety of satellite-captured data and images are used to understand underwater ecosystems and detect algal growth using various models, such as ResNet, VGGNet, and MobileNet [28].

iv. **Classification vs. Regression in Water Quality**: It helps in classifying water samples based on sensor data, and regression models help in anticipating various other attributes of water such as pH, turbidity, dissolved oxygen levels, and biochemical oxygen demand [29].

v. **Example**: A CNN+ regression layer enables prediction of the amount of nitrate present in the water samples collected simply by observing the color variations from the images.

13.6 INTEGRATION OF CV AND ML FOR ENVIRONMENTAL MONITORING

When CV and ML are combined, a powerful framework for automated, real-time environmental monitoring is formed. When these technologies increase the capacity for capturing images or data automatically using drones or surveillance systems. Certain key elements can then be recognized and secluded using this process. Thus, the data is further investigated by skilled ML models to identify or forecast appropriate water quality indicators.

In a case study conducted in Kerala, India, researchers utilized real-time evaluation and drone-captured information to evaluate riverine water quality applying CV and ML. The system substantially lowered response time by correctly recognizing and categorizing visible pollutants, such as plastic waste and eutrophication, with 92% accuracy [30].

In a different situation, researchers in Shanghai employed a CNN classifier in conjunction with drone-based data to monitor severe algal blooms. Within minutes, the device was able to recognize chlorophyll concentrations in images and could help notify environmental bureaus, reducing the customary response time by almost 60% [31].

13.7 ADVANTAGES AND CHALLENGES

There are many significant advantages to combining CV and ML for water quality monitoring. These systems are optimal for monitoring large and distant bodies of water because of their high agility. Once deployed, they can operate constantly without direct human involvement and have low operating expenses. They are especially useful for timely environmental management because of their remote operation capabilities and capacity to provide real-time alerts or forecast insights. Nevertheless, there are some difficulties with the strategy as well. Large, varied, and high-quality training datasets are essential for efficient performance. Environmental factors such as weather, reflections, and variations in illumination can significantly impact model accuracy and image quality.

Furthermore, one of the main drawbacks of DL models is their incapacity to identify pollutants that do not have obvious indicators, such as colorless or odorless toxins, solely from picture data.

13.8 LIMITATIONS

After all, knowing the truth that CV and ML are advanced tools in detecting water quality, many advantages and challenges still exist with their use in actual environmental settings. One important problem is environmental variability, where weather, climate, and turbidity of water can greatly impact the collection of images and information, resulting in the quality of the images and their accurate prediction. The lack of extensive, annotated datasets, unequal class distributions in infrequent pollution occurrences, and the high costs of human labeling for supervised learning models are all equally important data-related concerns.

Technically speaking, other ML models are naturally challenging to understand, which raises questions regarding their trust and interpretability. The different algorithms can lead to more overfitting, especially when applied to untrained representative data while fitting it into policymaking [32]. Besides these, infrastructure constraints such as poor wavelength, high energy consumption, deployment and maintenance costs of sensor networks and imaging equipment must be considered. AI-related policies and legal implementations in the ecological system must carefully consider issues such as data privacy and governance of the environmental data system while complying with national and international rules [33].

13.9 FUTURE DIRECTIONS AND EMERGING TRENDS

Accurate, flexible, and quick methods are needed to monitor and maintain the criteria for drinking water, and the combination of ML and cutting-edge technology is ushering in a new era of increasingly sophisticated solutions. Explainable model development, multimodal data integration, remote sensing, autonomous edge systems, and transparency enabled by blockchain are some of the major trends.

Explainable AI (XAI): The existing AI models' black-box nature, which restricts openness and stakeholder trust, is one of their main problems. This is the goal of XAI, which offers insights into model judgments that are understandable to humans.

XAI in water quality modeling is becoming increasingly important for risk communication and regulatory compliance. By combining SHAP (SHapley Additive exPlanations) or LIME (Local Interpretable Model-agnostic Explanations) methodologies, for example, environmental scientists can determine which sensor data or visual cues were most important in making a given prediction. This promotes better environmental management decision-making in addition to increasing model accountability [34].

Multimodal Integration: Multimodal data fusion, which combines spectral, hyperspectral, and IoT sensor data with CV image data, is the direction of future water quality systems. This hybrid method improves environmental interpretation and prediction accuracy. An example where RGB imaging is integrated with hyperspectral reflectance data can increase the analysis of chemical pollutants that are not apparent in the actual data [35]. Different reliable models using images and photos can be created with the combination of information such as historical data about water quality and climatic conditions [36].

Remote Sensing and Satellite-Based Forecasting: Remote sensing is also an increasing trend used for water quality monitoring. Nowadays, ML-controlled satellite-captured images are used to forecast vital parameters such as turbidity, sea surface temperature, and chlorophyll-a.

These parameters can also be estimated using local scales such as by using recurrent neural networks (RNNs) or long short-term memory models [37].

Edge and Robotic Systems: Automated technologies such as underwater robots, pond-based IoT sensor networks, and edge computing-enabled water drones are revolutionizing real-time surveillance, which is highly useful in minimizing dependence on cloud connectivity, where collecting and processing data locally lowers expectancy [38]. Such ML and CV-controlled robotic technologies with microcontrollers and underwater cameras can conduct on-site analysis, making them especially suitable in settings with limited resources, such as rural or agricultural areas.

Blockchain for Data Security and Transparency: Blockchain technology provides a reorganized and solid way to store and share water quality data, which is vital for environmental monitoring as data accuracy becomes a major problem. Although the blockchain record guarantees the traceability of sensor readings and model predictions, smart contracts can automate compliance and alarm systems [39].

13.10 ML IMPLEMENTATION

This section describes the implementation of three ML models and analyzes the best model for water quality monitoring. The dataset used in this work is publicly available on the Kaggle website. The dataset attributes are shown in Figure 13.2. In Figure 13.2, it depicts the water quality dataset with 2011 complete entries, demonstrated high data integrity with no missing values.

In Figure 13.3, (potability = 0) is notably higher than that of safe water samples (potability = 1), with unsafe samples more than 1200 and safe samples only reaching the margin of 800. This signifies for that water classified as unsafe is more widespread than potable, indicating a probable alarm for water quality or safety in the population.

The pie chart (Figure 13.4) reveals that a majority of the water samples (59.7%) are classified as not potable, and only 40.3% are considered potable. This highlights a significant inequity, underscoring

```
Data columns (total 10 columns):
 #   Column           Non-Null Count  Dtype
---  ------           --------------  -----
 0   ph               2011 non-null   float64
 1   Hardness         2011 non-null   float64
 2   Solids           2011 non-null   float64
 3   Chloramines      2011 non-null   float64
 4   Sulfate          2011 non-null   float64
 5   Conductivity     2011 non-null   float64
 6   Organic_carbon   2011 non-null   float64
 7   Trihalomethanes  2011 non-null   float64
 8   Turbidity        2011 non-null   float64
 9   Potability       2011 non-null   int64
```

FIGURE 13.2 Water quality parameters dataset.

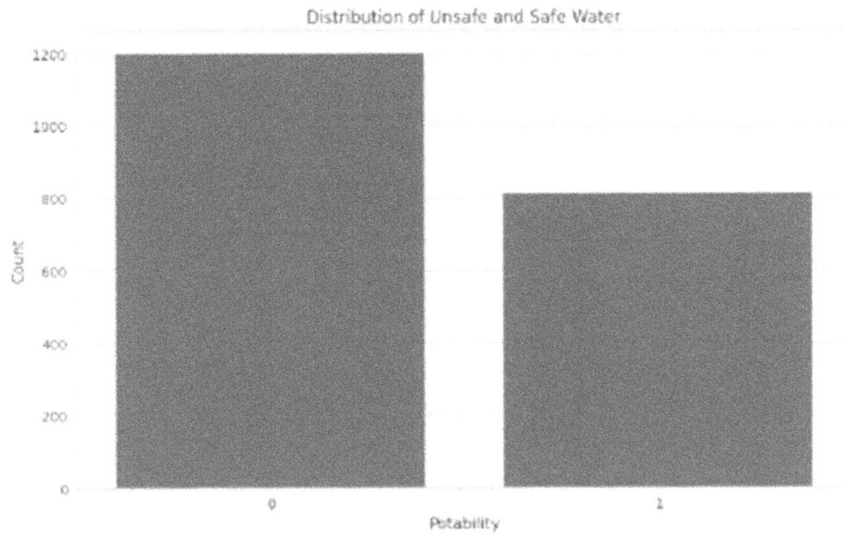

FIGURE 13.3 Bar chart: Comparison of safe and unsafe water.

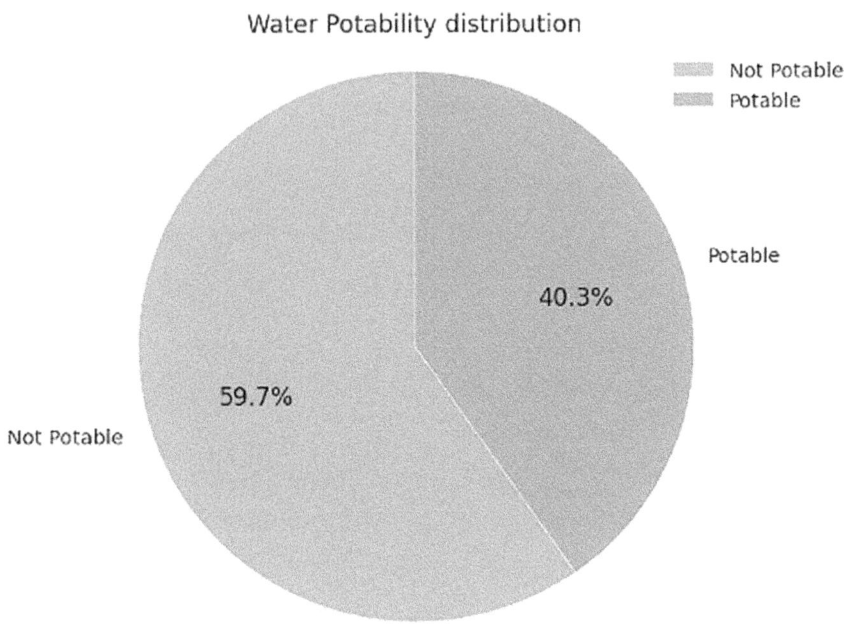

FIGURE 13.4 Pie chart showing water potability distribution.

that unsafe water is nearly 1.5 times more prevalent than safe water. This represents the urgency in addressing issues related to maintaining and retaining water quality for the public.

The correlation matrix in Figure 13.5 confirms that most of the water quality parameters have very weak correlations with each other. A limited negative correlation occurs (–0.163), denoting a weak inverse relationship. Overall, the dataset indicates that the parameters are largely independent, with no strong linear relationships.

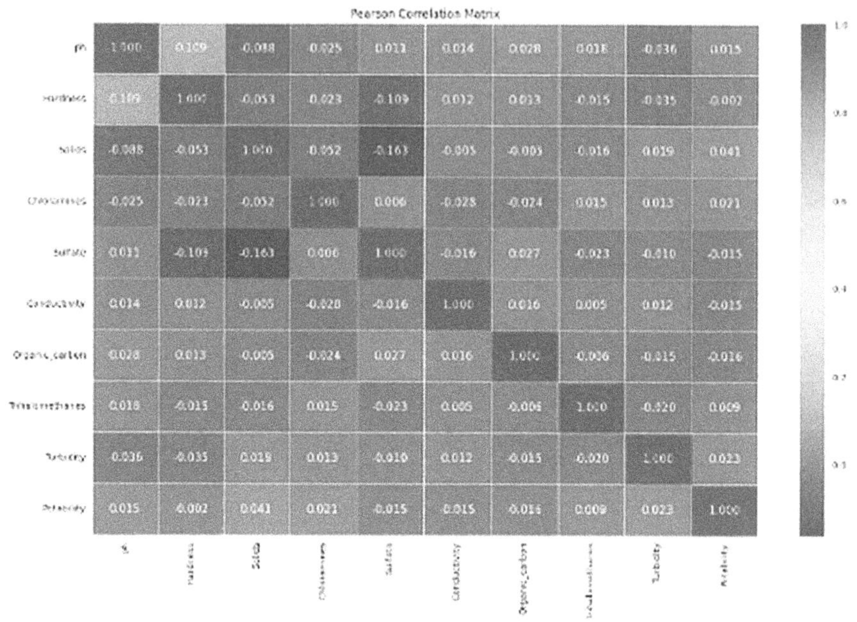

FIGURE 13.5 Pearson correlation matrix.

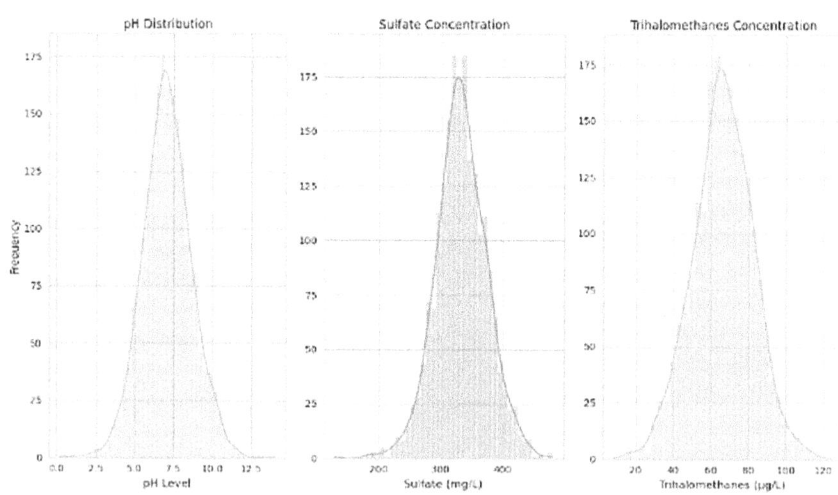

FIGURE 13.6 Histogram distribution.

The histograms in Figure 13.6 indicate that the three parameters exhibit an approximately normal distribution, indicative of stable and consistent patterns.

The confusion matrix (Figure 13.7) shows that the XGBoost Regressor model has some predictive capacity, as it shows both types of errors, particularly false negatives, which could influence reliability in critical applications such as water potability prediction.

Figure 13.8 shows an overall accuracy of 61%, with better performance on non-potable water compared to potable water. The lower recall (0.48) for potable water shows that the model often fails to correctly identify potable water, suggesting a bias toward predicting the non-potable class.

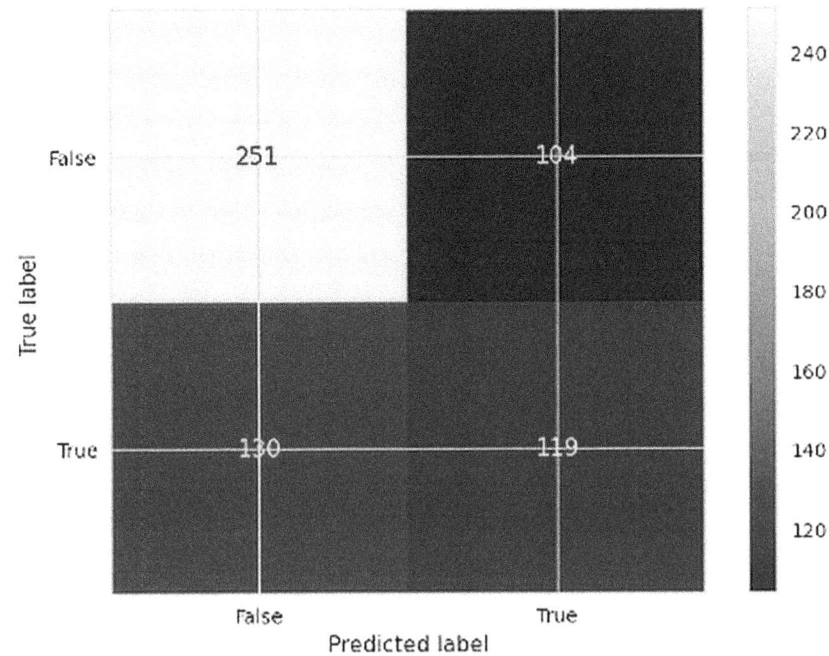

FIGURE 13.7 Confusion matrix: XGBoost Regressor.

	precision	recall	f1-score	support
0	0.66	0.71	0.68	355
1	0.53	0.48	0.50	249
accuracy			0.61	604
macro avg	0.60	0.59	0.59	604
weighted avg	0.61	0.61	0.61	604

FIGURE 13.8 Classification report: XGBoost Regressor.

The confusion matrix in Figure 13.9 shows that the Logistic Regression model makes many mistakes, especially by tagging unsafe water as safe, which makes it unsafe for real-world use without improvements, thus reducing its reliability for safe water quality detection.

The classification report shows a low overall accuracy of 47%, with both precision and recall values being weak, indicating that the model is not adequately predicting usable and non-usable water, rendering the model weak and underperforming, and in need of improvement for reliable estimations (Figure 13.10).

The confusion matrix in Figure 13.11 shows that the model misclassified 97 safe samples as unsafe (false positive) and 142 unsafe samples as safe (false negative). A high number of false negatives undermines the model's consistency, as unsafe water is projected as safe. Thus, this model will yield weak forecasting results as well.

The DTC model achieved 60% accuracy, performing well with a precision score of 0.65 and a recall of 0.73 but shows weak performance on unsafe water (Figure 13.12).

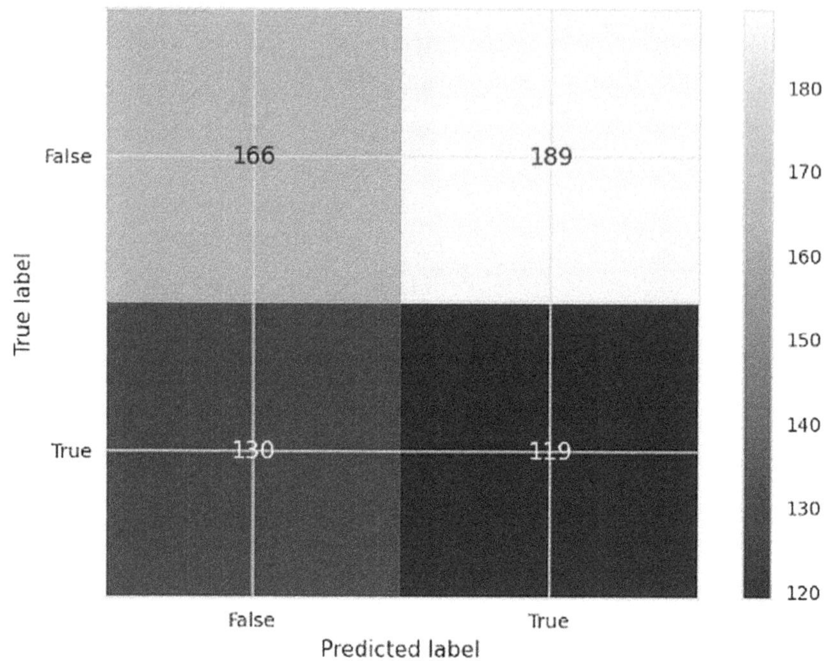

FIGURE 13.9 Confusion matrix: Logistic Regression.

	precision	recall	f1-score	support
0	0.56	0.47	0.51	355
1	0.39	0.48	0.43	249
accuracy			0.47	604
macro avg	0.47	0.47	0.47	604
weighted avg	0.49	0.47	0.48	604

FIGURE 13.10 Classification report: Logistic Regression.

13.11 CONCLUSION

Quality water is not only a basic requirement for human survival but also presents a challenge to provide due to the increasing pollution levels. There are many traditional methods to check and ensure the quality of water.

The use of artificial intelligence technology, especially CV and ML, is transforming water monitoring techniques. This chapter has studied how environmental investigation can be transformed by combining ML models with CV techniques. To achieve a highly effective high-tech distribution, it will also be essential to capitalize on open-access water quality information, array edge devices in underserved areas, and support environmental analysts through capacity-building initiatives. Societies may create vigorous, intelligent, and sustainable water resource management systems that are appropriate for the complexity of the 21st century by embracing AI as an addition to, rather than a replacement for, conventional techniques.

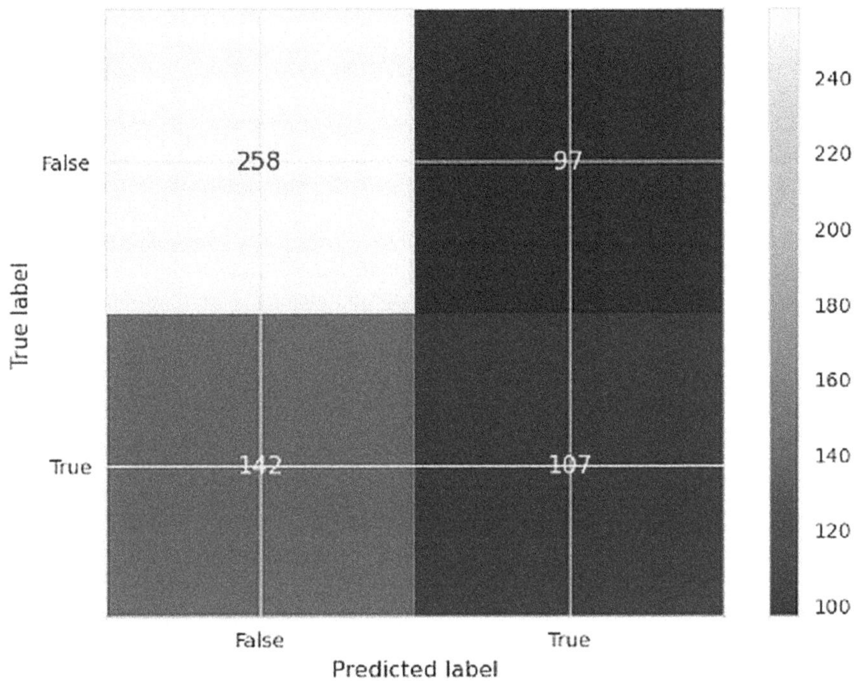

FIGURE 13.11 Confusion matrix: Decision Tree Classifier.

	precision	recall	f1-score	support
0	0.65	0.73	0.68	355
1	0.52	0.43	0.47	249
accuracy			0.60	604
macro avg	0.58	0.58	0.58	604
weighted avg	0.60	0.60	0.60	604

FIGURE 13.12 Classification report: Decision Tree Classifier.

REFERENCES

1. US Environmental Protection Agency. *The facts about nutrient pollution*. Washington, DC: US EPA; 2012 [cited 2025 Jul 26]. Available from: https://www.epa.gov/nutrientpollution
2. Bhuyan MH, Hassan MM, Almogren A. A novel machine learning approach for water quality prediction using IoT. *J Clean Prod*. 2021;287:125435.
3. Nallakaruppan MK, Gangadevi E, Shri ML, et al. Reliable water quality prediction and parametric analysis using explainable AI models. *Sci Rep*. 2024;14:7520. https://doi.org/10.1038/s41598-024-56775-y
4. Kaur A, et al. *Classification and analysis of water quality using machine learning algorithms*. Springer; 2021. https://doi.org/10.1007/978-981-33-4866-0_48
5. Koditala NK, Pandey PS. Water quality monitoring system using IoT and machine learning. In: *Proc Int Conf Res Intell Comput Eng (RICE)*, 2018, p. 1–5. https://doi.org/10.1109/RICE.2018.8509050

6. Kumar J, Sharma B, Gupta P. *IoT enabled advanced water quality monitoring system for smart pond management.* IEEE; 2024. Available from: https://ieeexplore.ieee.org/stamp/stamp.jsp?arnumber=10506512
7. World Health Organization. *Guidelines for drinking-water quality.* Geneva: WHO; 2017. Available from: https://www.who.int/publications/i/item/9789241549950
8. US Environmental Protection Agency. *Water quality standards handbook.* Washington, DC: US EPA; 2023. Available from: https://www.epa.gov/wqs-tech/water-quality-standards-handbook
9. Zhi W, et al. Deep learning for water quality. *Nat Water.* 2024. https://doi.org/10.1038/s44221-024-00202-z
10. World Health Organization. *Guidelines for drinking-water quality*, 4th ed. incorporating the 1st addendum [Internet]. Geneva: WHO; [cited 2025 Jul 26]. Available from: https://www.who.int/publications/m/item/guidelines-for-drinking-water-quality-4th-ed.-incorporating-the-1st-addendum-(chapters)
11. Sinay. What are the main water quality indicators and parameters? [Internet]. 2021 [cited 2025 Jul 26]. Available from: https://sinay.ai/en/what-are-the-main-indicators-of-water-quality/
12. Chapman D, editor. *Water quality assessments: A guide to use of biota, sediments and water in environmental monitoring.* 2nd ed. London: E&FN Spon, Chapman & Hall; 1996.
13. World Health Organization. *Radiation and health (RAD). Guidelines for drinking-water quality*, 4th ed., incorporating the 1st addendum [Internet]. 2017 [cited 2025 Jul 26]. Available from: https://www.who.int/publications/i/item/9789241549950
14. Lenntech. Drinking water standards [Internet]. [cited 2025 Jul 26]. Available from: https://www.lenntech.com/applications/drinking/standards/drinking-water-standards.htm
15. US Environmental Protection Agency. National Primary Drinking Water Regulations [Internet]. 2024 [cited 2025 Jul 26]. Available from: https://www.epa.gov/ground-water-and-drinking-water/national-primary-drinking-water-regulations
16. Sawyer CN, McCarty PL, Parkin GF. *Chemistry for environmental and engineering science.* 5th ed. New York: McGraw-Hill Inc.; 2003, pp. 587–590. Available from: https://www.scirp.org/reference/referencespapers?referenceid=1458612
17. Manahan S. *Environmental chemistry* [Internet]. CRC Press; 2017 [cited 2025 Jul 26]. Available from: https://doi.org/10.1201/9781315160474
18. UN-Water, Damania R. Progress on ambient water quality: Global Indicator 6.3.2 updates and acceleration needs. 2021 Executive Summary [Internet]. 2021 [cited 2025 Jul 26]. Available from: https://www.unwater.org/sites/default/files/app/uploads/2021/09/SDG6_Indicator_Report_632_Progress-on-Ambient-Water-Quality_2021_Executive-Summary_EN.pdf
19. Bhandari S, Bisht KS. Water quality assessment of Alaknanda River in the Srinagar Valley, Garhwal Himalaya, India. *J Mountain Res.* 2023;18(2). https://doi.org/10.51220/jmr.v18i2.20
20. Bureau of Indian Standards. *Indian Standard IS 10500:2012 – Drinking water – Specification* (Second Revision). New Delhi: BIS; 2012.
21. Fu L, You W, Zhang H. Deep learning-based computer vision for environment monitoring. *Nat Mach Intell.* 2020;2(5):245–252. https://doi.org/10.1038/s42256-020-0172-7
22. Zhao Y, Zhang Y, Han Y. Real-time water quality monitoring with AI: Progress, challenges, and opportunities. *Water Res.* 2021;196:116977. https://doi.org/10.1016/j.watres.2021.116977
23. Szeliski R. *Computer vision: Algorithms and applications.* Springer; 2010.
24. Redmon J, Divvala S, Girshick R, Farhadi A. You only look once: Unified, real-time object detection. *Proc IEEE Conf Comput Vis Pattern Recognit.* 2016;779–788. doi:10.1109/CVPR.2016.91.
25. Long J, Shelhamer E, Darrell T. Fully convolutional networks for semantic segmentation. *Proc IEEE Conf Comput Vis Pattern Recognit.* 2015;3431–3440. doi:10.1109/CVPR.2015.7298965.
26. Giri UK, Saxena M, Bordoloi D, Tripathy R, Mohapatra SK, Kumar Sarangi P. Machine learning model for heart disease detection: A comparative analysis of SVM vs KNN. *2024 International Conference on Electrical Electronics and Computing Technologies (ICEECT)*, Greater Noida, India, 2024, pp. 1–5, https://doi.org/10.1109/ICEECT61758.2024.10739169.
27. Grewal MK, Kaur N, Giri UK, Tripathy R, Sarangi PK, Agrawal AK, Medical data analysis and machine learning models for kidney disease prediction: Exploratory study, *2025 International Conference on Ambient Intelligence in Health Care (ICAIHC)*, Raipur, Chattisgarh, India, 2025, pp. 1–6, https://doi.org/10.1109/ICAIHC64101.2025.10957273.
28. Dhiman R, Giri UK, Tiwary A, Tripathy R, Sahoo AK, Sarangi PK, Machine learning based predictive models for lung cancer detection and classification, *2024 International Conference on Cybernation and Computation (CYBERCOM)*, Dehradun, India, 2024, pp. 157–161, https://doi.org/10.1109/CYBERCOM63683.2024.10803196.

29. Giri UK, Goel N, Sabitha P, Sahoo AK, Tripathy R, Sarangi PK, Using machine learning to analyze voice disorders for Parkinson's disease detection, *2024 4th International Conference on Technological Advancements in Computational Sciences (ICTACS)*, Tashkent, Uzbekistan, 2024, pp. 1395–1401, https://doi.org/10.1109/ICTACS62700.2024.10840915.
30. Sharma M, George J, Pillai A. Real-time water quality monitoring using drone imagery and AI models in Indian rivers. *J Environ Inform*. 2022;39(4):56–67.
31. Li X, Xu H, Zhang C. Deep learning and multispectral imaging in river water quality classification. *Remote Sens*. 2021;13(15):3003. https://doi.org/10.3390/rs13153003
32. Doshi-Velez F, Kim B. Towards a rigorous science of interpretable machine learning. arXiv [Preprint]. 2017. https://doi.org/10.48550/arXiv.1702.08608
33. Floridi L, Cowls J, Beltrametti M, Chatila R, Chazerand P, Dignum V, et al. AI4People—An ethical framework for a good AI society: Opportunities, risks, principles, and recommendations. *Minds Mach*. 2018;28(4):689–707. https://doi.org/10.1007/s11023-018-9482-5
34. Arrieta AB, Díaz-Rodríguez N, Del Ser J, Bennetot A, Tabik S, Barbado A, et al. Explainable artificial intelligence (XAI): Concepts, taxonomies, opportunities and challenges toward responsible AI. *Inf Fusion*. 2020;58:82–115. https://doi.org/10.1016/j.inffus.2019.12.012
35. Huang W, Ma Y, Hu W, Song K. Integrating hyperspectral imaging and machine learning for improved water quality assessment. *Remote Sens Environ*. 2021;262:112519. https://doi.org/10.1016/j.rse.2021.112519
36. Zhou J, Hu J, Zhang Y. Image-based water pollution monitoring using machine learning. *J Environ Inform*. 2020;35(2):1–12.
37. Pahlevan N, Sarkar S, Franz BA, Balasubramanian SV, Schott JR. Machine learning and remote sensing for chlorophyll-a retrieval in inland and coastal waters. *Remote Sens Environ*. 2022;276:113011. https://doi.org/10.1016/j.rse.2022.113011
38. Agarwal N, Gupta S, Mehta R. IoT and edge intelligence for water quality monitoring: Opportunities and challenges. *IEEE Internet Things J*. 2022;9(8):6123–6135.
39. Sharma R, Pathak M, Rao S. A blockchain-enabled data provenance framework for trustworthy water quality monitoring. *Environ Model Softw*. 2022;150:105325. https://doi.org/10.1016/j.envsoft.2022.105325

14 Cybersecurity in Smart Water Systems

Abhishek Sundi and Priyakanksha Mishra

14.1 INTRODUCTION

In this modern era of digitalization and automation, we can see intelligence models handling human tasks in every sector of occupation; it has a similar effect in water management systems as well. Hence, we can now witness a revolutionized form of water management systems that gives rise to smart water systems, combine advanced technologies such as the Internet of Things (IoT), Supervisory Control and Data Acquisition (SCADA) systems, and artificial intelligence (AI). It is also evident that wherever these new technology models are integrated, we have only seen advancement. Similarly, this integration gives advantages in optimizing operations, improving sustainability, and, most importantly, enabling predictive maintenance of the systems (Liu et al., 2022). Moreover, we can see advancements on a large scale by supporting facilities such as efficient water distribution to the public, real-time monitoring of water quality, and rapid fault detection upon any complaints, thereby addressing growing urbanization and climate-related challenges. However, the same interconnectedness introduces critical cybersecurity risks; for example, legacy SCADA systems are often designed without modern security features, and poorly secured IoT endpoints create vulnerable entry points for malicious actors (Chin et al., 2020).

To back up these talks, many real-world scenarios are already present which show the gravity of these risks, just like in the cyberattack case of Oldsmar, Florida, in the year 2021, in which we can see how adversaries can manipulate factors of grave concern such as the chemical dosing in a treatment facility, which directly threatens the health of the public (ICS-CERT, 2021). In some similar incidents, we can witness that water utilities worldwide continue to face escalating risks including ransomware, insider threats, and targeted attacks against critical control systems (Kumar et al., 2019). We are all already aware that sometimes, even a common necessity such as water, which is present everywhere, can create havoc when not present in the right condition or right amount, as water remains an essential service; even small disruptions can have cascading effects on health, safety, and economic stability. Therefore, developing robust cybersecurity frameworks, supported by AI-driven intrusion detection and global policy alignment, is crucial to safeguarding water utilities in the digital era (Liu et al., 2022).

DOI: 10.1201/9781003714217-14

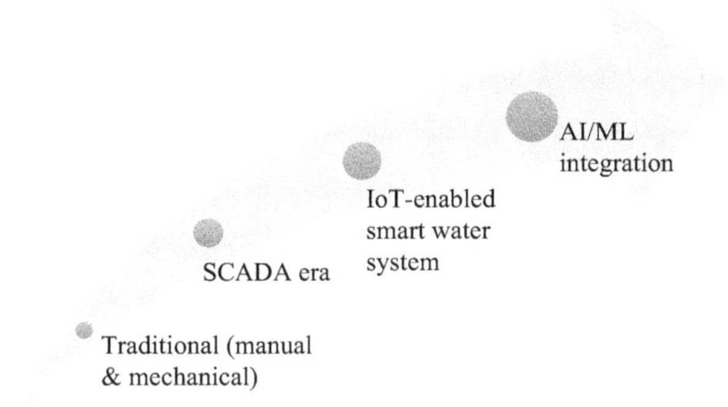

FIGURE 14.1 Progressive evolution of water systems highlighting key transitions from traditional mechanical setups to SCADA, IoT-enabled smart water, and AI/ML integration.

14.2 EVOLUTION OF SMART WATER SYSTEMS

We know that, like anything else, water management has also traditionally relied on manual and mechanically controlled infrastructure, but increasing urban demand and climate variability have accelerated the adoption of digital solutions. If one were to talk about the history of smart water systems, it should be kept in mind that the whole concept of smart water systems has emerged with the integration of SCADA platforms, which enabled centralized control and remote monitoring of distribution and treatment processes (Amin & Wollenberg, 2005). Over time, these systems were further enhanced through IoT sensors, cloud-based analytics, and advanced wireless networks, enabling utilities to achieve real-time visibility of leaks, contamination, and pressure fluctuations (Pereira et al., 2020).

When we look at the renowned idea of integrating smart water systems as a solution to tackle cyberattacks or maintain the purity of water systems, we can understand that the transition was shaped by global trends in smart cities, where water infrastructure is treated as a cyber-physical system and includes many advanced techniques such as IoT-enabled smart meters. For instance, it allows for precise consumption tracking, predictive maintenance, and user engagement, thereby improving efficiency and reducing non-revenue water losses (Hadjiefthymiades et al., 2019). In past studies, it has also been seen that the successful integration of AI and machine learning (ML) has further strengthened facilities such as forecasting, anomaly detection, and automated decision-making in water utilities (Liu et al., 2022). However, while these developments mark a significant leap in water management, they also establish a foundation where cybersecurity must be prioritized, as the interconnected nature of digital utilities increases vulnerability to malicious exploitation (Figure 14.1).

14.3 CYBERSECURITY THREAT LANDSCAPE

With the growing field of digital updates, our water systems are also evolving into an interconnected cyber-physical infrastructure. Simultaneously, with the integration of cyber advancements, there are several cybersecurity threats which are increasingly exposed at a wide range. We also know that the traditional SCADA systems were originally designed for greater reliability and process control rather than focusing on security measures, which is why water management systems were vulnerable majorly,

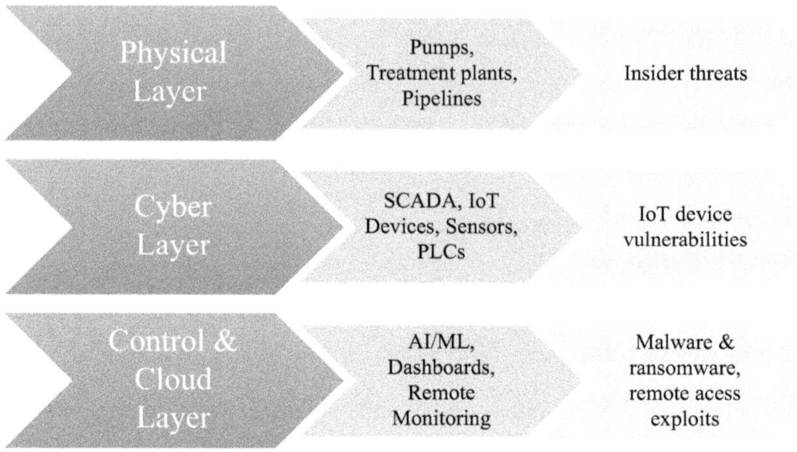

FIGURE 14.2 Cybersecurity threat landscape in smart water systems, illustrating how malware, IoT vulnerabilities, remote exploits, and insider threats target different infrastructure layers.

due to weak authentication, outdated protocols, and inadequate patching (Cárdenas et al., 2008). Moreover, in past few research studies, it was seen that the integration of IoT devices further expands the attack surface in water systems because of using low-cost hardware such as sensors and controllers with limited encryption and poor access control (Kumar et al., 2019).

The severity and importance of these cyberattacks and proper security from them can be accurately measured by reviewing many of the past real-world attacks, such as the one that happened in Oldsmar, Florida, in the year 2021, in which a cyberattack demonstrated how adversaries can exploit remote access vulnerabilities to manipulate chemical dosing in a treatment plant, posing direct threats to public health (ICS-CERT, 2021). Another example can be set by studying the reports published by the United Kingdom and Israel, which show how targeted ransomware and disruption attempts emphasize the fact that both state-sponsored and criminal actors view critical water infrastructure as high-value targets (Ziade et al., 2021). Moreover, beyond these external threats, some insider risks such as disgruntled employees or negligent operators remain a persistent concern in utility cybersecurity (Chin et al., 2020).

Till date, we have seen that collectively these challenges illustrate a rapidly evolving threat landscape where water systems are exposed not only to conventional malware but also to sophisticated, multi-vector attacks. This is why it is very important to address these vulnerabilities in time, which requires a layered defence strategy and proactive monitoring tailored to the unique cyber-physical nature of water utilities (Figure 14.2).

14.4 CASE STUDY: OLDSMAR WATER TREATMENT PLANT ATTACK 2021

14.4.1 Overview of the Incident

One of the most famous cases in history related to a cyberattack in a water management system is the attack that occurred on Oldsmar, Florida's water treatment facility in February, 2021. This attack underscored the vulnerabilities of small-scale water utilities. The basic overview of this case is that an attacker gained unauthorized remote access to the plant's SCADA system through TeamViewer, a legitimate remote

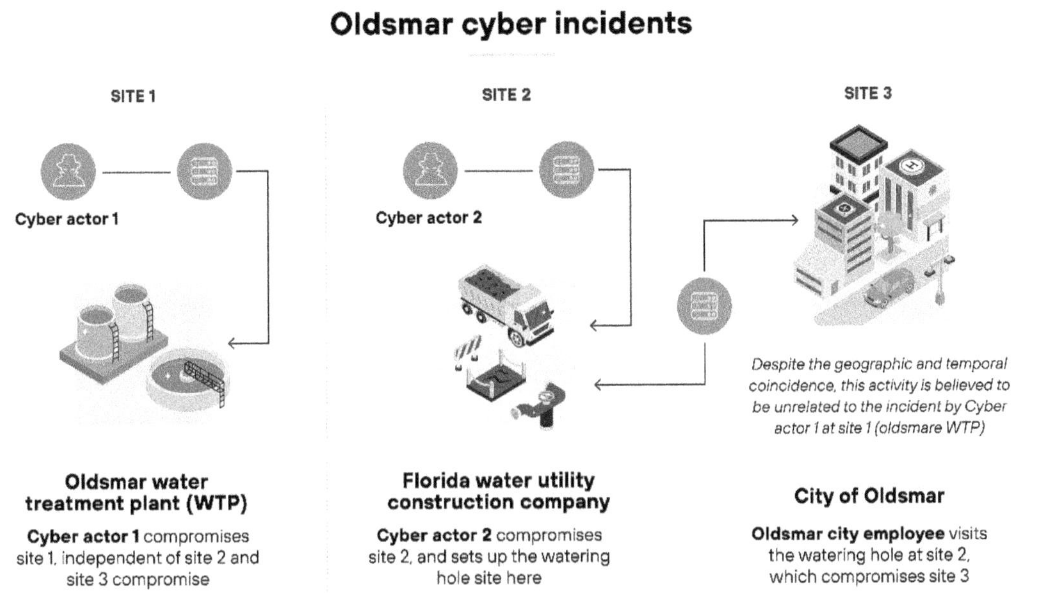

FIGURE 14.3 A diagrammatic overview of the Oldsmar cyberattack incident.

desktop tool used for maintenance. Once the attacker was inside the system, the adversary attempted to alter the chemical dosing by increasing the level of sodium hydroxide (lye) from 100 parts per million (ppm) to 11,100 ppm. It is noteworthy that such manipulation, if undetected, could have caused serious public health risks, including chemical burns and water contamination (CISA, 2021). Fortunately, operators observed the activity in real time and immediately reversed the changes before any harm occurred.

14.4.2 Lessons Learned and Implications

From this case, we can see how important it is to keep a highly secured system if it is integrated with the cyber world or internet. The Oldsmar incident highlights the dangers of insecure remote access, reliance on outdated software, and inadequate authentication protocols in water utilities (Pipes et al., 2022). On a wider glance, we can understand that many facilities continue to depend on consumer-grade remote tools rather than hardened industrial solutions. Furthermore, the event demonstrated how even unsophisticated attackers can exploit weak cyber hygiene to threaten critical infrastructure. As a result, U.S. agencies such as the EPA and CISA emphasized the need for stronger cybersecurity frameworks, improved operator training, and the adoption of layered defence mechanisms tailored for the water sector (Pipes et al., 2022) (Figure 14.3).

14.5 VULNERABILITIES IN SMART WATER SYSTEMS

14.5.1 IoT Device Vulnerabilities

Everyone wants to save money by buying cheap stuff in places where more advanced and expensive tools are needed, but sometimes this habit can make us pay a hefty sum amount afterward. Similarly, the rapid deployment of low-cost IoT devices in water utilities has created significant attack surfaces,

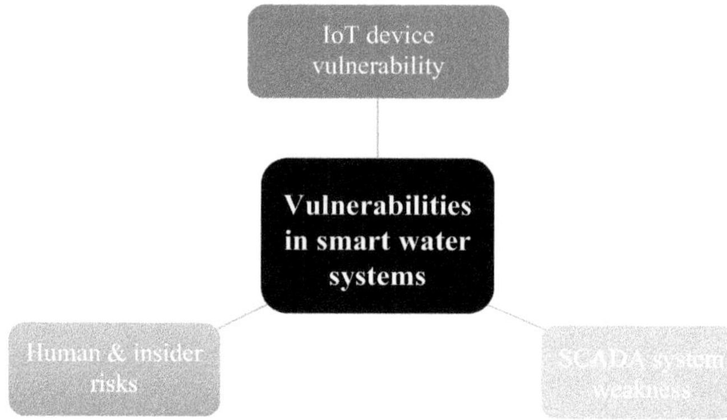

FIGURE 14.4 Classification of vulnerabilities in smart water systems, illustrating risks associated with IoT devices, SCADA infrastructures, and human or insider factors.

resulting in many sensors and controllers operating with default credentials; they also lack strong encryption and are not regularly updated, leaving them exposed to unauthorized access (Sadeghi et al., 2015). Compromised IoT nodes can serve as entry points for larger cyber-physical attacks, enabling adversaries to manipulate data streams or disrupt operational continuity.

14.5.2 SCADA System Weaknesses

If we look at the SCADA systems closely, we will understand that they were originally designed with the prior motive of availability and efficiency in mind; however, the purpose of cybersecurity was not considered at that time, even though it is central to automation and process control. To support this fact about SCADA systems we can look at their reliance on legacy protocols, unpatched software, and unsecured remote access solutions, which ultimately make them vulnerable to intrusion (Knapp & Langill, 2014). Moreover, once compromised, attackers can alter chemical dosing, disable alarms, or cause service outages, as demonstrated in incidents such as Oldsmar (ICS-CERT, 2021).

14.5.3 Human and Insider Risks

However, the most common and persistent vulnerability is human error and insider threats, which depend on some basic facts about human understanding and vigilance regarding cybersecurity and cyberattacks. Some common practices in the public that make them prone to cyberattacks include weak password practices, insufficient cybersecurity training, and disgruntled employees; all these can compromise water system security (Chin et al., 2020). Moreover, some social engineering attacks, such as phishing, further exploit human weaknesses to bypass technical safeguards.

Collectively, these vulnerabilities highlight the urgent need for layered security strategies, integrating technical, organizational, and human-focused defences to protect critical water infrastructure (Figure 14.4).

14.5.4 Cybersecurity Solutions and Defence Mechanisms

An advanced proactive cybersecurity approach is what the complexity and interconnectivity of smart water systems demand, which can address both technical and human vulnerabilities, because traditional perimeter-based defences are no longer sufficient in protecting these infrastructures from evolving threats such as ransomware, zero-day exploits, and insider attacks. In place of still using traditional methods, utilities must adopt a holistic strategy that integrates AI/ML-based detection, layered architecture, and operational best practices.

14.5.5 AI/ML-Based Intrusion Detection

In the past few decades, we have already seen that the integration of AI and ML into traditional methods only brings endless advancement and enhancement in the techniques by providing advanced capabilities in detecting anomalies and identifying cyber intrusions within SCADA and IoT environments. However, we already know that, unlike rule-based systems that rely on predefined signatures, AI/ML algorithms analyse large volumes of real-time data, which includes network traffic, command signals, and sensor readings to identify unusual patterns (Liu et al., 2022). For the purpose of advanced detection of unauthorized control commands and abnormal water treatment operations, the successful application of techniques such as supervised learning, clustering, and deep neural networks in regular processes is much needed (Khan et al., 2021); this adaptability is critical for recognizing zero-day attacks and minimizing false positives. Moreover, AI-driven predictive analytics can forecast potential system failures, which enables preventive interventions before disruptions occur.

14.6 LAYERED CYBERSECURITY ARCHITECTURE

Many of the cyberattacks that have happened in the past have taught us one important fact: a defence-in-depth strategy is required for safeguarding water utilities. At the device level, IoT sensors and controllers should be secured with strong authentication, regular firmware updates, and end-to-end encryption (Knapp & Langill, 2014). Then, at the network level, SCADA systems must be isolated from public-facing networks through segmentation, firewalls, and virtual private networks. At the enterprise level, Security Information and Event Management platforms can consolidate logs and detect anomalies across the infrastructure (Chin et al., 2020). Finally, redundancy and fail-safe mechanisms ensure continuity of operations even during successful breaches. Ultimately, we will notice that this layered approach distributes defensive barriers, ensuring that a single point of compromise does not endanger the entire system.

14.6.1 Operational and Policy Measures

With the evolving field of cyber vigilance and cybersecurity, it is evident that beyond technical defences, organizational practices are equally critical. This is why utilities should implement role-based access controls, enforce strong password and multi-factor authentication policies, and regularly train staff to recognize phishing and social engineering tactics (Ziade et al., 2021). In addition to this, if we introduce some more cybersecurity drills and incident response exercises into the weekly routine, it will enhance preparedness for real-world scenarios. At a broader level, alignment with frameworks such as the NIST Cybersecurity Framework and national cybersecurity strategies ensures consistency and compliance with global best practices (NIST, 2018). Moreover, public–private collaboration also plays a key role in information sharing and rapid response to emerging threats (Figure 14.5).

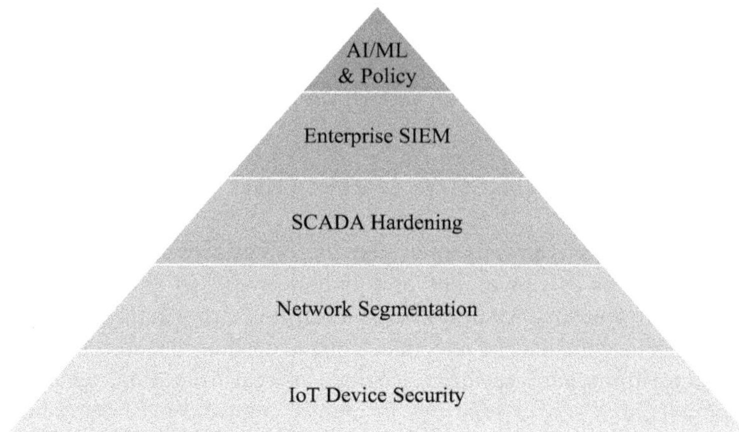

FIGURE 14.5 Layered cybersecurity defence model for smart water systems, illustrating security measures from IoT device protection to AI/ML-based monitoring and policy enforcement.

TABLE 14.1 Comparative overview of national and international cybersecurity frameworks relevant to smart water systems

FRAMEWORK	*FOCUS AREA*	*APPLICATION TO WATER SYSTEMS*
India – NCSS (2020)	Risk assessment, incident response, workforce training	Implementation gaps in the water sector
US – NIST Framework (2018)	Identify, Protect, Detect, Respond, Recover	Widely adopted in utilities, adaptable to water
EU – NIS Directive	Incident reporting, operator responsibility	Legal obligation for essential services

14.6.2 Policy and Regulatory Frameworks

Till date, we have already seen that smart water systems are a very critical infrastructure that requires strong regulatory oversight and policy frameworks to mitigate cybersecurity risks. National and international agencies have recognized the vulnerability of utilities and developed guidelines to strengthen resilience through compliance, monitoring, and risk-based approaches (Figure 14.6). Table 14.1 compiles all these policies in a comparative format.

14.7 CASE STUDIES AND GLOBAL PERSPECTIVES

14.7.1 Real-World Case Studies

See Table 14.2.

14.7.2 Global Perspective

From these global incidents, we can understand that facts such as the reliance of water management systems on legacy infrastructure, poor access controls, and insufficient workforce training were the

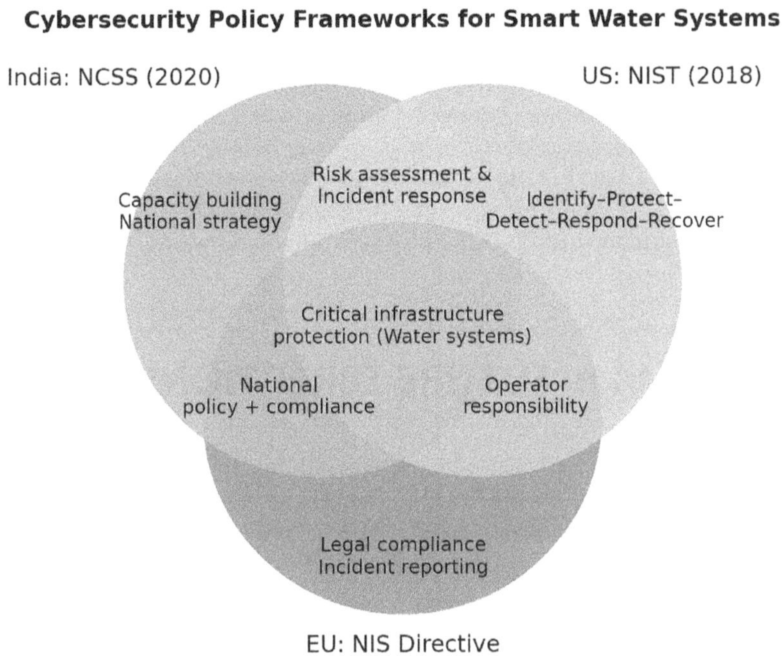

FIGURE 14.6 Comparative view of cybersecurity policy frameworks (NCSS, NIST, and NIS Directive) showing their shared focus on securing critical water infrastructure (MeitY, 2020).

TABLE 14.2 Documented cyber incidents targeting water systems, arranged chronologically, highlighting diverse threat actors and vulnerabilities

S. NO.	*CASE SUMMARY*	*REFERENCE YEAR*
1	Maroochy Shire Wastewater Attack, Australia (2000) – A disgruntled contractor hacked SCADA controls, releasing ~800,000 litres of untreated sewage.	Slay and Miller (2007)
2	Harrisburg, USA (2006) – Attackers gained access to a water utility's control system through the internet, exposing weak remote access protocols.	U.S. GAO (2007)
3	Canal Pumping Station, USA (2011) – A SCADA hack disrupted water pumps in Illinois; initially suspected as foreign intrusion, later attributed to insider negligence.	ICS-CERT (2011)
4	Kemuri Water Company Breach, USA (2016) – Hackers manipulated water flow and chemical treatment data, exposing SCADA vulnerabilities.	Verizon DBIR (2016)
5	Onslow Water and Sewer Authority (ONWASA), USA (2018) – Ransomware attack disrupted billing and operational systems; recovery costs exceeded $2 million.	ONWASA (2018)
6	Israel Water Facilities Attacks (2020) – Suspected state-sponsored cyberattacks targeted water pumps and wastewater plants via PLCs.	Ziade et al. (2021)
7	Oldsmar Water Treatment Plant, USA (2021) – Attacker remotely accessed SCADA via TeamViewer, attempted to increase sodium hydroxide dosing 100-fold.	ICS-CERT (2021)
8	UK Water Utility Ransomware Attack (2021) – A ransomware infection hit IT systems of a water supplier, disrupting internal operations.	Water UK (2021)
9	South Staffordshire Water Hack, UK (2022) – Cybercrime group claimed access to IT systems of water utility serving 1.6 million customers.	NCSC UK (2022)
10	Clop Ransomware on Veolia North America (2023) – Hackers exfiltrated customer data from water services giant, highlighting rising data theft risks in utilities.	Veolia (2023)

main points that led to severe attacks on them. This is why it is very important to establish resilience through layered defences by combining technical measures such as segmentation and intrusion detection with organizational practices such as training and incident response drills. Moreover, on an international level, collaboration between governments, utilities, and research bodies is crucial for sharing threat intelligence and best practices (IWA, 2020).

Therefore, we can see from these facts that the global perspectives emphasize that while technological solutions such as AI/ML intrusion detection are promising, governance, awareness, and cross-border cooperation remain equally critical for securing digital water utilities.

14.8 FUTURE DIRECTIONS

After stating all these challenges and problems, we learned that cyber threats to smart water systems grow in frequency and sophistication, which led to the conclusion that the future of water infrastructure security will rely on proactive innovation and long-term resilience planning. It will need emerging technologies coupled with governance reforms and workforce development to shape the next phase of protection for critical water resources.

14.8.1 Emerging Technologies

If we specifically talk about the emerging technologies for smart water systems, then it is impossible to avoid AI and ML models as advancements. Moreover, technologies such as blockchains and digital twins are expected to redefine cybersecurity in water systems as well. In addition to this, we will see some AI-driven anomaly detection continue to evolve, employing deep learning and reinforcement learning to improve accuracy in identifying malicious behaviour within SCADA and IoT networks (Liu et al., 2022). With the high demand for privacy and security tools, blockchains can also be used for enhancing data integrity by ensuring tamper-proof transaction logs, while digital twins, virtual replicas of water infrastructure, can simulate cyberattack scenarios to test defence mechanisms before deployment (Rathore et al., 2021).

14.8.2 Resilience and Adaptive Infrastructure

Cyber resilience is a concept that should be embraced by future water utilities, emphasizing the fact that not only prevention but also rapid recovery from incidents is important. This concept includes embedding redundancy into system design, creating self-healing networks, and developing automated failover mechanisms to maintain essential services during disruptions (Cherdantseva et al., 2016). Moreover, it is already known that adaptive infrastructure will rely on predictive analytics and real-time monitoring to detect threats early, reducing the risk of cascading failures across interconnected networks.

14.8.3 Capacity-Building and Collaboration

For the betterment of smart water systems, the best thing future research can integrate is applying human and institutional capacity-building, and training programmes for operators, engineers, and cybersecurity professionals will be crucial to address the current skill gap (Ziade et al., 2021).

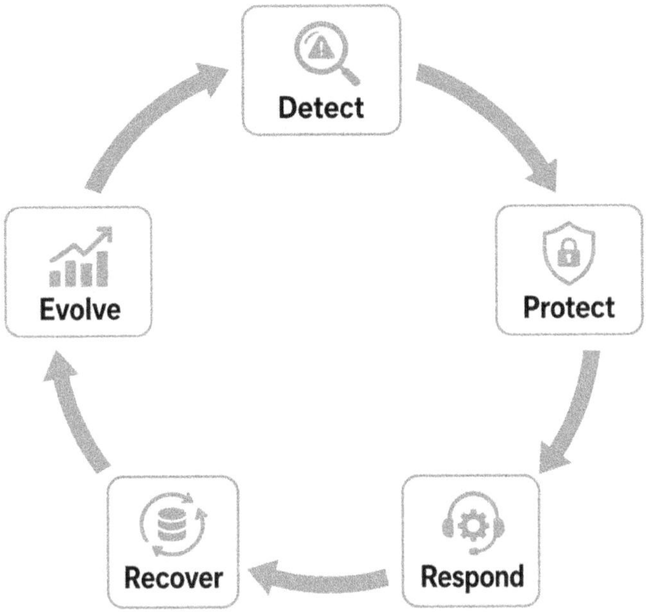

FIGURE 14.7 Cybersecurity resilience cycle for future smart water systems.

Moreover, governments and utilities should strengthen public–private partnerships, fostering collaboration across sectors and borders for effective threat intelligence sharing. As we already know, some global organizations, such as the International Water Association (IWA), will play a critical role in aligning best practices and ensuring that even resource-constrained utilities can adopt modern defences (IWA, 2020) (Figure 14.7).

14.9 CONCLUSIONS

After all this discussion on the cybersecurity measures for water systems, we have come to the conclusion that the digital transformation of water utilities, driven by IoT, SCADA, and AI integration, has brought significant benefits in operational efficiency, sustainability, and service reliability. However, this same evolution has created new cyber-physical vulnerabilities that adversaries are increasingly exploiting. Case studies, such as the Oldsmar incident in the USA and the state-sponsored attacks on Israel's water infrastructure as well as ransomware incidents, teach us about the importance of integrating more technical and layered security measures into our water management systems. Moreover, the UK and USA demonstrate that cyber threats to water utilities are not hypothetical but real, recurring, and escalating.

This chapter highlighted the threat landscape and pinpointed vulnerabilities in IoT devices, legacy SCADA systems, and human factors, while emphasizing the importance of layered cybersecurity architectures and AI/ML-based intrusion detection systems as defence mechanisms. Policy frameworks such as India's National Cyber Security Strategy, the U.S. NIST Cybersecurity Framework, and the EU NIS Directive illustrate how nations are beginning to address these risks, but global case studies underscore that gaps in implementation, awareness, and resources remain pressing challenges.

REFERENCES

Amin, M., & Wollenberg, B. (2005). Toward a smart grid: Power delivery for the 21st century. *IEEE Power and Energy Magazine, 3*(5), 34–41. https://doi.org/10.1109/MPAE.2005.1507024

Cárdenas, A. A., Amin, S., & Sastry, S. (2008). Research challenges for the security of control systems. *Proceedings of the 3rd Conference on Hot Topics in Security*, San Jose, CA, 1–6. USENIX Association.

Cherdantseva, Y., Burnap, P., Blyth, A., Eden, P., Jones, K., Soulsby, H., & Stoddart, K. (2016). A review of cyber security risk assessment methods for SCADA systems. *Computers & Security, 56,* 1–27. https://doi.org/10.1016/j.cose.2015.09.009

Chin, T., Huang, C., & Xu, H. (2020). Cybersecurity for critical infrastructures: A review of attacks and defense in SCADA systems. *Journal of Information Security and Applications, 54,* 102596. https://doi.org/10.1016/j.jisa.2020.102596

CISA. (2021). *Unprotected remote access and SCADA vulnerabilities: Lessons from Oldsmar water treatment attack.* U.S. Cybersecurity and Infrastructure Security Agency. https://www.cisa.gov

Government Accountability Office (GAO). (2007). *Critical infrastructure protection: Multiple efforts to secure control systems are under way, but challenges remain* (GAO-07–1036). U.S. Government Accountability Office. https://www.gao.gov/products/gao-07-1036

Hadjiefthymiades, S., Kolomvatsos, K., Anagnostopoulos, C., & Zafeiropoulos, A. (2019). Smart water management in smart cities: A cyber-physical approach. *Sensors, 19*(1), 62. https://doi.org/10.3390/s19010062

ICS-CERT. (2011). *Public water district cyber intrusion.* Industrial Control Systems Cyber Emergency Response Team. https://www.cisa.gov

ICS-CERT. (2021). *Advisory on Florida water treatment facility intrusion.* U.S. Department of Homeland Security. https://www.cisa.gov/news-events/cybersecurity-advisories/aa21-042a

International Water Association (IWA). (2020). *Cybersecurity principles for digital water infrastructure.* IWA Publishing.

Khan, R., Gani, A., Wahab, A. W. A., & Guizani, M. (2021). Industrial control system security: Requirements, threats, and countermeasures. *Computer Networks, 190,* 107950. https://doi.org/10.1016/j.comnet.2021.107950

Knapp, E. D., & Langill, J. T. (2014). *Industrial network security: Securing critical infrastructure networks for smart grid, SCADA, and other industrial control systems* (2nd ed.). Syngress.

Kumar, R., Khan, R. A., & Alenezi, M. (2019). Evaluating the cybersecurity of smart water management systems. *Journal of King Saud University – Computer and Information Sciences, 31*(1), 10–16. https://doi.org/10.1016/j.jksuci.2017.06.001

Liu, Y., Li, Y., & Zhang, C. (2022). AI-enabled anomaly detection in SCADA systems: A review. *Journal of Water Resources Planning and Management, 148*(7), 04022045. https://doi.org/10.1061/(ASCE)WR.1943-5452.0001584

Ministry of Electronics and Information Technology (MeitY). (2020). *National Cyber Security Strategy.* Government of India. https://www.meity.gov.in/ministry/our-groups/details/cyber-security-division-gNxIDOtQWa

National Cyber Security Centre (NCSC). (2022). *South Staffordshire water: Cyber incident.* UK NCSC. https://www.ncsc.gov.uk

National Institute of Standards and Technology (NIST). (2018). *Framework for improving critical infrastructure cybersecurity* (Version 1.1). U.S. Department of Commerce. https://nvlpubs.nist.gov/nistpubs/CSWP/NIST.CSWP.04162018.pdf

ONWASA. (2018). *Onslow Water and Sewer Authority cyberattack report.* Onslow Water and Sewer Authority, North Carolina. https://www.onwasa.com

Pereira, R. S., Alves, D., & Martins, A. (2020). Smart water grid: A cyber-physical system perspective. *Journal of Cleaner Production, 263,* 121202. https://doi.org/10.1016/j.jclepro.2020.121202

Pipes, L., Paliath, S., & Lambert, D. (2022). Cybersecurity challenges in water utilities: Insights from the Oldsmar incident. *Utilities Policy, 76,* 101356. https://doi.org/10.1016/j.jup.2022.101356

Rathore, H., Mohamed, A., & Guizani, M. (2021). A survey of blockchain applications in smart grids and smart water systems. *IEEE Access, 9,* 125360–125376.

Sadeghi, A. R., Wachsmann, C., & Waidner, M. (2015). Security and privacy challenges in industrial Internet of Things. *Proceedings of the 52nd Annual Design Automation Conference (DAC),* San Francisco, CA, 1–6. ACM. https://doi.org/10.1145/2744769.2747942

Slay, J., & Miller, M. (2007). Lessons learned from the Maroochy water breach. In E. Goetz & S. Shenoi (Eds.), *Critical infrastructure protection* (pp. 73–82). Springer. https://doi.org/10.1007/978-0-387-75462-8_6

Veolia. (2023). *Veolia North America confirms data breach following Clop ransomware attack.* Veolia Water Technologies. https://www.veolia.com

Verizon. (2016). *Data Breach Investigations Report (DBIR).* Verizon Enterprise Solutions. https://www.verizon.com/business/resources/reports/dbir

Water UK. (2021). *Cybersecurity in the water sector: Annual review.* Water UK Publications. https://www.water.org.uk

Ziade, J., Chotard, L., & Halgand, J. (2021). Cybersecurity challenges for water utilities: Lessons from recent attacks. *Utilities Policy, 73,* 101299. https://doi.org/10.1016/j.jup.2021.101299

Water Quality Monitoring with Computer Vision and ML

15

Kapil Sethi, Anu Thakur, Vishal, Mayank Attri, Kavita Dhiman, and Abhineet Anand

15.1 INTRODUCTION

Water is the basis of life, a requirement for consumption, agriculture, industry, and ecological equilibrium. However, the availability of clean and safe water has become a global challenge as the health of water supplies declines. Increased population, urban development, industrialization, and climate change attributes have severely deteriorated the chemical and physical quality of surface water (rivers, lakes, and reservoirs), and also groundwater sources (Barzegar et al., 2020). According to estimates made by the World Health Organization (WHO), billions of people around the globe drink water that does not meet safety standards and is susceptible to waterborne diseases and ecological collapse. Solutions to tackle the problem will require new monitoring methods that do not merely rely on laboratory tests. Water quality monitoring (WQM) is the repeated appraisal of the physical, chemical, or biological composition of water in order to evaluate characteristics to find specific conditions of use. Traditionally, these assessments have relied on extensive sample collection and laboratory analysis, where the accuracy of the assessment, oftentimes, has come at the cost of efficiency (Sanya et al., 2022). For some developing regions where laboratory and staff available are already limited, the capacity and frequency of assessment is further hampered (Li et al., 2021). In addition, by the time laboratory results are available, the pollution event may have dispersed or increased in severity, limiting the possible options for remediative response. These weaknesses in traditional approaches to water assessment and monitoring have highlighted the need for perhaps more agile, affordable, and accurate ways to monitor and understand (Sethi et al., 2017).

DOI: 10.1201/9781003714217-15

Advancements in AI, computer vision, and machine learning (ML) have opened up new options regarding water quality assessments. Prior to the availability of these more facile technological instruments, previous water quality measurement tools would have likely been reliant primarily on laboratory analysis (which exposed the environment, overall health, or food product to considerable delays), through the analysis of either video or still images. Along with the improvement in timing, we can act in a much more timely manner, if not in real time, allowing us to monitor, confront, and deal proactively with changing water conditions instead of responding to confirmation and the much-too-late act of reframing that separates us from our own exposure (Panwar & Sethi, 2022). Not only that, but these image analysis systems are in a significant push towards automating, thinking of them as traditional screening systems where, in essence, your limit is your own capacity, the conditions where we could be fatigued, and the inconsistency of your conclusions (Gupta et al., 2022). Automation is a particularly useful advantage when activity is remote or when there is a potential risk of accessing a physical location for inspections when it could be risky (e.g., in floodplains or near a pollution input point, or hazardous shorelines). With access to advanced imaging systems and AI automation for analysis, this presents opportunities for scaled, ongoing monitoring of sites in near-real-time through the entire river basin, in urban reservoir settings, or across extensive irrigation systems, and with automated systems, there is no increased cost to the operation.

Computer vision, which is a distinct form of AI, is part of the new monitoring approaches described above. Computer vision's goal is to perceive, analyse, and remember visual information in ways that can be similar to, but likely more accurate than, human perception (Mohan et al., 2025). When used for WQM, computer vision algorithms can analyse pictures or video from a wide variety of sources, including stationary bank-side cameras, mobile sensors on autonomous surface vehicles, aerial images or video captured by drones, or satellite images (Sethi & Jaiswal, 2022). A computer vision algorithm can find, quantify, and measure visual indicators of water impairment, such as colour changes from chemical pollutants, turbidity from sedimentation runoff, oil sheen from industrial spills, or colour changes, such as green-tinted water from algal blooms. Computer vision algorithms can find visual impairments both in the visible spectrum and sometimes in the multispectral or hyperspectral light signatures, the latter of which can identify light signatures outside human eyesight perception (Thakur & Sethi, 2025). For example, near-infrared imaging can determine if aquatic vegetation is healthy or stressed, while thermal imaging can identify temperature changes associated with industrial discharge or thermal pollution of freshwater. Unlike human observers, who might not notice slow and/or minor changes in visual perception, computer vision algorithms can be trained to measure changes using historical images.

Once trained, the historical images act as comparisons so that change can be detected, anomalies can be identified, and predictive models can be made (Meng et al., 2022). This improves accuracy and enables water resources management decisions to be carried out with data in mind. Much of the time, ML enhances the performance of computer vision or image recognition because it offers a technology wherein systems can learn and enhance their accuracy over time. For example, many different types of models, and in this case convolutional neural networks (CNNs), are good at recognizing complex sequential visual cues associated with certain water quality parameters. With enough previous samples and experience, CNNs may be able to classify a water sample as clean, grossly polluted, or heavily polluted, while random forest (RF) algorithms may have been able to classify a water sample more reliably using pictures of the sample along with readings from environmental sensors.

The possibilities of synergy between AI-supported visual WQM and Internet of Things (IoT) sensors are likely one of the most exciting opportunities in engaging WQM with AI development. Not only can visual data help to visualize surface-level changes, IoT sensors can measure specific physicochemical parameters that are aligned with the observed changes, such as pH, dissolved oxygen (DO), and conductivity (Sethi et al., 2020a). While integrating all of these data streams together into hybrid AI–IoT models may ultimately represent the real-time nature reflecting the situation of the water quality, it will be equally important that the potential systems are scalable. Drones equipped with cameras and multispectral imaging devices have the ability to monitor much larger water bodies than manual surveying, in only a fraction of the time. Multispectral and thermal imaging can detect pollution indicators that may not be visible in the regular red–green–blue (RGB) spectrum, such as chemical spills or specific temperature changes that may affect aquatic life (Yao et al., 2021). In remote or dangerous environments, this is essential, and we may not have the capability for human access.

The possibility of the real-world deployment of computer vision and ML in WQM is no longer just an idea or conjecture. The documentation and case studies of rivers, reservoirs, and industrial facility pollution created by maqamat show that systems designed using computer vision and ML can detect and classify pollution events, with very accurate ML/classification methods that are usually faster in tempo and greater in spatial coverage compared to regular/traditional monitoring of WQM. The predictive capability of ML models can support alerting systems. With such systems, we can act on issues before the pollution reaches a state of intolerable contamination. Not to say, however, that using AI in WQM will be without its obstacles. Complications such as inconsistent quality of images, noise in the environment, minimal amounts of data, and problems with site-specific calibration can all negatively affect accuracy. Ethical issues such as the use of surveillance programming or drones in populated areas by governments are essential aspects to consider. It's also important for AI-based decisions to have the greatest level of transparency, and there is a heavy burden of accountability when using AI. This chapter aims to create a basic guide for designing computer vision and ML systems for monitoring water quality. In addition to demonstrating how to create systems that will monitor water quality, this guide will give details about how to collect data, what to do with collected data, and how to design models that can be used with data and IoT devices. Each of the case studies included in this document illustrates that this outline is effective. Finally, we will discuss some of the issues that need to be addressed before deploying these systems, such as ethical considerations and the future development of systems. The ultimate goal of this chapter is to demonstrate how these technologies may be utilized in a continuous and sustainable manner for managing water systems in a way that supports global environmental resource management.

15.2 TRADITIONAL WQM

Conventional WQM practices have traditionally served as a foundation for many environmental evaluation processes over many years. The routine methods rely on manually sampling water for subsequent testing and chemical analyses in laboratories for key demographic indicators (physicochemical characteristics) such as pH levels, turbidity, DO, biological oxygen demand, total dissolved solids, and concentrations of heavy metals or toxins (see Figure 15.1). These procedures are accepted globally in view of their accuracy and precision. For example, spectrophotometric analysis can provide quantification of contaminant concentrations at micro-levels, while microbiological culture tests can identify harmful pathogens.

Despite these considerations, the limits of traditional methods are important. Firstly, due to the need to transport samples to certified laboratories and the complexity of testing, the duration from sample to test can take hours to days. Secondly, in regard to coverage of sampling points, sampling is often spatially limited, which means sampling only includes certain locations periodically and can miss localized contamination events. Thirdly, operational costs (including lab equipment, trained personnel, and logistics in the field) can be prohibitively high, especially for continuous assessments or for developing countries with a lack of available infrastructure. Furthermore, in terms of real-time interventions, by the time results are observed, the opportunity may be lost—allowing the pollution incident

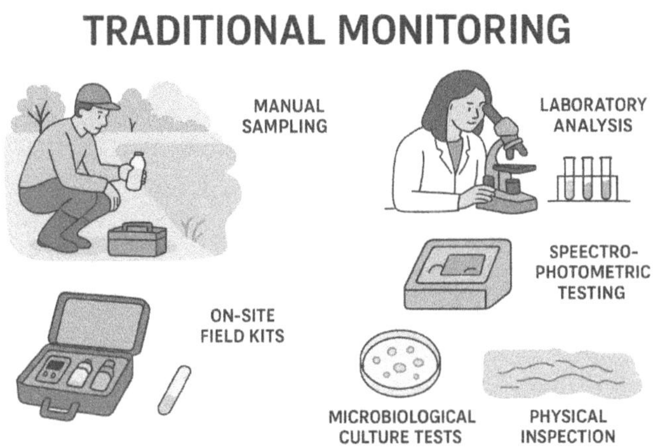

FIGURE 15.1 Conventional techniques for water analysis from on-site sampling to laboratory-based assessments.

to be exacerbated. As reported by Guo et al. (2023), laboratory prospective assessments (using spectrophotometry, chromatography, and microbial culture tests) achieved over 95% measurement accuracy for parameters including chlorophyll-a, total nitrogen, and total phosphorus in freshwater lakes. Chen et al. (2022) illustrated that for baseline water quality assessments, chemical oxygen demand (COD) and BOD measurements using normal wet chemistry methods were consistently produced with variation coefficients below 3% and confirmed their reliability. Singha et al. (2021) evaluated water quality indices (WQIs) for Indian river systems and found that conventional sampling gave WQI values from 42 (poor) to 78 (good) over seasonal changes, also showing the inherent sensitivity of the process in capturing environmental disturbance. However, Podgorski et al. (2020), when looking at a groundwater contamination survey across Southeast Asia, showed that while traditional testing allowed accurate mapping of arsenic concentration, there were problems associated with cost and labour, and traditional testing takes weeks to survey very large areas. If taken together, their results suggest conventional methods are scientific meta-analysis methods, but they are slow, area-restricted, and expensive. Therefore, as methods of continuous, large-scale monitoring, conventional methods are not very effective, and especially not effective for real-time analysis.

15.3 EMERGENCE OF AI IN ENVIRONMENTAL MONITORING

AI has proven to be an extremely effective tool for improving environmental monitoring in the last decade. The technological advances of AI will become a basis for informing environmental monitoring, and the use of images in environmental monitoring will likely soar as monitoring becomes more rapid, flexible, and accurate (Figure 15.2). For example, Meng et al. (2022) used a deep CNN designed with 5 convolutional layers, 1 pooling layer, and 1 fully connected layer that used images from Sentinel-2 to monitor water quality in Lake Dianchi, China, which produced mean absolute errors (MAEs) of 32.57%–42.58% when predicting chlorophyll-a concentrations. Similarly, Liu et al. (2021) used a long short-term memory (LSTM) neural network to predict drinking water quality in the Yangtze River

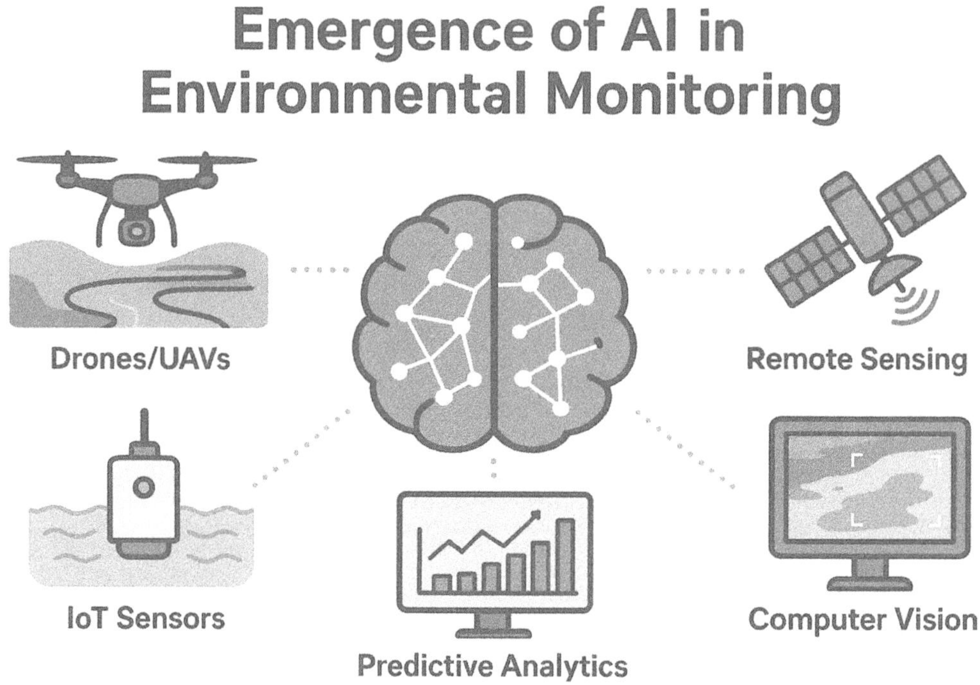

FIGURE 15.2 Emergence of AI in water and environmental monitoring systems.

basin, and in their study, they predicted more than just chlorophyll-a, including pH, turbidity, DO, and conductivity. Liu et al. (2021) acquired a mean squared error (MSE) of only 0.0017, which is indicative of great predictive ability regarding time-series environmental data.

Barzegar et al. (2020) professed a hybrid CNN–LSTM model to estimate chlorophyll-a and DO from coastal waters and also noted a correlation coefficient of $R = 0.97$, which is impressive when compared to a standard CNN or LSTM model alone. In the authors' study, Khullar and Singh (2021) applied a bidirectional LSTM model for predicting multiple water quality parameters, noting accuracy (root MSE (RMSE) = 0.117), which considerably validated it to be relied upon when using complex environmental datasets from the Yamuna River, India.

AI applications have also extended to marine water bodies. A recent study (Yao et al., 2021, and all references listed in the original study) applied Landsat-8 imagery to a gradient boosting decision tree algorithm to predict chlorophyll-a in coastal waters at an MAE of 0.998 μg/L and an RMSE of 1.626 μg/L. Regardless of the environment, these studies indicate that AI-based models could process a broad range and large quantity of heterogeneous environmental data; they can also adapt to changing conditions, make predictions at equal or better accuracies than standard registries, and operate for near-real-time and broad-scale observational WQM.

15.4 COMPUTER VISION FOR WATER ANALYSIS

Computer vision has emerged as one of the most promising AI approaches for WQM because it can directly analyse images and videos to detect pollution indicators without physical contact with the water. Guo et al. (2023) developed a CNN-based image classification model for water quality assessment using multispectral aerial imagery, achieving an overall classification accuracy of 94% for categories such as clean, moderately polluted, and heavily polluted water. Similarly, Chen et al. (2022) integrated a CNN with decision tree ensembles to estimate COD from surface water images, reaching a coefficient of determination (R^2) of 0.92 and an RMSE of 0.82 mg/L, demonstrating that visual analysis can accurately predict key chemical parameters.

Zhou et al. (2022) applied object detection algorithms on a growing dataset of microalgae from the oceans (YOLOv5, TOOD, and Faster R-CNN) with a mean average precision score of greater than 0.90 for identifying multiple types of algae. These findings collectively confirm that computer vision, especially when paired with advanced deep learning architectures and multispectral or hyperspectral imaging, can deliver highly accurate, scalable, and non-invasive WQM solutions that rival laboratory-based methods.

15.5 HYBRID AI–IoT INTEGRATION IN WATER MONITORING

The combination of AI and IoT technologies is an impactful approach to improve the monitoring of water quality by adding visual analysis when paired with real-time physicochemical sensing. Surya et al. (2024) implemented a hybrid CNN–LSTM model and integrated IoT-enabled sensors to demonstrate real-time monitoring of the Ganga River in India (Table 15.1). In their study, the system demonstrated a pollution event detection accuracy of 93% and was able to predict parameter variation (i.e., pH, turbidity, and DO) as far as 24 hours ahead of an event, thus improving an early-warning system. Following the same premise, Lu and Ma (2023) applied empirical mode decomposition with RF and extreme gradient boosting (XGBoost) algorithms (CEEMDAN-RF for RF and CEEMDAN-XGBoost for XGBoost) to IoT sensor data streams from urban water bodies. They noted prediction accuracies of at least 90% for temperature, pH, turbidity, and dissolved organic matter.

Podgorski et al. (2020) demonstrated a large-scale groundwater monitoring network in Southeast Asia that is made up of IoT sensors connected via ML models. They successfully produced a balanced

TABLE 15.1 Hybrid AI–IoT integration framework for water quality monitoring, showing the complementary roles of computer vision, IoT sensors, and data fusion in enabling accurate, real-time, and predictive assessments

COMPONENT	*FUNCTION*	*EXAMPLES OF PARAMETERS/FEATURES*	*BENEFITS*
Computer vision (AI)	Detects visible surface indicators using images/videos from cameras, UAVs, or satellites.	Turbidity, discolouration, algal blooms, floating debris	Rapid surface anomaly detection; wide-area coverage
IoT sensors	Measures real-time physicochemical properties directly from water.	pH, dissolved oxygen (DO), conductivity, temperature, COD	Accurate subsurface and chemical monitoring
Data fusion layer	Integrates AI-based visual outputs with sensor data for cross-validation.	Correlation between discolouration and pH/turbidity changes	Reduces false positives, enhances reliability
Alerts and predictions	Provides real-time notifications and forecasts using historical + live data.	Early warning for algal bloom, industrial effluent, or oxygen depletion	Faster intervention; predictive water management

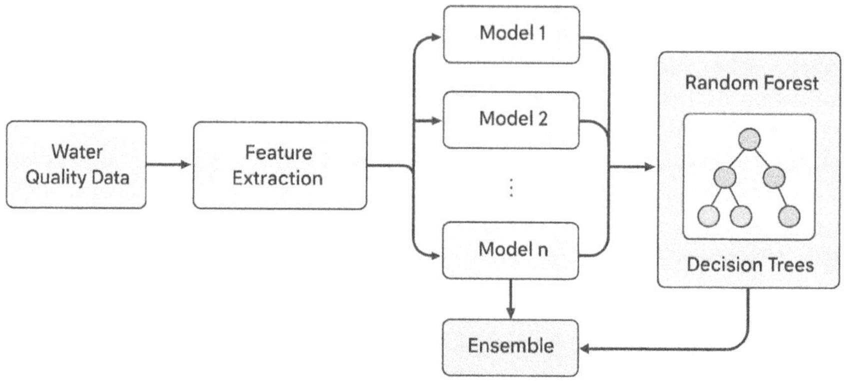

FIGURE 15.3 Architecture of random forest and ensemble methods in water quality monitoring, illustrating data flow from feature extraction to individual models, random forest classifiers, and final ensemble-based decision-making.

accuracy score of all predictive models greater than 85% in predicting arsenic contamination, while providing significant cost and time savings over a fully manual sampling approach. Another study by Samantaray et al. (2018) conveyed that IoT-connected autonomous vehicles with cameras coupled with AI algorithms for algal bloom detection can produce continuous surface monitoring and can monitor freshwater and marine surfaces with a resolution of 95% detection accuracy.

These studies present the idea that hybrid A-IoT frameworks (2018) can potentially solve the gap between precision laboratory-based work and real-time operational-based functions with continuous, automated, and specific monitoring of water quality. By correlating visual data from drones or stationary cameras with physicochemical sensor data, such systems provide a more comprehensive view of water health and are especially well-suited for large-scale deployments where both surface and subsurface conditions must be assessed simultaneously (Figure 15.3).

15.6 RF AND ENSEMBLE METHODS IN WQM

Ensemble learning methods such as RF, gradient boosting machines (GBMs), and hybrid tree-based models have proven highly effective for predicting and classifying water quality parameters due to their robustness against overfitting and ability to handle heterogeneous datasets. Li et al. (2021) applied RF to Sentinel-2 satellite imagery for salinity estimation in estuarine waters, achieving an R^2 of 0.89 and RMSE of 0.37 PSU, outperforming linear regression and support vector regression models. Chen et al. (2022) compared decision trees, RF, support vector machine, and naive Bayes classifiers to classify the WQI from 12 rivers and lakes and presented the highest weighted F1-score (0.94) for the RF model, with the highest predictive performance consistency. Singha et al. (2021) experimented with deep neural networks, multilayer perceptrons, RFs, and XGBoost that could predict WQI in India's river systems. RF provided the best compromise between accuracy and speed, with RMSE values <3.5 and mean absolute percentage errors <4%. Podgorski et al. (2020) also used RF and GBM with over 50 ancillary spatial predictors for a greater study of groundwater in Southeast Asia, reporting AUC scores >0.90 in predicting arsenic contamination and classifying contamination status. Additionally, Lu and Ma (2023) reported that predictive hybrid ensemble models (i.e., CEEMDAN-RF and CEEMDAN-XGBoost) improved the accuracy of physicochemical parameter prediction by 5%–7% compared to stand-alone models on noisy, real-world IoT datasets. Collectively, there is enough evidence to suggest that RF and ensemble learning techniques offer computational efficiency; they are also adaptable when used with a variety of data sources and make a perfect fit for fully incorporated, large-scale, real-time monitoring of water quality.

15.7 REAL-TIME AND RIVER-BASED AI MONITORING

Real-time WQM using AI models is becoming increasingly important as it allows continuous evaluation of water quality and immediate action in the case of pollution events, especially with larger river systems. Surya et al. (2024) created a hybrid CNN–LSTM model with IoT-based sensors to monitor the Ganga River in India (Table 15.2). Their model detected pollution events with 93% accuracy and was able to predict changes to the parameters (e.g., the associated changing pH, turbidity, and DO) with a forecast of 24 hours ahead with an MAE of 0.12, thus providing an excellent early-warning tool. A similar study by Khullar and Singh (2021) made use of a bidirectional LSTM network to predict water quality parameters of the Yamuna River and reported RMSE values as low as 0.117, which outperformed normal univariate time-series forecasting tasks.

As an example of being able to achieve wider geographic coverage, Podgorski et al. (2020) applied a real-time IoT sensor network and ML algorithms to provide a spatial mapping of groundwater contamination in Southeast Asia. The system generated near-real-time estimates of arsenic concentrations, achieving balanced accuracy scores > 85%, all in much less time than a typical manual sampling campaign. Also, Lu and Ma (2023) provided an example of the use of streaming IoT data from urban rivers and RF and XGBoost algorithms, which forecasted aspects of key parameters such as temperature, pH bioindicators and turbidity with > 90% accuracy, while being exposed to changing environmental conditions.

These examples illustrate not only that there is potential for real-time AI-based monitoring systems, including AI systems using deep learning models coupled with IoT infrastructure, which can

TABLE 15.2 Real-time and river-based AI monitoring framework integrating cameras, IoT sensors, drones, and AI models for continuous and predictive water quality assessment

COMPONENT	*FUNCTION*	*EXAMPLES OF PARAMETERS*	*BENEFITS*
Fixed cameras	Continuous capture of surface water conditions from riverbanks	Colour changes, turbidity, floating debris	24/7 localized monitoring; immediate anomaly detection
IoT sensor network	In situ measurement of physicochemical parameters	pH, DO, turbidity, conductivity, temperature	High-frequency, real-time water quality readings
Drones/UAVs	Periodic aerial surveys for wide-area coverage	Algal blooms, sediment plumes, effluent tracking	Extends monitoring to remote or inaccessible zones
AI models (CNN, LSTM, RF)	Analyse visual and sensor time-series data to detect and predict anomalies	Pollution events, oxygen depletion, seasonal shifts	Accurate classification and predictive forecasting
Alert system	Generates automated notifications and predictive warnings	Early alerts for contamination events	Rapid response, reduced damage, improved safety

generate both temporal resolution (continuous updates) and spatial scalability (multiple site coverage) (Rohit et al., 2023). Together these attributes are particularly well-suited for phasing complex, highly dynamic river systems where pollution and discharges occur as a consequence of rainfall, industrial discharges and seasonal conditions.

15.8 ALGAE DETECTION USING COMPUTER VISION AND RESIDUAL LEARNING

The identification and classification of algae in water bodies is an important aspect of WQM because excessive algal growth can develop into harmful algal blooms (HABs) that can threaten both aquatic ecosystems and human health. Samantaray et al. (2018) developed a real-time detection system for algae that can be deployed using unmanned surface vehicles as well as unmanned aerial vehicles (UAVs), with computer vision algorithms capable of detecting algal blooms in field conditions with different levels of light and weather. Their detection system recorded over 95% accuracy in field conditions, showing the practicality of using a mobile algae monitoring vision-based system.

Going beyond the detection of blooms, Deglint et al. (2018) developed the Spectral Absorption Microscopy for Species Observation and Naming system that fused spectral and morphological features in a deep residual CNN for the classification of algae, including cyanobacteria and green algae. The model was classified with 96% accuracy using microscopy images, demonstrating the potential to incorporate spectral analysis with deep learning for species-level identification.

For large-scale marine applications, Zhou et al. (2022) tested the object detection architectures YOLOv5, TOOD, and Faster R-CNN on their manually annotated marine microalgae dataset. In these tests, they achieved mean average precision scores of over 0.90 for multi-class algae detection. Hence, their results indicate that modern deep learning object detection architectures are capable of locating and identifying algae in many aquatic contexts. All of these studies demonstrate that computer vision technology, particularly when coupled with spectral imaging and residual learning, may be a highly effective non-invasive approach for the real-time identification and classification of algae and is extremely important in mitigating the potential negative environmental and financial consequences of HAB occurrences.

15.9 PROPOSED FRAMEWORK FOR WQM

The suggested framework combines multiple technologies: IoT-based sensing with computer vision and ML to build one complete automation system for scalable and close to real-time water quality assessment. The framework has four distinct components to achieve this goal: (1) data acquisition, (2) data preprocessing, (3) model training and classification, and (4) hybrid AI–IoT integration (as illustrated in Figure 15.4).

15.9.1 Data Acquisition

The stage of gathering data is the foundation of the proposed WQM framework, as the quality of the data collected during this phase ultimately determines the ability for the data to be accurately analysed later in the process. The data gathered during this stage consists of both visual and physicochemical

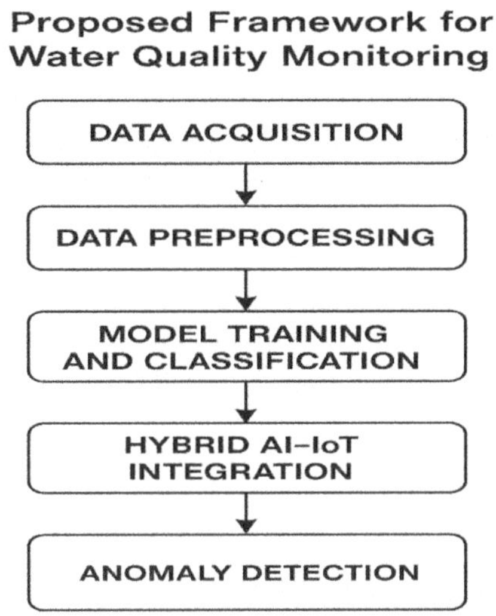

FIGURE 15.4 Proposed framework for water quality monitoring, illustrating the sequential process from data acquisition to anomaly detection through AI–IoT integration.

data gathered from a variety of complementary sources. The sources of visual data are generally split into three main types.

The first source of visual data is fixed-position ground cameras that are placed along the shorelines of streams, rivers, lakes, or reservoirs to continuously monitor a specific location or area from a fixed location. The goal of fixed-position ground cameras is to monitor a specific area 24 hours a day through still images or video streams to provide real-time data to the monitoring agencies.

The second source of visual data is aerial (drone or UAV) imagery, which allows for a greater field of view and speed in deployment after pollution events. Aerial imagery can be taken with RGB cameras, multispectral sensors (to view chlorophyll levels, sediment concentration, and the health of the surrounding vegetation), or thermal sensors (for identifying changes in temperature due to industrial discharge).

The third source of visual data is satellite imagery, where high-resolution satellites are capable of taking pictures of water bodies and identifying seasonal and long-term changes to the bodies of water; this wide-area view complements localized monitoring of waterbodies and is particularly useful for trend analysis of large, inaccessible areas.

IoT-enabled sensors provide real-time monitoring of water physicochemical characteristics by being placed directly into the water for consistent and continuous recording of these measurements at specific points throughout the body of water. Key parameters being measured with IoT-enabled sensors include pH level; turbidity; electrical conductivity; DO level; temperature; and concentration of nutrients. A central monitoring platform will receive the data from the sensors through a wireless transmission system as the data is continuously recorded for these parameters. A hybrid approach of using fixed/stationary sensor nodes alongside mobile sensor nodes (such as those used on boats and autonomous surface vehicles (ASVs)) enables both localized (or "point") and distributed measurements to be captured through the use of this combined method (Sood & Zhu, 2024).

Using a combination of both fixed/stationary and mobile sensor nodes enables capturing all visible indicators of pollution, such as colour change, the presence of floating debris or algae blooms (surface water), and all other invisible (non-visible) chemical and biological changes that occur below the surface of the water (Dennison et al., 2005). The use of multiple sensor nodes positioned on multiple platforms (air, land, satellite, and in situ) also allows the sensor network to provide total spatial coverage, timeliness of data acquisition, and flexibility for operational use, even in very remote or unsafe areas.

15.9.2 Data Preprocessing

Collecting raw data from cameras, drones, satellites, and IoT sensors is the first step to preprocessing the data into an accurate, consistent format that is ready for training the models. Preprocessing must take into consideration the environmental and technical factors that impact both the visual and numeric data used to train the models. Failure to correct for these environmental and technical factors will lead to decreased reliability of the models for analysis. For visual data, common challenges include lighting variations caused by changing weather conditions or time of day, surface reflections from sunlight or artificial lighting, motion blur from camera movement, and occlusions from floating debris or passing objects. To remedy these problems, many preprocessing methods are used:

1. **Image Enhancement**: Image enhancement manipulates the brightness, contrast, and sharpness to improve the visibility of visual features.
2. **Colour Normalization**: Standardizing colour values between images in an attempt to minimize the impact of changing illumination, allowing consistent measurement of water colour over time and across camera devices.
3. **Noise Reduction**: Filtering methods such as Gaussian blur, median filtering, or wavelet denoising are all options for reducing noise at the pixel level without introducing important detail loss for features of interest.
4. **Segmentation**: An automatic exercise in which the water region has been identified from other surrounding information (land, vegetation, buildings/infrastructure, sky) in such a way that is not considered in the analysis of the image. Some examples of segmentation include thresholding, edge detection, or automatic segmentation using a deep learning-based segmentation model.

For sensor data, anomalies can occur because of temporary interference with the device or errors caused during the transmission of data, but they can also originate from across-the-board changes in the environment unrelated to actual water quality trends. When preparing sensor data for potential analysis:

1. **Outlier Detection and Removal**: Clearly identify readings that are significantly outside normal ranges and filter the readings out, so they don't disproportionately bias a model during training.
2. **Signal Smoothing**: Use methods such as moving averages or exponential smoothing to minimize random short-term fluctuations but still allow for long-term trends to remain.
3. **Statistical Normalization**: Normalize values to allow more consistent scaling and therefore be more appropriate to combine with visual data during multi-modal analysis.
4. **Signal Decomposition**: Use methods such as empirical mode decomposition or methods from Fourier analysis to decompose a more complex signal into simpler components, allowing for more targeted analysis.

Utilising effective preprocessing techniques for sensor data has two roles: (1) to normalise and/or clean the data and (2) the degree to which signal processing can improve the characteristics of the signal, thus creating clearer patterns in our ML model(s) when we get to that stage. Overall, the goal is to create a clean, clear, error-free input to the framework for the purpose of producing clean signals for either a predictive purpose or as autonomous signals for real-time applications. Therefore, developing the sensor data will not only produce valuable results, but it will also increase the predictive capability and reliability of the sensor data for making automated real-time decisions (Zhou et al., 2022).

15.9.3 Model Training and Classification

What happens next, after collecting and preparing your data for analysis, is using either ML or deep learning models to help identify the condition of the water quality and predict certain physical characteristics. This is where the data is converted into something "actionable". When it comes to identifying physical characteristics of the water based on visual imagery, CNNs are the technology of choice due to their effectiveness at extracting spatial features and texture characteristics. CNNs use sophisticated algorithms that automatically match visual patterns with physical characteristics associated with those patterns (such as colour change, debris floating on the surface, algal blooms, or an increase in the turbidity of the water). To teach a CNN how to associate visual patterns to a water quality category, it needs to be trained using large databases of water photographs, with each photograph being classified according to one of three categories: clean; moderately polluted; or heavily polluted. Advanced CNN versions, such as ResNet, DenseNet, and EfficientNets, can also be applied to enhance accuracy while minimising overfitting, particularly in situations where the size of the dataset is limited. For sensor data, models such as RF (Sethi et al., 2018), GBM (Sethi et al., 2020b), XGBoost, and support vector regression (Lu & Ma, 2023) are commonly employed. These algorithms are robust against noise and can handle heterogeneous inputs from different types of IoT sensors. They are particularly effective for predicting numerical values of water quality parameters, such as pH, DO, or turbidity levels.

Hybrid models, which use both visual and sensor information, are used in multi-modal systems. For example, CNNs would create features from an image, RF or GBMs would create features from the numerical sensor input, and the features would be passed into a fusion layer in order to build a fused prediction (Khullar & Singh, 2021). Each type of model has its advantages. CNNs can create spatial recognitions, and the ensemble models are more reliable in making decisions.

Model training is typically as follows:

1. **Dataset Preparation**: This includes separating the data into training, validation, and, of course, test data so that performance can be computed with unbiased data.
2. **Model Selection and Tuning**: Model selection is determining what architecture will give you the best prediction, whereas tuning hyperparameters such as learning rate, tree depth, and kernel size are simply optimization steps.
3. **Training and Validation**: During this phase, the data is fed into the model iteratively, and performance metrics such as accuracy, R^2 score, RMSE, and F1-score metrics are monitored carefully in order to avoid overfitting (Pan et al., 2025).
4. **Testing and Deployment**: The final model is run on data that has never been seen by the model, and then the model use case can be evaluated in practice before the data can be monitored in the system.

At the conclusion of this stage, the framework will produce models that will detect early signs of contamination, classify water quality into defined categories, and predict future changes. All of this information will provide the basis for the last hybrid AI–IoT stage, where the predictions can be updated in continuous stages with real-time measures, allowing model inference to remain accurate within dynamic environmental conditions.

15.9.4 Hybrid AI–IoT Integration

The last stage of the proposed framework would bring together the advantages of computer vision with IoT-based sensing in an integrated framework for real-time monitoring (Ambe et al., 2024). Computer vision can identify surface observable changes, such as discolouration, flow, turbidity, algal blooms, and debris on the surface, whereas IoT-based sensor systems can provide continuous measurable indicators of subsurface and chemical condition, including pH, DO, conductivity, and levels of nutrients. Integrating these two data types brings a multi-dimensional understanding of water quality, composited visually, and in physicochemical state (Glasgow et al., 2004).

The integration process would begin by collecting data concurrently. Because the monitoring uses fixed time intervals for all data collection, it is possible to collect visual feed from both cameras and drones simultaneously. All visual feeds from monitoring systems would have time stamps to identify which set of IoT sensors monitoring at the same site were recording in the same period. This will allow the framework to compare visual observation of contamination with measured changes in water chemistry (Lowe et al., 2022; Rohit et al., 2024). For instance, if the vision model detects discolouration over a large area, the IoT-based sensors will capture whether the discolouration was accompanied by abnormal turbidity or pH fluctuation in the water column, thus lowering the likelihood of false alarms.

In terms of the computational architecture, an overall data fusion layer will integrate the final outputs of each model. The visual data is handled by CNNs or other image-based networks (Sethi et al., 2021). The numerical sensor-based data is handled by ensemble learning or regression models. Ultimately, the fusion layer will combine the two outputs in some fashion: simple average; simple or weighted voting; or, even more advanced deep learning data fusion sequences will output a single comprehensive prediction of water quality.

This hybrid method is advantageous because of the following:

1. **Improved Accuracy**: The cross-validated predictions between image-based prediction and sensor-based prediction are more robust.
2. **Immediate Detection**: Visual observations can provide surface change detection at the moment of observation, while sensor data can assess chemical changes that may occur before the visible signs of surface change are observed.
3. **Scalability**: IoT sensors could be deployed over a large area while drones or satellites could provide spatial coverage with periodic imagery as needed.
4. **Uncommon First Response**: While the data can be streamed in continuously, authorities could receive instant alerts and act before the appearance of confirmed surface change.

This type of approach in practice is based on a predictive WQM design that functions on a framework of analysing past and current data simultaneously to forecast predicted conditions. Simultaneously, over time, the integrative system will build relationships based on the pattern associations of visual cues and sensor readings, thus improving predictive capacity (Mukonza & Chiang, 2023). Hybrid AI–IoT plans like this example are a perfect fit for smart city asset (water) management, industrial effluent management, and river basin monitoring at scale to manage surface appearance and subsurface chemistry at the same time.

15.10 DIFFERENT CASE STUDIES

15.10.1 River Water Monitoring

Rivers are moving waterways that can be contaminated with pollutants from multiple sources such as chemical discharges, pollutants from agriculture, wastewater from settled urban developments, and natural seasonal flooding. It is difficult to monitor the water quality conditions of rivers as water quality can change rapidly over time and/or vary between locations along the river (Mukonza & Chiang, 2023). Developing an AI computer vision IoT monitoring system will remove barriers to monitoring rivers, as both a continuous stream and sites along the river can be monitored spatially.

In this case study example, fixed-position cameras are placed along the river of interest at strategic locations along the banks, for example, upstream, at industrial discharge locations, and downstream of urban developments. The fixed-position cameras capture images or videos with high resolutions that can be streamed into a neural network image classifier (Singha et al., 2021). The deep learning system will be trained to detect visual parameters, or indicators, such as turbidity changes, colour changes, evidence of oil films on the water surface, and the presence of visible debris. IoT-enabled sensors can be placed in the water to measure the pH, DO, turbidity, conductivity, and temperature of the water in real time, which will also be transmitted via wireless capability. The sensors will transmit their readings to a central server that will also process images or videos and capture, synchronize, and record the stream of images and/or videos in its database.

Drones with RGB and multispectral cameras are used periodically to survey longer segments of the river, particularly after heavy rains or after suspected pollution events. The aerial point of view allows the system to identify any upstream contamination sources, as well as track the contaminants along the river, which can take place across kilometres. The multispectral imagery can provide information on vegetation stress on the banks of the river and detect algal blooms or sediment plumes that cannot be reliably observed in the low spectral RGB visible spectrum (Samantaray et al., 2018).

Since we take both datasets—imagery and physicochemical baseline data—we can cross-validate anomalies. For instance, if the CNN identifies a sudden brown colour on the surface of the water, we can check the IoT sensors to see if there is a concurrent spike in turbidity (Guo et al., 2023). Conversely, it is also possible, for example, that if one of the sensors logs a drop in DO concentration and the surface is clear, a subsurface contamination event or organic matter decomposition may be taking place, so additional information would be warranted. The exposure of a system like this demonstrates several benefits over traditional manual monitoring (Yan et al., 2024). First, response time to an alert is improved from weeks to minutes. Alerts can be generated within minutes of a contamination event. Second, the spatial coverage is enhanced as drones and multiple fixed stations can survey areas that would have taken several days to determine using traditional fieldwork. Third, the system establishes a historical database of images of the contamination event and sensor readings, allowing for the identification of trends and predictive modelling of seasonal pollution.

15.10.2 Reservoir Surveillance

Reservoirs are a vital source of water for their use in drinking, irrigation, and the generation of hydropower. Management of water quality is paramount when managing reservoirs. Reservoirs can be large, enclosed water bodies; they can receive and hold contaminants, unlike rivers; and they can accumulate contaminants over long time periods, which can lead to damage to the ecosystem (Rohit & Milan, 2023). One of the largest threats posed to the health of a reservoir is HABs that can deplete DO and release toxins, and vertical stratification can lead to serious alterations of native aquatic ecosystems.

In this case study, UAVs are used to perform periodic surveillance flights of the entire reservoir surface with multispectral imaging sensors. Contrary to RGB cameras, multispectral sensors collect data across multiple wavelengths (including regions of near-infrared), suggesting higher sensitivity to the levels of chlorophyll-a, which is useful when screening for algal presence in water bodies (Geetha & Gouthami, 2016). This will enable the detection of early-stage blooms, beyond visual appearance. Multispectral imagery captured by a UAV is then processed using an image segmentation and spectral analysis pipeline to isolate water bodies from other land areas. Historical records of DO, temperature, and nutrient level sensor readings from in-person sensors are used to help validate the presence of a bloom (and its severity) by comparison.

In addition to the detection of blooms, this process allows for the implementation of preventative measures through the use of controlled aeration of water, reduction of human-added nutrient sources, and/or application of targeted chemical treatment prior to the point of critical severity of a bloom. This proactive strategy is beneficial in preventing fish kills and depletion of DO, decreases the cost of treating an immediate area, and supports a healthy reservoir ecosystem over time (DuPont and Maher, 2018).

15.10.3 Industrial Effluent Detection

Often, industrial zones will dump wastewater into bodies of water, such as lakes or rivers, through drainage channels. If the treatment of this wastewater prior to discharge isn't effective, or if monitoring for contaminants isn't effective, an accidental release (or illegal dumping) could severely contaminate surrounding water resources. In this case study, we evaluated a hybrid AI/IoT model that allows for the continuous monitoring of industrial contaminants as one method of detecting industrial wastewater contamination (Dennison et al., 2005).

Static cameras were placed to capture images of the surface of the water upstream of the discharge point in order to provide baseline data prior to the installation of the hybrid model. Video captured by the camera was then processed using computer vision software that is specifically designed to search for irregularities in detection patterns or colour changes within the image captured (Podgorski et al., 2020). Aside from video analysis, we also utilised IoT-enabled monitoring systems to take measurements using advanced instruments and collect information related to critical water quality parameters such as pH, electrical conductivity, turbidity levels, as well as COD on a continuous basis using an IoT-connected data logger. Once the visual analysis identifies a probable discolouration event, it will check the visualization against the most recent sensor data reading (Liu et al., 2021). If both indicate there is a problem, it will automatically notify the environmental authorities of a polluted effluent event. By confirming the two sources used to identify the effluent anomaly, we can greatly reduce the number of false positives from shading, lighting, and sensor drift.

Since the implementation of this system, it has reduced the average response time of investigating an effluent enforcement event from several days to under 2 hours (Figure 15.5). The ability to quickly identify effluent contamination enables authorities to take preventative action to help contain the contamination, such as suspending industrial discharges, deploying absorbent barriers, or implementing emergency treatment regimens before the contamination is further advanced downstream (Maier & Keller, 2019). Over time, the system will build a historical record of effluent events that can be utilized to substantiate compliance audits, or enforcement proceedings, or be informational about the industrial waste diversion practices.

15.10.3.1 Advantages of the Approach

- **Real-Time Monitoring**: Typically, quality testing for aquatic environments can take weeks to complete. The proposed system can provide immediate insights as it continuously assesses real-time video feeds, drone imagery/hyperspectral image collection, and IoT sensor readings. A rapid, real-time sensor network will allow for immediate detection of events as they occur (e.g., rapid reduction in DO, fish kills, changes in colour of water, or the presence of trash) (Baird et al., 2024). Ultimately, if contamination is discovered during a real-time monitoring event, immediate detection allows authorities to recognize and respond accordingly before further contamination or negative ecological outcomes take place.
- **Scalability**: The proposed system is modular in design. This means it can be implemented in several different sizes of deployments. Small installations could consist of one or several

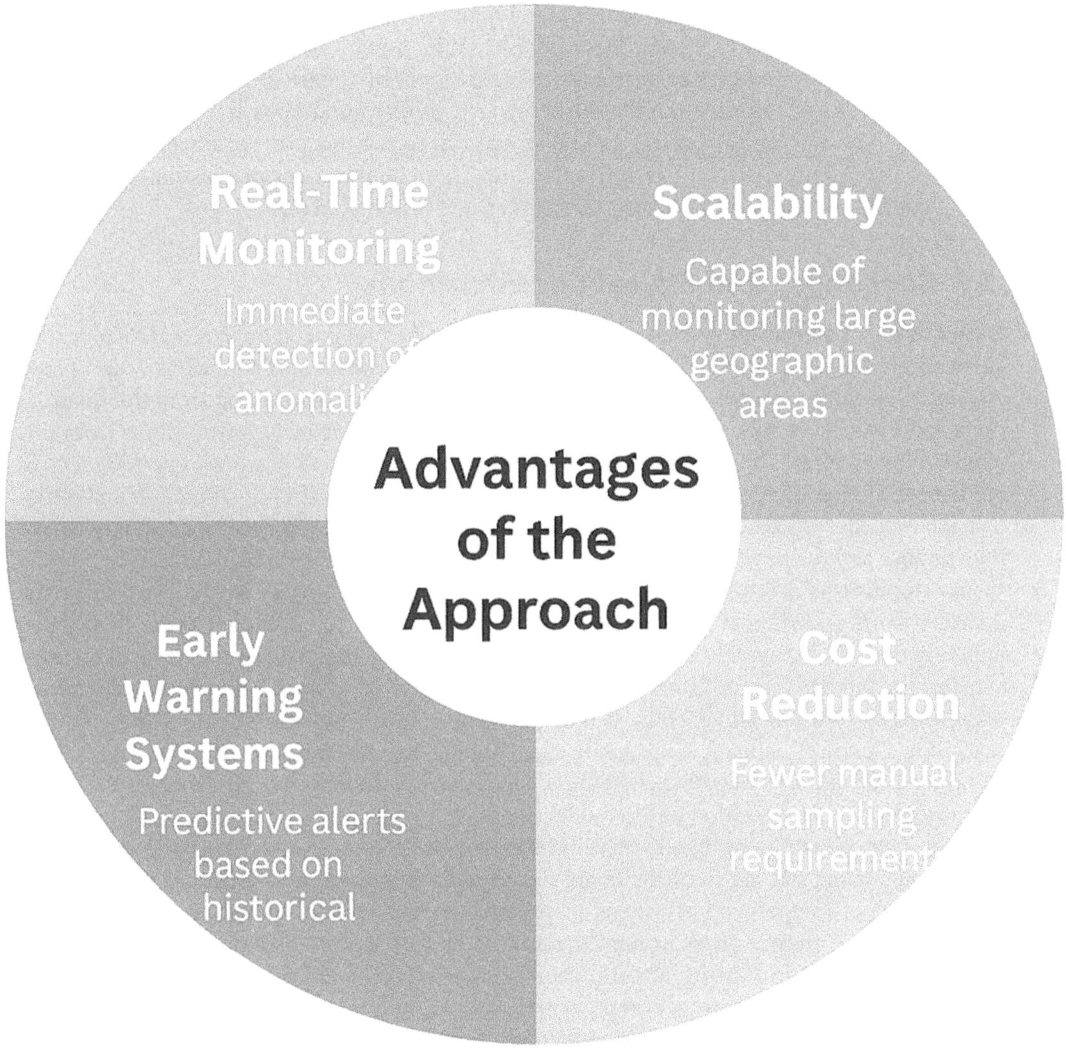

FIGURE 15.5 Circular infographic illustrating the four main advantages of the proposed approach: real-time monitoring, scalability, cost reduction, and early-warning systems.

fixed cameras and sensors monitoring and assessing a local reservoir. Large installations have the capability to fuse networks of IoT devices and other sources of data derived from drone patrol and satellite data to monitor hundreds of kilometres of rivers or coastlines (Surya et al., 2024). Special consideration is given to new technologies that will permit assessments in remote or difficult-to-access gaps with no human presence on-site.

- **Cost Savings**: Manual sampling and lab testing are inherently resource-intensive, with significant logistics and specialized equipment and trained personnel issues (Kumar et al., 2025). The net result is that the automated monitoring system can reduce the frequency of expensive field activities by automating much of the monitoring process. Although there will always be a place for lab testing to confirm anomalous results and regulatory compliance, lab sampling occurs only when unusual observations are encountered, providing a true long-term reduction in operational costs.

- **Early Warnings**: The combined real-time data and historical data allow for the predictive modelling of water quality trends. For example, if the system has identified environmental conditions that were in place prior to known algal bloom or chemical spill events, alerts can be issued based on incomplete data if the system thinks conditions have crossed certain thresholds. The warnings or alerts can be structured to not just protect aquatic life, if possible, but also to safeguard communities that are reliant on the water in question, for compliance purposes, and to lessen the cost of emergencies (Chen et al., 2022).

15.10.3.2 Challenges and Ethical Considerations

- **Data Quality Limitation**: Visual and sensor data can be influenced by external factors such as poor lighting, sun glare, cloud cover, heavy rain, fog, and surface reflections on water. These anomalies in image processing can change how much the variation found from the image is impacted by these factors and may also lead to incorrect positive or negative results from this analysis (Yong et al., 2021). Similarly, sensors may have different readings due to defective equipment or biofouling, which could necessitate establishing a regular calibration and maintenance procedure. These limitations must be addressed to achieve proper data acquisition of high quality and to enable accurate and dependable monitoring via AI methods.
- **Generalization of Models**: If ML models are developed with only a small subset of the datasets from a particular geographic region, time of year, or pollutant type, then when applying those same models in different environments, with much different physical circumstances, the models may not function correctly. For example, if an AI algorithm was trained using images of a particular river, it may perform poorly categorizing water in a tropical reservoir, a coastal estuary, or a newly morphologically altered river (Tang et al., 2019). Continued retraining using datasets with a variety of spatiotemporal information is needed to improve the generalization of the model.
- **Privacy Concerns**: Drone-based surveillance provides the most comprehensive options for water monitoring, but many of the factors used to facilitate the use of drones can unintentionally collect images and/or videos of private citizens, private property, and/or critical infrastructure. Considerations around the ethical and legal implications regarding the collection, retention, and sharing of data would then need to be considered (Boccadoro et al., 2020). Operators would have to comply with privacy regulations considering the flight paths and camera angles, and they also need to ensure that the data they collect can be stored securely to prevent misuse.
- **Regulatory Compliance**: AI and IoT technologies for environmental monitoring must comply with national and regional environmental regulations, national and regional regulations on drone operations, and data protection laws. Failure to comply with these regulations may result in legal actions or prohibition from carrying on any business. In some cases, regulatory systems are very detailed regarding how collected data through monitoring can be utilised in enforcement; therefore, it is necessary to work closely with applicable authorities (e.g., environmental) and other partners to achieve good governance in this area.

15.11 CONCLUSION

Water monitoring has historically relied upon laboratory sampling methods; however, rapidly evolving cities, industries, and climate are shifting this practice to include computer vision, ML, and IoT sensors as the future of environmental monitoring due to their abilities to allow continuous, scalable, and low-cost environmental monitoring. Combining real-time visual indicators and real-time chemical (physicochemical) measuring instrumentation would allow for the assessment of changes to water quality, with visible, but also less visible (or otherwise) changes associated with surfaces above and beyond the subsurface conditions relating to aquatic environments.

The individual case studies related to river/reservoir monitoring and industrial effluent provide examples of how these types of systems have provided the capacity to generate rapid detection/alerts in the context of anomaly detection. Importantly, these systems reduce response time and improve the accuracy of environmental assessments through cross-validation from different data sources and types. While the benefits are considerable, full implementation of this approach will also require addressing open questions of data quality, generalization of models to other forms of water, safeguarding privacy, and compliance with regulations on the environment and data protection. Essential responses to the aforementioned issues will require continuous research, robust governance systems, and extensive collaboration among technology developers, regulators, and community members where this and other digital technology will be used. Firstly, it was noted that developments in edge computing, satellite data, multi-modal AI models, and community-based monitoring could make water quality observation more responsive, inclusive, and predictive. Secondly, a continuous iteration cycle of evidence-based gathering and updating decision support systems is feasible and represents transformative potential that can fulfil future safeguarding of global water resources, progress on UN sustainable development goals, and benefit human and aquatic ecosystems for generations to come.

REFERENCES

Ambe, N. A., Sharma, V., Sethi, K., and Fallah, M. H. 2024. Regenerative braking systems in electric vehicles (EVs): A review. In *3rd Edition of IEEE Delhi Section Flagship Conference (DELCON)*, New Delhi, India. 1–5.

Baird, M., et al. 2024. A wireless sensor network IoT platform for consumption and quality monitoring of drinking water. *Discover Applied Sciences*. 7: 15.

Barzegar, R., Moghaddam, A. A., and Adamowski, J. 2020. A hybrid convolutional neural network–long short-term memory model for chlorophyll-a prediction in inland waters. *Environmental Monitoring and Assessment*. 192(11): 1–14.

Boccadoro, P., Daniele, V., Di Gennaro, P., Lofù, D., and Tedeschi, P. 2020. Water quality prediction on a Sigfox-compliant IoT device: The road ahead of WaterS. arXiv preprint arXiv:2007.13436.

Chen, J., Zhang, X., and Li, Y. 2022. Water quality classification based on machine learning algorithms: A comparative study. *Water*. 14(3): 451.

Deglint, J. L., Jin, C., and Wong, A. 2018. SAMSON: Spectral absorption microscopy for species observation and naming. arXiv preprint arXiv:1810.10889.

Dennison, P. E., Roberts, D. A., Peterson, S. H., and Rechel, J. 2005. Use of normalized difference water index for monitoring live fuel moisture. *International Journal of Remote Sensing*. 26(5): 1035–1042.

DuPont, M., and Maher, J. 2018. Smart system for water quality monitoring utilizing long-range-based Internet of Things. *Applied Water Science*. 8(5): 1–10.

Geetha, S., and Gouthami, S. 2016. Internet of Things enabled real-time water quality monitoring system. *Smart Water*. 2(1): 1.

Glasgow, H. B., Burkholder, J. M., Reed, R. E., Lewitus, A. J., and Kleinman, J. E. 2004. Real-time remote monitoring of water quality: A review of current applications, and advancements in sensor, telemetry, and computing technologies. *Journal of Experimental Marine Biology and Ecology*. 300(1–2): 409–448.

Guo, J., Wang, X., and Zhao, Q. 2023. Remote sensing-based estimation of lake water quality parameters using convolutional neural networks. *Sustainability*. 17(3): 998.

Gupta, S., D., U., and Hebbar, R. 2022. Analysis and application of multispectral data for water segmentation using machine learning. arXiv preprint arXiv:2212.08749.

Khullar, S., and Singh, N. 2021. Predictive modeling of river water quality using bidirectional LSTM networks. *Environmental Science and Pollution Research*. 28(42): 59735–59746.

Kumar, A., Sethi, K., and Verma, A. 2025. A comprehensive review of machine learning techniques in fraud detection. In *12th International Conference on Computing for Sustainable Global Development (INDIACom)*, New Delhi, India. 1–7. IEEE.

Li, H., Sun, D., and He, Q. 2021. Estimating salinity from Sentinel-2 imagery using random forest regression. *Water Supply*. 21(7): 3407–3419.

Liu, Y., Zhang, T., and Zhou, Y. 2021. Water quality prediction in the Yangtze River using long short-term memory networks. *Journal of Hydrology*. 603: 126884.

Lowe, M., Qin, R., and Mao, X. 2022. A review on machine learning, artificial intelligence, and smart technology in water treatment and monitoring. *Water*. 14(9): 1384.

Lu, J., and Ma, Y. 2023. Hybrid CEEMDAN-RF and CEEMDAN-XGBoost models for real-time water quality prediction using IoT data. *Heliyon*. 9(5): e15123.

Maier, P. M., and Keller, S. 2019. Estimating chlorophyll-a concentrations of several inland waters with hyperspectral data and machine learning models. arXiv preprint arXiv:1904.02052.

Mehta, A., and Sethi, K. 2025. Enhancing liver disease diagnosis using ensemble machine learning. In *12th International Conference on Computing for Sustainable Global Development (INDIACom)*, New Delhi, India. 1–5. IEEE.

Meng, X., Wu, C., and Lin, Y. 2022. Monitoring lake water quality with deep convolutional neural networks using Sentinel-2 imagery. *Remote Sensing*. 14(5): 1192.

Mohan, S., Kumar, B., and Nejadhashemi, A. P. 2025. Integration of machine learning and remote sensing for water quality monitoring and prediction: A review. *Sustainability*. 17(3): 998.

Mukonza, S. S., and Chiang, J. L. 2023. Meta-analysis of satellite observations for United Nations sustainable development goals. *Environments*. 10(10): 170.

Pan, D., Deng, Y., Yang, S. X., and Gharabaghi, B. 2025. Recent advances in remote sensing and artificial intelligence for river water quality forecasting: A review. *Environments*. 12(5): 158.

Panwar, N., and Sethi, K. 2022. Breast cancer: Early diagnosis using artificial intelligence. In *2nd International Conference on Innovative Sustainable Computational Technologies (CISCT)*, Dehradun, India. 23–24 December. 1–4.

Podgorski, J. E., et al. 2020. Extensive groundwater arsenic contamination in Pakistan. *Science of the Total Environment*. 740: 140181.

Rawat, N., and Sethi, K. 2023. Early detection of breast cancer using machine learning algorithms. In *3rd International Conference on Innovative Sustainable Computational Technologies (CISCT),* Dehradun, India, 8–9 September. 1–3.

Rohit, N. A., Bathla, G., Sethi, K., and Sharma, G. 2024. Harnessing AI for competitive edge in religious tourism. In Nadda, V., Tyagi, P. K., Singh, A. and Singh, V. (eds.), *AI Innovations in Service and Tourism Marketing*. Hershey, PA: IGI Global. 397–410. doi: 10.4018/979-8-3693-7909-7.ch020.

Rohit, S. K., and Milan, S. 2023. Epidemiology of occupational injuries and hazards in Indian hospitality. *European Chemical Bulletin*. 12: 2586–2589.

Rohit, M., Sethi, K., Khan, M., and Raina, A. 2023. Machine learning model for prediction of chemical harmfulness in hospitality. *Data and Metadata*. 2: 161.

Samantaray, S., Mohanty, S., and Rout, M. 2018. Real-time algae detection using computer vision on USV and UAV platforms. arXiv preprint arXiv:1811.10847.

Sanya, W. M., Alawi, M. A., and Eugenio, I. 2022. Design and development of Smart Water Quality Monitoring System using IoT. *International Journal of Advances in Scientific Research and Engineering*. 8(3): 1–13.

Sethi, K., Chauhan, S., and Jaiswal, V. 2021. Artificial intelligence in higher education: First attempt. In Verma, S. and Tomar, P. (eds.), *Impact of AI Technologies on Teaching, Learning, and Research in Higher Education*. Hershey, PA: IGI Global. 1–29. doi: 10.4018/978-1-7998-4763-2.ch001.

Sethi, K., Gupta, A., Gupta, G., and Jaiswal, V. 2017. Comparative analysis of machine learning algorithms on different datasets. In *Circulation in Computer Science International Conference on Innovations in Computing (ICIC 2017)*. 87–91

Sethi, K., Gupta, A., and Jaiswal, V. 2018. Machine learning based performance evaluation system based on multi-categorial factors. In *2018 Fifth International Conference on Parallel, Distributed and Grid Computing (PDGC)*, Solan, Himachal Pradesh, India, 20–22 December. 86–89. IEEE.

Sethi, K., and Jaiswal, V. 2022. PSU-CNN: Prediction of student understanding using CNN. *Materials Today: Proceedings*. 62: 4957–4964.

Sethi, K., Jaiswal, V., and Ansari, M. D. 2020a. Machine learning based support system for stream selection. *Recent Advances in Computer Science and Communications*. 13(3): 336–344.

Sethi, K., Sharma, A., Chauhan, S., and Jaiswal, V. 2020b. Impact of social and cultural challenges in education using AI. In Habib, M. K. (ed.), *Revolutionizing Education in the Age of AI and Machine Learning*. IGI Global. 130–151. doi: 10.4018/978-1-5225-7793-5.ch007.

Singha, S., Kumar, A., and Bera, A. 2021. Comparative analysis of machine learning models for water quality index prediction. *Water Resources Management*. 35(12): 4119–4134.

Sood, R., and Zhu, K. 2024. Improving water quality time-series prediction using Sentinel-2 MSI and GEE. arXiv preprint arXiv:2408.14010.

Surya, A., Singh, R., and Verma, A. 2024. Real-time water quality monitoring using hybrid deep learning and IoT. *Journal of Environmental Management*. 349: 118768.

Tang, T., Strokal, M., van Vliet, M. T. H., Seuntjens, P., and Burek, P. 2019. Bridging scale gaps in water quality modeling. *Current Opinion in Environmental Sustainability*. 39: 101–108.

Thakur, A., and Sethi, K. 2025. Advances in non-contact cardiovascular monitoring via face-based ML. In *2025 12th International Conference on Computing for Sustainable Global Development (INDIACom)*, New Delhi, India, 2–4 April. 1–4. IEEE.

Yan, X., Zhang, T., Du, W., Meng, Q., Xu, X., and Zhao, X. 2024. Review of machine learning for water quality prediction. *Journal of Marine Science and Engineering*. 12(1): 159.

Yao, H., Xu, X., and Li, M. 2021. Predicting chlorophyll-a using gradient boosting and Landsat-8. *Journal of Marine Science and Engineering*. 9(2): 151.

Yong, S. K., Kim, S., and Lee, J. 2021. Estimating chlorophyll-a using Landsat-8 and machine learning. *Journal of Marine Science and Engineering*. 9(4): 428.

Zhou, Y., Li, H., and Wang, M. 2022. Object detection for marine microalgae using YOLOv5, TOOD, and Faster R-CNN. arXiv preprint arXiv:2211.07546.

Scalable AI/ML Deployment Strategies

16

Case Study on Water Management, Print, and Logistics

Cigil Achenkunju and Ciril Achenkunju

FINDINGS

A. A consistent theme across the reviewed literature emphasizes monitoring, whether it be of the model, data, or deployment. Notably, incorporating a monitoring-centric approach early in the development lifecycle helps reduce key operational issues.
B. More than 50% of the projects and industry research reviewed utilize some form of fully managed cloud services AI/ML platform.
C. Industry studies highlight increased model maturity with a focus on reliability, scalability, and lifecycle management.

16.1 INTRODUCTION

The deployment of artificial intelligence (AI)/machine learning (ML) systems to address industry-specific problems in production environments varies significantly. These challenges stem from a range of operational and infrastructural issues that are not always evident during model development. This work aims to investigate proven strategies and best practices for identifying and addressing such challenges

DOI: 10.1201/9781003714217-16

in real-world scenarios. Focusing on use cases in manufacturing, retail, and logistics, we propose a scalable, monitoring-centric framework that aligns architectural choices with deployment-critical concerns, including data drift, data variance, resource constraints, and latency requirements. By doing so, we seek to inform system design decisions that are resilient to the complexities of production ML environments.

AI/ML systems have rapidly evolved to become essential components of modern business applications. However, the journey from development to production deployment of these systems presents complexities that many organizations struggle to navigate effectively. The hidden technical costs of deploying ML systems can accumulate rapidly when development and deployment are treated as separate concerns [1]. Our findings reveal that, despite sectoral differences, common patterns emerge in the architectural approaches and deployment strategies. Using case studies and best practices, the article presents system designs, ML components, architecture, and lessons learned from the successful implementation of MLOps practices, providing valuable references for other organizations to build and optimize their deployments. During our empirical review, we identified a gap in connecting architectural approaches with the key challenges that may arise during or after the implementation of AI/ML systems.

Deploying ML models into production environments introduces challenges that can significantly impact their performance. We will explore these pressing issues in the operational challenges section. To address these challenges, implementing a monitoring-centric architecture is crucial. A monitoring-centric architecture focuses on continuously overseeing the performance and behavior of AI/ML models in production. By integrating various monitoring tools and frameworks, organizations can promptly detect relevant issues and take corrective actions.

A. We identify key challenges in the deployment of AI/ML systems and aggregate metrics from different papers.
B. We present architectural patterns that address the identified challenges and validate them through case studies in diverse industry settings.
C. We propose a monitoring-centric decision-making approach to manage increased maintenance costs.

16.2 METHODOLOGY AND LITERATURE REVIEW

A multivocal literature review from formal and informal sources of industry leaders who have deployed most ML and AI models in production. The study identifies similarities and differences in their architectures and approaches for monitoring different issues faced during deployment and maintenance.

The review follows a systematic approach by defining research questions, identifying relevant keywords (MLOps, production deployment, experiment, architecture, etc.), and sourcing content from digital libraries (Google Scholar, ArXiv) as well as industry whitepapers, blogs, and technical reports. We also employed snowball sampling by analyzing related work sections to expand the review scope. Our criteria required each selected paper to include practical implementation details for production-level MLOps tools (e.g., MLflow, Kubernetes, and Docker). The literature review was conducted by systematically analyzing multiple industry professional papers and open-source projects related to AI/ML models, monitoring, and production deployment. Each paper was evaluated based on the following criteria: system design and architecture; types of issues/challenges detected (data, concept drift, bias, feature attribution); methodologies employed (e.g., deep learning, statistical analysis, uncertainty estimation); scalability ; and real-world application scenarios. A comparative analysis was performed to identify common approaches, unique methodologies, and the strengths and limitations of each framework. The findings were synthesized to provide a comprehensive understanding of various production practices in AI/ML model monitoring and deployment.

16.3 THE EVOLUTION OF ML DEPLOYMENT PRACTICES

Early works primarily focused on model-serving frameworks [2], while recent research has expanded to address the entire ML lifecycle management [3]. Notably, a study [4] surveyed the challenges of deploying ML systems in various industries, identifying monitoring and maintenance as particularly difficult.

Traditional ML workflows follow established methodologies such as CRISP-DM (Cross-Industry Standard Process for Data Mining), which emphasize data understanding, preparation, modeling, and evaluation [5]. However, these frameworks often treat deployment and monitoring as final steps rather than integral considerations throughout the development process. A study [6] documented Microsoft's experiences in integrating software engineering practices with ML development, highlighting the tensions between rapid experimentation cycles and production stability requirements. Similarly, another study [1] emphasized that when deployment and development concerns are separated, such ML systems incur significant technical debt.

16.4 OPERATIONAL CHALLENGES IN PRODUCTION ML

Several operational challenges specific to production ML systems were identified:

A. **Model Drift**: Monitoring (data drift, concept drift, etc.) and quality control have been cited as critically important for managing ML systems in production [7]. Data drift and concept drift are both types of model drift that can degrade ML performance. Data drift occurs when the statistical properties of the input features change over time, while concept drift occurs when the relationship between the input features and the target variable changes. Additionally, many types of concept drift have been identified [8], each of which can affect model performance. A paper [9] presents 28 monitoring needs and tests for detecting when production data diverges from training data distributions. One of the key challenges in model monitoring is distribution shift, which is also a key source of model performance degradation.

B. **Resource Constraints**: The rapid advancement of AI has propelled edge intelligence to the forefront of technological innovation, enabling real-time decision-making and localized data processing. However, deploying AI models on resource-constrained edge devices introduces multifaceted challenges stemming from the increasing complexity of AI models and the inherent computation and memory limitations of edge hardware [10]. The work [10] examines computational and memory constraints on model selection and optimization for edge deployment. Furthermore, once a deployed model is identified as deteriorating, it is retrained to maintain high performance. Known as continual learning [7], this process can be computationally expensive and place high demands on the software engineering system.

C. **Latency Requirements**: Latency is the duration between initiating a request and the moment the requester can utilize the returned result. Research [2] explores a set of novel techniques to reduce and bound latency while maximizing throughput that generalizes across different predictive ML frameworks, such as caching, batching, and adaptive model selection. An empirical study analyzing attributes from 791 production services revealed a significant trend: over 50% of cloud service owners monitor metrics such as success rate, capacity, availability, and latency for resources.

D. **System Complexity**: The paper [1] discusses how the tight coupling between ML components creates maintenance challenges that exceed those of traditional software systems. The lines between model components blur, making it hard to make small changes to data inputs or model parameters, leading to cascading and unpredictable outcomes. Dependencies may include other services, databases, messaging systems, and other infrastructure components [11].

16.5 SYSTEM DESIGN AND MLOPS WORKFLOW

We describe the prevalence of various architectural paradigms identified in the reviewed literature [12] and discuss the challenges associated with deploying these approaches in production. From a long-term perspective, industry practitioners can use these architectural insights to inform decisions when adopting different MLOps system practices. We extract and analyse key components that comprise a typical architecture for such systems and identify various challenges. For the benefit of readers, we focus on architectural components commonly referenced in the literature and production practices. The target audience for this research is AI/ML practitioners and leaders who can use this work to implement their systems. Researchers and practitioners can use this research to identify suitable architectural variants and associated challenges as per their requirements.

The research reported in [12] systematically reviewed 43 different papers to define key architecture paradigms, but without mapping to the challenges associated with such approaches. Key design principles are used to evaluate diverse models, such as accessibility, scalability, handling historical data, monitoring capacity, and incorporation of external data.

MLOps is an approach to managing and automating the development, deployment, and maintenance of models based on the implementation of such solutions in industry settings. We synthesize the workflow into a few main stages that multiple papers and research works [4,12,13] have detailed in the MLOps workflow and process.

A. **Data Management**: Data management typically involves the collection, processing, augmentation, and analysis of data. This stage aims to enrich the dataset and learn the business process for model training.
B. **Model Learning and Verification**: This stage involves selecting algorithms, developing the model architecture, training, and validating models. It includes evaluating the models on test data to assess performance, reliability, compliance, and security.
C. **Model Deployment and Monitoring**: The developed model is packaged with the required dependencies and deployed to production as a complex software system. Integrating ML systems often presents unique and shared challenges with other software systems. Monitoring and updating the deployed model is crucial, as ML systems are prone to inference degradation.
D. **Model Lifecycle Management**: This stage deals with coordinating all the stages of the workflow lifecycle while ensuring compliance with business, ethical, and regulatory requirements.

16.5.1 Frameworks, Architecture, and Tools

Many frameworks and architectural approaches facilitate the development of ML solutions. We list scalable, monitoring-centric approaches to support systematic development in organizations delving into ML practices. There can be multiple such architectural approaches, not limited to those listed here. The choice of frameworks and architecture is highly dependent on the specific requirements and constraints of organizations. A wide range of MLOps tools are used for pipelines and model operationalization. The choice depends on the business use case and compatibility with the existing technology stack.

MLflow is an open-source platform for managing the deployment and experimentation of ML models. Cloud platforms, such as AWS Sage Maker, Azure ML, and Google Cloud Vertex AI, are used in lifecycle management and model operationalization. Containerization tools, such as Docker, and orchestration tools, such as Kubernetes, are widely used by many ML operationalization teams. Other tools, such as Jenkins, Kubeflow, and Kafka-ML, are used for pipelines and model deployment across different practices. The choice of MLOps tools that offer various features to support the ML model lifecycle depends on specific needs and goals. Some tools provide specific features that solve certain issues in the previously listed MLOps workflow. A study examined different tools and platforms used to support the ML components in production [12]. Please note that Table 16.1 is not an exhaustive list. One observation from reviewing different papers and projects was that some tools are used in various ways and for multiple purposes (Figure 16.1).

16.5.2 Maturity Models

A maturity model is used to assess the level of automation across the development, deployment, testing, and monitoring of ML solutions. The maturity model approach outlined in in-depth research [14] uses three levels based on manual or automated processes in the ML workflow: the first level, fundamental maturity involves manual script-driven processes at each stage of the ML workflow. The expectation at this level is that the generated model does not change often, but in practice, models are regularly optimized. This level generally lacks active performance monitoring, and only a few CI/CD concepts are orchestrated. At the second level, experimentation steps for ML are orchestrated to automate continuous model training in production. At the third level, the automation of the CI/CD pipeline at each stage of the workflow enables rapid release to production systems. This level supports different types of monitoring, such as data drift, system latency, and concept drift.

TABLE 16.1 List of different tools identified in various research papers and projects reviewed

COMPONENT	*# OF TIMES IDENTIFIED IN [12]*	*ADDITIONAL IDENTIFICATION*
Jenkins	12	1 [26]
AWS SageMaker	11	1 [19]
Docker	6	3 [19, 27, 28]
MLflow	6	4 [25, 36, 29, 30]
AWS S3	4	1 [19]
Azure ML	4	2 [31, 30]
Google Vertex AI	4	1 [27]
Apache Kafka	3	3 [23,31,32]
Kubernetes	3	3 [28,29,31]

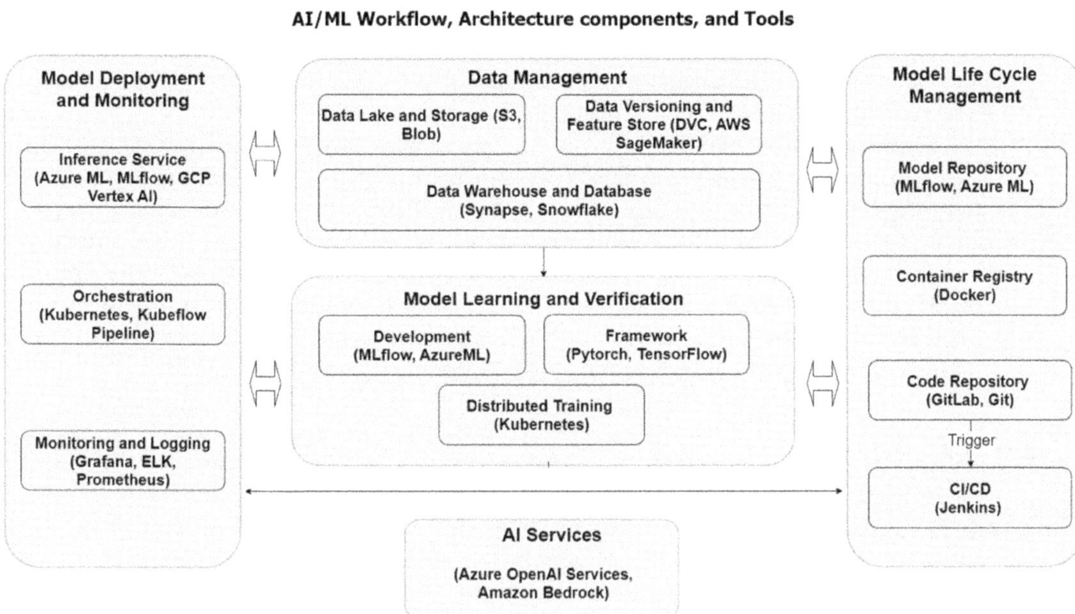

FIGURE 16.1 AI/ML workflow, architecture components, and tools.

The most common methods of deploying models to production are listed below. As previously stated, these choices are made in accordance with the goals, technology stack, and maturity models of different organizations.

A. **Microservices**: An empirical research [15] examined how microservice architectures can improve the maintainability and scalability of complex systems, including those with ML components. It is an architectural style that distributes applications into small, independently deployable services that communicate via lightweight mechanisms. The main benefit of this approach is that complexity is shifted from the application to the architecture level. Each service is usually autonomous, with its own CI/CD pipeline for testing and deployment. Changing one service implementation does not affect the others, as communication between them occurs via REST interfaces. Deploying models through a REST API or other protocols to provide real-time predictions helps avoid tight coupling between services. This practice helps with the continuous monitoring of not just performance but also facilitates the detection of operational anomalies.
B. **Containerization**: Containers are a popular mechanism used for deploying and running applications, as they can easily move across different computing environments without requiring any changes to the underlying code. Containers encapsulate resources, dependencies, and functionality and are the most granular independently hostable unit. They are typically monitored for availability [16]. Motivated by the difficulties in distributing modern high-performance software to a wide range of users and developers, we address the use of Linux containers as an approach to portable deployment of scientific software libraries, from laptop computers to supercomputers, with no noticeable performance penalty [16]. Netflix containerizes models and applications and leverages Kubernetes for deployment and management. Through containerization, they achieve rapid deployment, elastic scaling, and high model availability while ensuring consistency and portability of the environment [17].

However, a large-scale study [18] that analysed posts from the Stack Overflow community identified various challenges that arise when using Docker for application development, debugging, and resource management. Please note that Docker is a platform that uses containerization to package and run applications.

C. **Cloud Services**: A fully managed ML platform that enables developers and data scientists to quickly and easily build, train, and deploy ML models at each step of the ML workflow [19]. In an in-depth study [14] examining a range of ML platforms, it was found that many organizations, including Databricks, Toyota, Accenture, Facebook, and Microsoft, are adopting and contributing to MLflow, which provides the community with easy support for bug resolution. MLflow is fundamentally an open-source platform for managing the ML lifecycle. You can deploy and manage it yourself on various platforms, including your local machine, on-premises clusters, or cloud platforms (AWS, Azure, Google Cloud, etc.) [20].

16.5.3 Model Monitoring

Testing and monitoring are essential strategies for improving reliability, reducing technical debt, and lowering long-term maintenance costs [9]. Monitoring dependencies, such as data version changes, source system changes, and upgrades, throughout the ML deployment pipeline is crucial. Model monitoring is a continuous process; therefore, early identification of issues and implementation of strategies are essential [19]. Models need to be monitored to ensure that they continue to meet predictive quality requirements; this monitoring is a vital part of the MLOps lifecycle. Tracking the computational performance of ML systems helps with cost estimation, including GPU memory allocation, disk usage, and network traffic. The benefits of implementing a monitoring-centric architecture include proactive issue detection, enhanced model reliability, regulatory compliance, and better decision-making.

To include MLOps monitoring, the first step is to monitor data that clearly indicate problems in the model deployment lifecycle [21]. On the model input side, TFX deploys a data validation system that checks schema correctness and ensures that training code evolves along with data schema changes. Arize AI takes the monitoring definition one step further by monitoring symptoms of model degradation, including concept and feature drift, online –offline inconsistency, and outliers in the feature distribution. SageMaker [12] monitors four aspects of MLOps : data quality, model quality, bias drift, and feature attribution drift.

16.6 CASE STUDIES AND CHALLENGES

Drawing on case studies and best practices from manufacturing, logistics, and print forecasting, we present key strategies and lessons learned for various MLOps implementations, providing valuable insights for other organizations to build and optimize their own MLOps practices.

16.6.1 Smart Manufacturing

In the research, a digital platform was developed to detect failures in water pumps using a real-time kappa architecture proposed by Wingerath [22,23]. This architecture is a layered system comprising Data Sources, a Data Bus, a Processing Layer, and a Presentation Layer. As per industry standards, the Cross-Industry Standard Process for Data Mining is followed to deploy the model. The data source layer consists of a dataset comprising 52 sensors on a water pump with different parameters (temperature, pressure, and humidity). A big data platform is a significant part of the implementation of the data bus, processing layer, and presentation layer, composed of Zookeeper (coordination service), Kafka (queue manager), Spark (processing), Spring Framework (web service or API), and GUI (angular framework) to perform predictions in real time.

The classifiers used for detection were random forest, support vector machine, artificial neural network, and K-nearest neighbors. Variable selection methods, such as Principal Component Analysis (PCA) and chi-squared tests, were used to reduce dimensionality while maximizing explained variance. They validated the models using a 10-fold cross-validation method and evaluated them using accuracy, FPR, TPR, and response time (average prediction response time). For deployment, the research provisioned five AWS EC2 instances and developed and deployed a monitoring service. The result obtained multiple models with high accuracy of 99%, with just one model below the 100 ms mean response time.

During our research, we observed several solved and unsolved challenges that align with findings from [1,4]:

A. **Data Collection, Cleaning, and Missing Values**: In a large-scale setup, it becomes difficult to understand the services that extract and store industrial sensor data. Decentralization improves scalability; however, it also increases the complexity of tracking which services perform specific tasks and how data is stored across the system. Furthermore, production sensor data may contain intermittent missing values and occasional outliers that may not be present in the curated dataset.
B. **Monitoring Complexities**: The statistical properties of the target variable can change, affecting predictions. The model requires sensor data to be processed efficiently to enable timely alerts. Furthermore, the computing resources at different manufacturing sites may vary. Hence, monitoring of evolving input data, prediction bias, and physical resource performance should be an essential part of all ML lifecycles.
C. **Glue Code or Pipeline Jungles**: Extensive use of generic packages and complex data preparation pipelines can lead to integration issues, making systems brittle and expensive to maintain [1]. For example, the initial cost of implementing experimental code paths on conditional branches of production code may seem relatively low. However, over time, such code paths can create increasing difficulties in maintaining backward compatibility and result in an exponential increase in cyclomatic complexity.

16.6.2 Route Optimization

A system of self-autonomous connected vehicles, where Internet of Things (IoT) devices, data centers, and smart vehicles exist in a distributed architecture, significantly impacts the quality of urban environments and societal and economic welfare [24]. These systems need to collect, process, and broadcast large amounts of data effectively to succeed in real-world scenarios. While research in this context has been extensive from a theoretical perspective, works focusing on the implementation, deployment, and evaluation of the practical performance of graph algorithms for real-world autonomous vehicles have been scarce.

Edge Computing is a paradigm that extends Cloud Computing and services to the edge of the network, closer to data sources and end users, enabling faster processing, reduced latency, better end-user data security, and improved efficiency. The implementation of route optimization problems follows the Edge-Cloud Computing paradigm, with computational efforts distributed across three different modules. The first module (FM: services hosted on a set of elastic cloud servers) is responsible for route computation based on roadmap inputs, user requests, and the fleet vehicle structure. The second module (SM: multi-access edge computing) computes the distance between users' locations using Dijkstra's algorithm. Additionally, SM hosts an urban simulation tool and a toolbox to compute road parameters, enabling experimentation with different vehicle movement strategies before implementing them in the real world. In the third module (TM: human –computer interface), the optimized routes are encoded and displayed along with those of the users and other vehicles in the fleet.

FM instantiates the optimization problem on graphs based on different data (vehicle positions, users' locations, distances between users' locations, etc.) and then exploits a routine provided by OR-Tool to find a solution to the resulting problem. The output is a sequence of users' locations for each vehicle. To transform each sequence into a route, a subroutine (SM) is called on a dataset representing the road map to find the shortest path for each pair of consecutive users' locations in the sequence. The concatenation of all the shortest paths for a given vehicle forms the route that the vehicle must follow.

Different modules use a broker system running on each computer to coordinate data exchange. The broker system provides channels to which different modules can subscribe to send or read messages. More detailed information regarding each module is omitted as it is out of scope for our literature review; we refer the reader to [24] for further information.

As noted in our previous findings during research evaluations, we identify several challenges that align with findings from [1,4]:

A. **Data Latency**: As the number of vehicles increases, the execution time, scalability, and quality of computed solutions may worsen.
B. **Integration Complexity**: Each module contains different subroutines, tools, and services. The model closely integrates solutions from other systems. In a large-scale setup, such systems become cumbersome and difficult to maintain over time.
C. **Computational Constraints**: Route calculations may be forced to return lower quality solutions in a limited time interval to be actionable for drivers.

16.6.3 Print Forecasting

The MLOps architecture for print demand forecasting at HP Inc., as detailed by Harshvardhan et al. [25], focuses on building a scalable and automated system that integrates ML with human oversight. The core of this approach is the development and deployment of a tree-based ML model, specifically LightGBM, to predict demand for over 18,000 print products across more than 170 countries. This ML model leverages historical demand data alongside additional datasets, such as channel inventory and sell-through information, to improve forecasting accuracy.

The ML system primarily utilizes LightGBM, a gradient boosting framework that excels in speed and efficiency, making it suitable for large-scale datasets and complex patterns. For hyperparameter optimization, Hyperopt was employed to fine-tune the LightGBM model for optimal performance across various market conditions. To manage the ML lifecycle, including experiment tracking, reproducibility, and deployment, MLflow was a critical open-source tool, allowing for the storage of serialized models and parameters as artifacts for future reference and warm starting future training. Jupyter Notebooks, combined with the papermill library, were used to parameterize and automate script execution, ensuring reproducibility and streamlining routine tasks like monthly time-series forecasting. The system is designed to dynamically switch between ML and statistical forecasts on a product-by-product and geography-specific basis, ensuring the most accurate method is applied globally. Iterative design, experimental validation, and MLflow for tracking experiments and model versions address the challenge of maintaining model performance and reliability over time, preventing model degradation or drift.

Real-world challenges arise when we implement systems in production. Detailed below are some of these challenges mentioned in [1,4]:

A. **Hidden Feedback Loop and System Complexity**: If the ML forecasts influence inventory decisions, which in turn affect future demand signals (e.g., product availability), this could create a feedback loop. The paper focuses on the accuracy of the ML model and its integration, but managing system-level complexity and potential feedback loops that could degrade forecast quality over time —a form of ML-specific technical debt —poses an implicit challenge.
B. **Entanglement and Erosion of Abstraction Boundaries**: The tight integration of historical data, channel inventory, and other external factors into a single LightGBM model could make independent changes difficult in the future (entanglement). If a change is needed in how channel inventory data is processed, it might have unforeseen consequences on the ML forecasting model itself, increasing maintenance costs and hindering agility.
C. **Handling Extreme and Unforeseen Events**: The adaptability of current ML models in predicting and responding to truly unprecedented events, which are by definition outside historical data patterns, remains a potential area of concern. Robustness to extreme outliers or entirely novel market conditions might require specialized anomaly detection or scenario-planning techniques.

16.6.4 Conclusion

In summary, hosting a monitoring-centric architecture is essential for maintaining the performance and reliability of ML models in production. By integrating real-time monitoring, organizations can effectively address challenges and ensure their ML systems continue to deliver accurate and valuable insights. Our case studies provide real-world examples of large-scale implementation in different domains highly relevant to many business use cases: demand forecasting, water pump failure detection, and logistics optimization. The research enables industry practitioners to see how ML concepts and MLOps practices translate into tangible business value. This research highlights the critical role of MLOps in robust and sustainable AI/ML deployments. It explicitly connects technical implementation with the challenges outlined in broader AI/ML deployment literature, such as scalability, monitoring, and complexity management. By bridging the gap between specific industry applications and general deployment hurdles, it equips practitioners with a framework to anticipate and cost-effectively address similar issues in their own projects.

REFERENCES

1. D. Sculley, Gary Holt, Daniel Golovin, Eugene Davydov, and Todd Phillips (2015). Machine learning: The high-interest credit card of technical debt. https://proceedings.neurips.cc/paper_files/paper/2015/file/86df7dcfd896fcaf2674f757a2463eba-Paper.pdf
2. Daniel Crankshaw, Xin Wang, Giulio Zhou, Michael J. Franklin, Joseph E. Gonzalez, and Ion Stoica (2017). Clipper: A low-latency online prediction serving system. https://arxiv.org/abs/1612.03079
3. Monika Steidl, Michael Felderer, and Rudolf Ramler (2023). The pipeline for the continuous development of artificial intelligence models—Current state of research and practice. https://www.sciencedirect.com/science/article/pii/S0164121223000109
4. Andrei Paleyes, Raoul-Gabriel Urma, and Neil D. Lawrence. (2022). Challenges in deploying machine learning models: A survey of case studies. https://arxiv.org/pdf/2011.09926

5. Pete Chapman (NCR), Julian Clinton (SPSS), Randy Kerber (NCR), Thomas Khabaza (SPSS), Thomas Reinartz (DaimlerChrysler), Colin Shearer (SPSS), and Rüdiger Wirth (DaimlerChrysler). 2000. CRISP-DM 1.0—Step-by-step data mining guide. https://public.dhe.ibm.com/software/analytics/spss/documentation/modeler/14.2/es/CRISP-DM.pdf
6. Saleema Amershi, Dan Weld, Mihaela Vorvoreanu, Adam Fourney, Besmira Nushi, Penny Collisson, Jina Suh, Shamsi Iqbal, Paul N. Bennett, Kori Inkpen, Jaime Teevan, Ruth Kikin-Gil, and Eric Horvitz. 2019. Guidelines for continuous human-AI. https://www.microsoft.com/en-us/research/wp-content/uploads/2019/01/Guidelines-for-Human-AI-Interaction-camera-ready.pdf
7. T. Diethe, T. Borchert, E. Thereska, B. Balle, and N. Lawrence (2019). Continual learning in practice. *Stat*, 1050: 18
8. J. Quinonero-Candela, M. Sugiyama, A. Schwaighofer, and N. D. Lawrence. (2008). *Dataset shift in machine learning.* MIT Press.
9. E. Breck, Cai, S., Nielsen, E., Salib, M., and Sculley, D. (2017). The ML test score: A rubric for ML production readiness and technical debt reduction, in *Proceedings of IEEE Big Data 2017.* https://static.googleusercontent.com/media/research.google.com/en//pubs/archive/aad9f93b86b7addfea4c419b9100c6cdd26cacea.pdf
10. T. Wang, J. Guo, B. Zhang, G. Yang, and D. Li. (2025). Deploying AI on edge: Advancement and challenges in edge intelligence. *Mathematics*, 13, 1878. https://doi.org/10.3390/math13111878
11. P. Srinivas, F. Husain, A. Parayil, A. Choure, C. Bansal, and S. Rajmohan. (2024). Intelligent monitoring framework for cloud services: A data-driven approach. https://arxiv.org/html/2403.07927v1
12. F. Amou Najafabadi, J. Bogner, I. Gerostathopoulos, and P. Lago. (2024). An analysis of MLOps architectures: A systematic mapping study. [Data set], Zenodo. https://doi.org/10.1007/978-3-031-70797-1_5
13. Behrouz Derakhshan, Alireza Rezaei Mahdiraji, Tilmann Rabl, and Volker Markl. (2019). Continuous deployment of machine learning pipelines. https://openproceedings.org/2019/conf/edbt/EDBT19_paper_23.pdf
14. S. Garg, P. Pundi, G. Rathee, P. K. Gupta, S. Garg, and S. Ahlawat. (2022). On continuous integration/continuous delivery for automated deployment of machine learning models using MLOps. https://arxiv.org/pdf/2202.03541
15. Carlos M. Aderaldo, Nabor C. Mendonça, Claus Pahl, and Pooyan Jamshidi. (2017). Benchmark requirements for microservices architecture research. https://dl.acm.org/doi/10.5555/3101282.3101285
16. Jack S. Hale, Lizao Li, Chris N. Richardson, and Garth N. Wells. 2016. Containers for portable, productive and performant scientific computing. https://arxiv.org/pdf/1608.07573
17. P. Liang, B. Song, X. Zhan, Z. Chen, and J. Yuan. (2024). Automating the training and deployment of models in MLOps by integrating systems with machine learning. https://arxiv.org/pdf/2405.09819
18. M. Ul Haque, L. H. Iwaya, and M. Ali Babar. (2020). Challenges in docker development: A large-scale study using stack overflow. In *Proceedings of ACM Conference (Conference'17).* ACM, New York, 11 pages. https://arxiv.org/pdf/2008.04467
19. D. Nigenda, Z. Karnin, M. B. Zafar, R. Ramesha, A. Tan, M. Donini, and K. Kenthapadi. (2022). Amazon SageMaker model monitor: A system for real-time insights into deployed machine learning models. https://arxiv.org/pdf/2111.13657
20. MLflow: A tool for managing the machine learning lifecycle. https://mlflow.org/docs/latest/ml
21. Zhentao Xu, Ruoying Wang, Girish Balaji, Manas Bundele, Xiaofei Liu, Leo Liu, and Tie Wang. (2023). AlerTiger: Deep learning for AI model health monitoring at LinkedIn. https://arxiv.org/pdf/2306.01977
22. W. Wingerath, F. Gessert, S. Friedrich, and N. Ritter. (2016). Real-time stream processing for big data. *It.—Information Technology*, 58, 186–194. https://doi.org/10.1515/itit-2016-0002
23. Ricardo Dintén Herrero, and Marta Zorrilla. (2022). An I4.0 data intensive platform suitable for the deployment of machine learning models: A predictive maintenance service case study. https://doi.org/10.1016/j.procs.2022.01.300
24. Mattia D'Emidio, Esmaeil Delfaraz, Gabriele Di Stefano, Giannantonio Frittella, and Edgardo Vittoria. (2024). Route planning algorithms for fleets of connected vehicles: State of the art, implementation, and deployment—Route optimization in Italy. https://www.mdpi.com/2076-3417/14/7/2884
25. M. Harshvardhan, Cara Curtland, Jerry Hwang, Chuck VanDam, Adam Ghozeil, Pedro A. Neto, Frederic Marie, and Chuanren Liu. (2025). Print demand forecasting with machine learning at HP Inc. https://pubsonline.informs.org/doi/10.1287/inte.2024.0126
26. Rakshith Subramanya, Seppo Sierla, and Valeriy Vyatkin. (2022). From DevOps to MLOps: Overview and application to electricity market forecasting. https://www.mdpi.com/2076-3417/12/19/9851

27. Matteo Testi, Luca Clissa, Matteo Ballabio, Salvatore Ricciardi, Federico Baldo, Emanuele Frontoni, Sara Moccia, and Gennaro Vessio. (2025). Enhancing cell counting through MLOps: A structured approach for automated cell analysis. https://arxiv.org/pdf/2504.20126
28. Robson S. Santos, Robesvania Araújo, Paulo A. L. Rego, Jose M. da S. M. Filho, Jarelio G. da S. Filho, Jose D. C. Neto, Nicksson C. A. de Freitas, Emanuel B. Rodrigues, Francisco Gomes, and Fernando Trinta. (2025). A microservice-based architecture for real-time credit card fraud detection with observability. https://www.scitepress.org/Papers/2025/136500/136500.pdf
29. Sergio Moreschini, David Hästbacka, and Davide Taibi. (2023). MLOps pipeline development: The OSSARA use case. https://dl.acm.org/doi/10.1145/3599957.3606211
30. Matti Linna. (2025). MLOps architecture for machine vision: A practical architecture for industry application. https://urn.fi/URN:NBN:fi:tuni-202506086926
31. Hansheng Ren, Bixiong Xu, Yujing Wang, Chao Yi, Congrui Huang, Xiaoyu Kou, Tony Xing, Mao Yang, Jie Tong, and Qi Zhang. (2019).Time-series anomaly detection service at Microsoft. https://arxiv.org/abs/1906.03821
32. Mikel Canizo, Enrique Onieva, Angel Conde, Santiago Charramendieta, and Salvador Trujillo. (2017). Real-time predictive maintenance for wind turbines using big data frameworks. https://arxiv.org/pdf/1709.07250

Index

For Product Safety Concerns and Information please contact our EU
representative GPSR@taylorandfrancis.com
Taylor & Francis Verlag GmbH, Kaufingerstraße 24, 80331 München, Germany

www.ingramcontent.com/pod-product-compliance
Ingram Content Group UK Ltd.
Pitfield, Milton Keynes, MK11 3LW, UK
UKHW052212190926
475702UK00006B/987
* 9 7 8 1 0 4 1 1 9 9 3 1 1 *